Crina Grosan and Ajith Abraham

Intelligent Systems

Intelligent Systems Reference Library, Volume 17

Editors-in-Chief

Prof. Janusz Kacprzyk
Systems Research Institute
Polish Academy of Sciences
ul. Newelska 6
01-447 Warsaw
Poland
E-mail: kacprzyk@ibspan.waw.pl

Prof. Lakhmi C. Jain
University of South Australia
Adelaide
Mawson Lakes Campus
South Australia 5095
Australia
E-mail: Lakhmi.jain@unisa.edu.au

Further volumes of this series can be found on our homepage: springer.com

Vol. 1. Christine L. Mumford and Lakhmi C. Jain (Eds.)
Computational Intelligence: Collaboration, Fusion and Emergence, 2009
ISBN 978-3-642-01798-8

Vol. 2. Yuehui Chen and Ajith Abraham
Tree-Structure Based Hybrid Computational Intelligence, 2009
ISBN 978-3-642-04738-1

Vol. 3. Anthony Finn and Steve Scheding
Developments and Challenges for Autonomous Unmanned Vehicles, 2010
ISBN 978-3-642-10703-0

Vol. 4. Lakhmi C. Jain and Chee Peng Lim (Eds.)
Handbook on Decision Making: Techniques and Applications, 2010
ISBN 978-3-642-13638-2

Vol. 5. George A. Anastassiou
Intelligent Mathematics: Computational Analysis, 2010
ISBN 978-3-642-17097-3

Vol. 6. Ludmila Dymowa
Soft Computing in Economics and Finance, 2011
ISBN 978-3-642-17718-7

Vol. 7. Gerasimos G. Rigatos
Modelling and Control for Intelligent Industrial Systems, 2011
ISBN 978-3-642-17874-0

Vol. 8. Edward H.Y. Lim, James N.K. Liu, and Raymond S.T. Lee
Knowledge Seeker – Ontology Modelling for Information Search and Management, 2011
ISBN 978-3-642-17915-0

Vol. 9. Menahem Friedman and Abraham Kandel
Calculus Light, 2011
ISBN 978-3-642-17847-4

Vol. 10. Andreas Tolk and Lakhmi C. Jain
Intelligence-Based Systems Engineering, 2011
ISBN 978-3-642-17930-3

Vol. 11. Samuli Niiranen and Andre Ribeiro (Eds.)
Information Processing and Biological Systems, 2011
ISBN 978-3-642-19620-1

Vol. 12. Florin Gorunescu
Data Mining, 2011
ISBN 978-3-642-19720-8

Vol. 13. Witold Pedrycz and Shyi-Ming Chen (Eds.)
Granular Computing and Intelligent Systems, 2011
ISBN 978-3-642-19819-9

Vol. 14. George A. Anastassiou and Oktay Duman
Towards Intelligent Modeling: Statistical Approximation Theory, 2011
ISBN 978-3-642-19825-0

Vol. 15. Antonino Freno and Edmondo Trentin
Hybrid Random Fields, 2011
ISBN 978-3-642-20307-7

Vol. 16. Alexiei Dingli
Knowledge Annotation: Making Implicit Knowledge Explicit, 2011
ISBN 978-3-642-20322-0

Vol. 17. Crina Grosan and Ajith Abraham
Intelligent Systems, 2011
ISBN 978-3-642-21003-7

Crina Grosan and Ajith Abraham

Intelligent Systems

A Modern Approach

Springer

Dr. Crina Grosan
Department of Computer Science,
Faculty of Mathematics and Computer Science
Babes-Bolyai University, Cluj-Napoca,
Kogalniceanu 1, 400084 Cluj - Napoca
Romania
E-mail: cgrosan@cs.ubbcluj.ro

Prof. Ajith Abraham
Machine Intelligence Research Labs
(MIR Labs)
Scientific Network for Innovation and Research Excellence
P.O. Box 2259
Auburn, Washington 98071-2259
USA
Email: ajith.abraham@ieee.org

ISBN 978-3-642-26939-4 ISBN 978-3-642-21004-4 (eBook)

DOI 10.1007/978-3-642-21004-4

Intelligent Systems Reference Library ISSN 1868-4394

Softcover reprint of the hardcover 1st edition 2011

Typeset & Cover Design: Scientific Publishing Services Pvt. Ltd., Chennai, India.

Printed on acid-free paper

9 8 7 6 5 4 3 2 1

springer.com

Preface

Recent advances in the information and communication technologies are leading to an exponential growth in the amount of data stored in databases. It has been estimated that this amount doubles every 20 years. For some applications, this increase is even steeper. Databases storing DNA sequence, for example, are doubling their size every 10 months. This growth is occurring in several applications areas besides bioinformatics, like financial transactions, government data, environmental monitoring, satellite and medical images, security data and web. As large organizations recognize the high value of data stored in their databases and the importance of their data collection to support decision-making, there is a clear demand for sophisticated computational intelligence tools. Most of the current real world problems involve global optimization (or a search for optimal solutions), function approximation / machine learning and approximate reasoning.

Global optimization deals with the task of finding the absolutely best set of admissible conditions to satisfy certain criteria / objective function(s), formulated in mathematical terms. Global optimization includes nonlinear, stochastic and combinatorial programming, multiobjective programming, control, games, geometry, approximation, algorithms for parallel architectures and so on. Due to its wide usage and applications, it has gained the attention of researchers and practitioners from a plethora of scientific domains. Global Optimization algorithms may be categorized into several types: Deterministic (example: branch and bound methods), Stochastic optimization (example: simulated annealing), Heuristics and meta-heuristics (example: evolutionary algorithms) etc.

Learning methods and approximation algorithms are fundamental tools that deal with computationally hard problems and problems in which the input is gradually disclosed over time. Both kinds of problems have a large number of applications arising from a variety of fields. Machine Learning is concerned with the study of building computer programs that automatically improve and/or adapt their performance through experience. Machine learning can be thought of as "programming by example". Decision trees are suitable for scientific problems entail labeling data items with one of a given, finite set of classes based on features of the data items. A decision-tree learning algorithm approximates a target concept using a tree representation, where each internal node corresponds to an attribute, and every terminal node corresponds to a class. Artificial Neural Networks are inspired by the way biological neural system works, such as the brain process information. The information processing system is composed of a large number of highly interconnected processing elements (neurons) working together to solve specific problems. Neural networks, just like people, learn by example.

Similar to learning in biological systems, neural network learning involves adjustments to the synaptic connections that exist between the neurons.

Probabilistic models and fuzzy logic offer a very powerful framework for approximate reasoning as it attempts to model the human reasoning process at a cognitive level. Fuzzy systems acquire knowledge from domain experts and this is encoded within the algorithm in terms of the set of if-then rules. Fuzzy systems employ this rule based approach and interpolative reasoning to respond to new inputs. The incorporation and interpretation of knowledge is straight forward, whereas learning and adaptation constitute major problems.

During the last two decades, several adaptive hybrid computational intelligence frameworks have been developed. Many of these approaches use a combination of different knowledge representation schemes, decision making models and learning strategies to solve a computational task. This integration aims at overcoming the limitations of individual techniques through hybridization or the fusion of various techniques.

This book offers a step-by-step introduction (in a chronological order) to the various modern computational intelligence tools used in practical problem solving. Chapters 2-5 and 14-16 deal with different search techniques including informed and uninformed search, heuristic search, minmax, alpha-beta pruning methods, evolutionary algorithms and swarm intelligent techniques. Chapters 6-9 introduce knowledge-based systems and advanced expert systems, which incorporate uncertainty and fuzziness. Chapters 10-13 illustrate different machine learning algorithms including decision trees and artificial neural networks. Finally Chapter 17 presents the fundamentals of hybrid intelligent systems with a focus on neuro-fuzzy systems, evolutionary fuzzy systems and evolutionary neural networks.

The authors would like to thank Springer - Verlag, Germany for the editorial assistance and excellent cooperative collaboration to produce this important scientific work. Last but not the least, we would like to express our gratitude to our scientific collaborators, colleagues and students for their wholehearted support during the last several years. We hope that the reader will share our excitement to present this book and will find this very useful.

Crina Grosan and Ajith Abraham* March 15, 2011

Department of Computer Science,
Faculty of Mathematics and Computer Science
Babes-Bolyai University, Cluj-Napoca,
Kogalniceanu 1, 400084 Cluj – Napoca, Romania
Email: cgrosan@cs.ubbcluj.ro

*Machine Intelligence Research Labs (MIR Labs)
Scientific Network for Innovation and Research Excellence
P.O. Box 2259, Auburn, Washington 98071, USA
Email: ajith.abraham@ieee.org / WWW: http://www.softcomputing.net

Contents

Chapter 1
Evolution of Modern Computational Intelligence

1.1 Introduction

A conventional computational intelligence book introduction starts, many a times, with a history of Artificial Intelligence (AI) and what has been done up to date. This book introduction will try to start with what can be envisaged as future Computer Intelligence.

It is important to have a common definition of AI. Although it might be possible to find many (somehow similar) definitions for *artificial intelligence*, probably the most appropriate one can be the one stating: *creating machines which solve problems in a way which, done by humans, require intelligence.*

A question arises, which can be the interactive question open in any AI course: do we have artificial intelligence? The answer is not a simple one. There are at least two ways of seeing things.

If we look around at the existing *intelligent machines*, we can tell (just to enumerate a few examples) that we have machines, which can interpret handwriting better than humans, we have machines which take decision better than humans do, we have machines which make calculation millions of times faster than humans, we have machines that interpret data, huge amount of data, much faster and accurate than humans, machines which understand language and interpret and transcript it at least at the same level as humans and examples can continue. All these are just natural nowadays, but were hard to believe two decades ago.

On the other hand, if we look at the existing *machines* from a human level intelligence point of view, it is hard to admit that we have a *human level intelligent machine*. Intelligence, on its own, has a broad interpretation sense. If we look at a very intelligent man (usually Einstein is given as reference for an intelligent man) and we look at a person from a remote place, with less or no contact with the civilized world, we see a huge difference (in terms of intelligence) between the two. But if we look at the same person from the remote mountain and at a cockroach, we think that difference between Einstein and our mountain man is nothing compared to difference between mountain man and cockroach. From the evolution of human habilis (first human-like ancestors) 2 million years ago, to homo sapiens 100 000 years ago, to agricultural revolution 10 000 years ago and then to the

C. Grosan and A. Abraham: Intelligent Systems, ISRL 17, pp. 1–11.
springerlink.com

industrial revolution, the human intelligence undergoes significant improvements. But nowadays scientists clearly state that, in a natural way, no improvement can be further performed to the human intelligence; thus, the need of an artificial, more powerful intelligence.

Human level intelligence has not yet been reached if we measure this achievement as passing the Turing test. Turing test - proposed by Alan Turing, a British mathematician – also known as "imitation game" is a simulation in which a judge attempts to distinguish which of two agents, in two separate rooms, is a human and which a computer imitating human from their responses in a wide-ranging conversation of any topic. As a note to the Turing test, there is an annual competition known as Leobner Prize, which awards the best instantiation of the Turing Test.

Nick Bostrom (Future of Humanity Institute, Oxford University) noted that for achieving artificial intelligence, three things are required: hardware, software, and input/output mechanisms. The input/output mechanisms refer to the technology required by a machine to interact with its environment. This is already available in the form of cameras and sensors. We already see robots performing several human like tasks, etc. Thus, this is the simplest required part.

For hardware, we really need to have human level speed and high memory machines. In terms of memory, things are promising. For speed, we still have to wait. Human brain processing power ranges between 100 million MIPS to 100 billion MIPS. (1 MIPS = 1 Million Instructions Per Second). Fastest supercomputer today (as of 2010) is Jaguar, built by the Cray Company and housed at the Oak Ridge National Laboratory in Tennessee; it has a top speed of 1.75 petaflops per second. This means we don't have yet human level computer power even at the range of supercomputers.

The other remaining problem is software. Once we get human intelligence level machines, software will be required. In doing so, one has to understand how human brain works. This is part of the current research these days and at least two main directions follow from there: computational neuroscience and molecular nanotechnologies. Neuroscience is concerned with how the individual components of the human brain work. Research up to date reports good computational models of primary visual cortex. But simulating the whole brain requires enormous computing power. Molecular nanotechnologies work at nanoscale level which is 1 to 100 nanometers – from 1/1,000,000 to 1/10,000 of the thickness of an American dime. Many of the key structures of human nervous systems exist at nanoscale. The major challenge is to use nanomachines to disassemble a frozen or a vitrified human brain.

In parallel with getting the artificial intelligence or human level artificial intelligence, small steps have been performed in terms of algorithms and methodologies, which can be applied to solve simple or more challenging real-world problems. Much of the current research focuses on the principles, theoretical aspects, and design methodology of algorithms gleaned from nature. Examples are artificial neural networks inspired by mammalian neural systems, evolutionary computation inspired by natural selection in biology, simulated annealing inspired by thermodynamics principles and swarm intelligence inspired by collective behavior of insects or micro-organisms etc. interacting locally with their environment causing coherent functional global patterns to emerge. These techniques have found their way in solving real world problems in science, business, technology and commerce. Computational intelligence is a well-established paradigm, where new theories with a sound biological understanding have been evolving. The current experimental systems have many of the characteristics of biological computers (brains in other words) and are beginning to be built to perform a variety of tasks that are difficult or impossible to do with conventional computers.

Although most of the AI related publications consider the birth of AI 15 years after the development of the first electronic computer (in 1941) and 7 years after the development of the invention of the first stored program computer (in 1949), evidences of artificial intelligence can be traced back in ancient Egypt and Greece. Most of AI scientists consider that the Dartmouth summer research project, organized in 1956 by John McCarthy (regarded as father of AI) at Dartmouth College in Hanover, New Hampshire, where the "artificial intelligence term has been coined" was the actual start of the AI as a science.

1.2 Roots of Artificial Intelligence

Logic is considered as being one of the main roots of AI. AI has been heavily influenced by logical ideas. Most members of the AI community would agree that logic has an important role to play in at least some central areas of AI research, and an influential minority considers logic to be the most important factor in developing strategic, fundamental advances. It started as long ago as in 5th century B.C. when Aristotle invented syllogistic logic, the first formal deductive reasoning system. The advances continued with small steps, with famous inventions of this millennium, examples like printing using movable type in the 15th century, invention of clocks as measuring machines in the 15th – 16th century, extension of this mechanism for the creation of other moving objects in the 16th century and so on. Pascal has invented the first mechanical digital calculating machine in 1642. This machine was an adding machine only, but later, in 1671, the German mathematician - philosopher Leibniz designed an improvement of the adding machine such as to incorporate multiplication and division. The machine – known as Step Reckoner – was built in 1973. The 19th century brings the ingenious project of the first computing machine. Looking for a method, which can overcome the high error rate in the calculation of mathematical tables, English mathematician Charles Babbage wished to find a way by which they could be calculated mechanically, removing human sources of error. He began to build Difference Engine, a mechanical device that can perform simple mathematical calculations in 1820 and then the Analytical Engine, which was designed to carry out more complicated calculations. Both devices finally remain just as prototype computing machines. Babbage's work has been later continued by Ada Augusta Lovelace, which remains as the world's first programmer. Babbage's Difference Engine was the first successful automatic calculator.

Another important contribution of 19th century is George Boole's logic theory, also known as Boolean logic or Boolean algebra. Even thought not much appreciated at the time it has been proposed, later after the publication of Boole's ideas, an American logician Charles Sanders Peirce spent more than 20 years modifying and expanding them, realizing the potential for use in electronic circuitry and eventually designing a fundamental electrical logic circuit. Pierce never actually built his theoretical logic circuit, being himself more of a logician than an electrician, but he did introduce Boolean algebra into his university logic philosophy courses.

It was later when one of his students – Claude Shannon, one of the organizers of the Dartmouth conference and one of the pioneers of AI, a Nobel Prize winner among others – made full use of all these ideas.

Gottlob Frege, a German mathematician, essentially reconceived the discipline of logic by constructing a formal system, which in effect, constituted the first predicate calculus (1893-1903). Frege's logic calculus system consisted of a language and an apparatus for proving statements. Predicate calculus system consisted of a set of logical axioms (statements considered to be truths of logic) and a set of rules of inference that lay out the conditions under which certain statements of the language may be correctly inferred from others.

The 20th century brings the most significant contributions to the AI field. If the first half of the century is not that remarkable, starting with the second half results will come in an impressive rhythm. Bertrand Russell, the British logician who pointed out some of the contradictions of Frege's logic during their correspondence and who refined the predicate calculus, revolutionizes formal logic with his three-volume work he co-authored with Alfred North Whitehead, *Principia Mathematica* (1910, 1912, 1913). The mathematical logician Emil Post had his important contributions to computer science in the beginning of the 20th century. In his later work during the early 1920s, Post developed his notion of production systems, developed a unification algorithm, and anticipated the later findings of Gödel, Church, and Turing. Post developed a programming language without thinking of a machine on which it could be implemented. Another important logician of the 20th century is Kurt Gödel, who proved the incompleteness of axioms for arithmetic, as well as the relative consistency of the axiom of choice and continuum hypothesis with the other axioms of set theory.

One of the most significant figures in the development of mathematical logic is Alonzo Church, A Princeton professor and Alan's Turing's supervisor. His book – *Introduction to Mathematical Logic* – published in 1944 comprises some of his earlier remarkable results. The Church-Turing Thesis, a controversial work, came to solve one of the important problems for logicians formulated in the 1930s by David Hilbert: Entscheidungsproblem. The problem asks if there was a mechanical procedure for separating mathematical truths from mathematical falsehoods. Probably the most controversial figure among the mathematicians of the 20th century, the British mathematician Alan Turing is well known as the founder of some fundamental principles, which are required to prove the evidence of artificial intelligence. The famous Turing test remains until today the biggest challenge for the existence of artificial intelligence. His famous work Computing Machinery and Intelligence has been published in 1950, soon after the development of the first electronic digital computer and the first stored computer program.

Built in 1943-1945 at the Moore School of the University of Pennsylvania for the War effort by John Mauchly and J. Presper Eckert, the Electronic Numerical Integrator And Computer (ENIAC) was the first general-purpose electronic digital computer. It was 150 feet wide with 20 banks of flashing lights. Even thought it was meant to help in the WWII, ENIAC has not been delivered to the Army until just after the end of the war.

The ENIAC was not a stored-program computer; it is described by David Alan Grier as a collection of electronic adding machines and other arithmetic units, which were originally controlled by a web of large electrical cables. EDVAC (Electronic Discrete Variable Automatic Computer) was the earliest electronic computer. Unlike its predecessor the ENIAC, it was binary rather than decimal, and was a stored program machine.

The paper by Warren McCulloch, a neuroscientist, and Walter Pitts, a logician, "*A logical calculus of the ideas immanent in nervous activity*" published in 1943 is regarded as the start point of two fields of research: the theory of finite-state machines as a model of computation and the field of artificial neural networks. McCulloch and Pitts tried to understand how the brain could produce highly complex patterns by using many basic cells that are connected together. They gave a highly simplified model of a brain cell – a neuron – in their paper. The McCulloch and Pitts model of a neuron has made an important contribution to the development of artificial neural networks. But their neuron model had limitations. Additional features were added, which allowed the neuron to learn and one of the next major development in neural networks was the concept of a perceptron, which was introduced by Frank Rosenblatt in 1958. Another paper published in the same 1943 – "Behavior, Purpose and Teleology" – by Arturo Rosenblueth, Norbert Wiener and Julian Bigelow set the bases for the new science of Cybernetics.

The problem solving has been a central challenge for computer scientists and for the AI community too. AI scientists came with their own problems and with their own methods of solving them. George Polya, a Hungarian born American mathematician, suggests in his very famous book – *How to solve it* – four main steps to approach a problem: understand the problem, devise a plan, carry on with the plan and look back. Problem solving remains a central idea of AI and a *How to solve it* modern version using heuristics has been published in 2004 by Zbigniew Michalewicz and David Fogel.

A few important scientific results preceded the Dartmouth Conference. Among them are the following: Norbert Wiener's results in cybernetics (he is among the first scientists who coined the term cybernetics) and also in the feedback theory as if all intelligent behavior is the results of feedback mechanisms. This discovery had a huge influence on the initial development of AI. *The logic theorist* developed between 1955-1956 by Allen Newell (researcher in computer science and cognitive psychology at Carnegie Mellon University), J. Clifford Shaw (a system programmer who is considered the father of the JOSS language) and Herbert Simon (originally a political scientist who also won the Nobel Prize in economics in 1978 and has been awarded the Turing Award along with Allen Newel in 1975 for their basic contributions to artificial intelligence and the psychology of human cognition) is considered as being the first AI program. The theorem proving can be reduced to search. The problem is represented as a tree model and the program will attempt to find a proof by searching the tree and by selecting the branch that will result in the correct proof. The program succeeded in proving thirty-eight of the first fifty-two theorems presented there, but much more importantly, the program found a proof for one theorem which was more elegant than the one provided by Russell and Whitehead (in *Principia Mathematica*). The impact of *The Logic Theorist* had in the development of AI made it a stepping-stone in the evolution of the AI field.

Although the enthusiasm of organizing the school at Dartmouth College was really huge and the expectations were great, as McCarthy he noted in the 1955 announcement of the conference:

"We propose that a 2 month, 10 man study of artificial intelligence be carried out during the summer of 1956 at Dartmouth College in Hanover, New Hampshire. The study is to proceed on the basis of the conjecture that every aspect of learning or any other feature of intelligence can in principle be so precisely described that a machine can be made to simulate it. An attempt will be made to find how to make machines use language, form abstractions and concepts, solve kinds of problems now reserved for humans, and improve themselves. We think that a significant advance can be made in one or more of these problems if a carefully selected group of scientists work on it together for a summer."

the results of the meeting were not really spectacular. The conference was organized by John McCarthy and formally proposed by John McCarthy, Marvin Minsky, Nathaniel Rochester and Claude Shannon with the scope of bringing together American scientists working on artificial intelligence. There were a total of 10 participants at the Dartmouth Summer Research Conference on Artificial Intelligence. John McCarthy (who was teaching at Dartmouth at that time and after moved to Stanford University; also won Turing Award in 1971), Marvin Minsky (who also won the Turing award in 1969), Trenchard More (from Princeton), Ray Solomonoff (the inventor of algorithmic probability and an originator of the branch of artificial intelligence based on machine learning, prediction and probability), Oliver Selfridge (graduate student of Norbert Wiener's at MIT, (but did not write up his doctoral research and never earned a Ph.D.) and a supervisor of Marvin Minsky), Claude Shannon (known for his contributions in information theory and cryptography during the World War II while he was at Bell Labs; among other contributions he made a chess playing computer program and made a fortune by applying game theory in Las Vegas games and in stock market), Nathaniel Rochester (who designed the IBM 701 the first general purpose, mass produced computer and wrote the first symbolic assembler), Arthur Samuel (who developed the alpha-beta tree idea and proposed a Checkers-playing program (on IBM's first commercial computer, the IBM 701) that appears to be the world's first self-learning program; 1962 his program beat a state champion), Herbert Simon and Allen Newell.

1.3 Modern Artificial Intelligence

The Dartmouth Conference opened the era of new and most significant advances in the AI field. Advances continued in a much faster rhythm than before. Technology was also advancing and this gave more room to more difficult and ambitious projects. AI research centers began forming at MIT and Carnegie Mellon University. The challenges were to create systems that could efficiently solve problems by limiting the search such as The Logic Theorist, and making systems that could learn by themselves. Newel, Shaw and Simon, the authors of The Logic Theorist, wanted programs that solved problems in the same ways as humans do. They developed the General Problem Solver (GPS) in 1957, which is basically a computer program intended to work as a universal problem solver machine. Any formalized symbolic problem can be solved, in principle, by GPS, for instance theorems proof, geometric problems and chess playing.

Using a means-end-analysis approach, GPS would divide the overall goal into sub-goals and attempt to solve each of those. The program was implemented in the low-level IPL programming language. While GPS solved simple problems such as the Towers of Hanoi that could be sufficiently formalized, it could not solve any real-world problems because search was easily lost in the combinatorial explosion of intermediate states.

McCulloch and Pitts' neuron was further developed in 1957 by Frank Rosenblatt at the Cornell Aeronautical Laboratory. Rosenblatt's perceptron was able to recognize patterns of similarity between new data and data it has already seen in a feed-forward model that demonstrated a primitive type of learning or trainability. His work was highly influential in the development of later multi-layered neural networks. Soon after the development of the perceptron, many research groups in the United States were studying perceptrons. Essentially the perceptron is a McCulloch and Pitts neuron where the inputs are first passed through some "preprocessors," which are called association units. These association units detect the presence of certain specific features in the inputs. In fact, as the name suggests, a perceptron was intended to be a pattern recognition device, and the association units correspond to feature or pattern detectors.

In 1958, John McCarthy showed how, given a handful of simple operators and a notation for functions, someone can build a whole programming language. He called this language LISP, for "List Processing," because one of his key ideas was to use a simple data structure called a *list* for both code and data. LISP is the second-oldest high-level programming language in widespread use today; only Fortran is older. LISP was heavy on computer power and it became more useful in 1970s with the existing technology.

In the late 50's and early 60's Margaret Masterman and colleagues from Cambridge design semantic nets for machine translation. A semantic net is a graph, which represents semantic relations among concepts. Silvio Ceccato also developed in 1961 correlational nets, which were based on 56 different relations, including subtype, instance, part-whole, case relations, kinship relations, and various kinds of attributes. He used the correlations as patterns for guiding a parser and resolving syntactic ambiguities. Masterman and her team developed a list of 100 primitive concept types, such as Folk, Stuff, Thing, Do, and Be. In terms of those primitives, her group defined a conceptual dictionary of 15,000 entries. She organized the concept types into a lattice, which permits inheritance from multiple supertypes.

The first industrial robot was installed at General Motors in 1961. It has been developed at Unimation Inc., the first robotic company founded in 1956 by Joseph F. Engelberger (a physicist, engineer and entrepreneur who is referred to as the "Father of Robotics"). Over the next two decades, the Japanese took the lead by investing heavily in robots to replace people performing certain tasks.

In 1963, John Alan Robinson, philosopher, mathematician and computer scientist invented resolution, a single inference method for first order logic. Resolution is a refutation method operating on clauses containing function symbols, universally quantified variables and constants. The essence of the resolution method is that it searches for local evidence of unsatisfiability in the form of a pair of clauses, one containing a literal and the other its complement (negation). Resolution and unification have since been incorporated in many automated theorem-proving systems and are the basis for the inference mechanisms used in logic programming and the programming language Prolog.

In 1963, DARPA (Defense Advanced Research Project Agency) and MIT signed a 2.2 million dollar grant to be used in researching artificial intelligence (to ensure that the US will stay ahead of the Soviet Union in technological advancements).

In 1966, Joseph Weizenbaum form MIT described in *Communications of the ACM*, ELIZA, one of the first programs that attempted to communicate in natural language. In only about 200 lines of computer code, Eliza models the behavior of a psychiatrist (the Rogerian therapist). ELIZA has almost no intelligence; it uses tricks like string substitution and canned responses based on keywords. The illusion of intelligence works best, however, if you limit your conversation to talking about yourself and your life.

Some of the more well-known AI projects that followed the General Problem Solver in the late 60's included: STUDENT, by Daniel G. Bobrow, which could solve algebra word problems and reportedly did well on high school mach tests, ANALOGY, by Thomas G. Evans (written as part of his PhD work at MIT), which solved IQ-test geometric analogy problems, Bert Raphael's MIT dissertation on the SIT program that demonstrates the power of logical representation of knowledge for question-answering systems and Terry Winograd's SHRDLU, which demonstrated the ability of computers to understand English sentences in a restricted world of children's blocks (such as a limited number of geometric shapes).

Another advancement in the 1970's was the advent of the expert system. Expert systems predict the probability of a solution under set conditions. Due to the large storage capacity of computers at the time, expert systems had the potential to interpret statistics, to formulate rules. The applications for real practical problems were extensive, and over the course of ten years, expert systems had been introduced to forecast the stock market, medicine and pharmacy, aiding doctors with the ability to diagnose disease, and instruct miners to promising mineral locations.

This was made possible because of the systems ability to store conditional rules, and storage of information.

One of the earliest expert systems was DENDRAL, developed at Stanford University. DENDRAL was designed to analyze mass spectra. DENDRAL did contain rules and consists of two sub-programs, Heuristic Dendral and Meta-Dendral and its developers believed that it can compete and experienced chemist (was marketed commercially in the United States). The program was used both in industry and academia. MYCIN, another expert system developed at Stanford University too has been used to diagnose blood infections and recommend treatments, given lab data about tests on cultures taken from the patient. Although never put to practical use, MYCIN showed to the world the possibility of replacement of a medical professional by an expert system. PROSPECTOR has been developed by NASA. It takes geological information about rock formations, chemical content, etc, and advises on whether there were likely to be exploitable mineral deposits nearby. Popular accounts of AI say that Prospector (in 1978-ish) discovered a hundred-million-dollar deposit of molybdenum.

These are only some of the first expert systems. Many more have been proposed, including applications in all major domains such as medicine, agriculture, engineering, etc. Rule-based systems are a relatively simple model that can be adapted to any number of problems. A general form of expert systems is an expert system shell. An expert system shell is actually an expert system whose knowledge is removed. Thus, the user can just add its own knowledge in the form of rules and provide information to solve the problem. Expert system shells are commercial versions of the expert systems.

The programming language PROLOG was born of a project aimed not at producing a programming language but at processing natural languages; in this case, French. The project gave rise to a preliminary version of PROLOG at the end of 1971 and a more definitive version at the end of 1972 at Marseille by Alain Colmerauer and Philippe Roussel. The name Prolog stands for Progammation en Logique in French and was coined by Philippe Roussel. It can be said that Prolog was the result of a combination between natural language processing and automated theorem-proving.

It was in 1964 when the new theory of *fuzzy logic*, a different kind of logic, has been proposed by Lotfi Zadeh at University of California (Berkeley). The concept was not much used at that time in the United States, but in the 70's the Japanese started using fuzzy ideas incorporated in electronic devices. The fuzzy mechanisms were first developed for years in Japan before the rest of the world started using them. It took a long time until fuzzy logic got accepted even though it fascinated some people right from the beginning. Besides engineers, philosophers, psychologists, and sociologists soon became interested in applying fuzzy logic into their sciences. In the year 1987, the first subway system was built which worked with a fuzzy logic-based automatic train operation control system in Japan. It was a big success and resulted in a fuzzy boom. Universities as well as industries got interested in developing the new ideas. Today, almost every intelligent machine has fuzzy logic technology inside it.

Neural networks remained for years only at the stage of a single neuron (perceptron) since their discoveries in the 60's. Due to lack of machine power required for their computational tasks, neural network research didn't progress much until in the mid 80's.

In the early 1980's, researchers showed renewed interest in neural networks. Recent work includes Boltzmann machines, Hopfield nets, competitive learning models, multilayer networks, and adaptive resonance theory models. With the backpropagation learning algorithm (and later on with other learning algorithms) neural networks became widely used. Neural networks are adequately used for data classification and modeling through a learning process.

Another important milestone in the field of AI is the development of Evolutionary Computation. Under the name evolutionary computation, four major domains are covered: genetic and evolutionary algorithms, evolution strategies, evolutionary programming and genetic programming. Although some work in this field can be traced back to the late 1950's, the field remained relatively unknown to the broader scientific community for almost three decades. This was largely due to the lack of available powerful computer platforms at that time. The fundamental work of John Holland, Ingo Rechenberg, Hans-Paul Schwefel, Laurence Fogel and John Koza represents the base of the evolutionary computation, as we know it today. Holland introduced genetic algorithms, probably the most studied and further developed branch of evolutionary computation, with remarkable application in optimization and search problems. Ingo Rechenberg and Hans-Paul Schwefel contributed to the development of evolution strategies. Fogel proposed evolutionary programming and Koza is known for his contributions to the genetic programming methods. All these methods have been (and still continue to be) further developed and improved, with hundreds of thousands of publications related to this subject.

Swarm intelligence is a method, which allows decentralized, self-organized systems with relative simple single software agents to solve complex problems and tasks together, which neither agent could do alone. Examples include ants (from which the Ant Colony Optimization system has derived), which leave pheromone trails for others to follow, and go as far as swarm-robots being able to symbiotically share computing resources, birds and fish (from which the Particle Swarm Optimization algorithm developed), bacteria (which gave birth to Bacterial foraging optimization algorithm), Multi-Agent Systems are systems of similar, possibly specialized entities, which are able to collectively solve problems and so on. Swarm robotics is a comparative young field of science, focusing on the development of limited single robots which are able to perform direct and indirect communication with each other and to create dynamic horizontal systems with a collective behavior.

Apart from all these, some progress has been registered in computer games playing. For checkers game, there exist Chinook. After 40-year-reign of human world champion Marion Tinsley, Chinook defeated it in 1994. Chinook used a pre-computed end game database defining perfect play for all positions involving 8 or fewer pieces on the board, a total of 444 billion positions. For Chess game, there exists Deep Blue. Deep Blue defeated human world champion Garry

Kasparov in a six-game match in 1997. Deep Blue searches 200 million positions per second, uses very sophisticated evaluation, and undisclosed methods for extending some lines of search up to 40 ply. In Othello game, human champions refuse to compete against computers, who are too good. Various kind of robots have been developed in the last century (and many continue to be developed today) to help and replace human work in hard and improper conditions.

1.4 Metamodern AI

In the new millennium, the trends in AI remain almost the same but more courageous, with more enthusiasm and by far with more advanced technologies. Apart from the new developments in terms of concepts and methods, ensembles of existing paradigms and hybrid intelligent approaches play an important role. Interdisciplinary approaches towards problem solving are another key idea. But involving experts from multiple domains such as engineering, biology, computer science and cognitive sciences, the progress is much faster. For example, there is a specific interdisciplinary trend, NBIC, which stands for Nano-Bio-Info-Cogno, whose ideas and research plans sound very promising. Universal Artificial Intelligence, idea proposed by Juergen Schmidhuber, comes with universal reinforcement learners and decision makers.

A more general Idea is that of Singularity, a concept originally coined by Vernor Vinge and sustained by Ray Kurzweil and other researchers of the Singularity Institute for Artificial Intelligence. The Singularity is the technological creation of smarter-than-human intelligence and it is most likely to happen next the machine will reach human level artificial intelligence.

The book offers a gentle introduction to modern computational intelligence field starting with the first and most simple ways to approach problem solving (some standard search techniques) and then continues with other methods in a chronological order of their development. The contents of this book would be beneficial for various disciplines and is structured for a larger audience, from medical doctors, researchers / scientists / students / academicians and engineers from the industry.

Chapter 2
Problem Solving by Search

2.1 Introduction

An important aspect of intelligence is *goal-based* problem solving. Several problems can be formulated as finding a sequence of actions that lead to a desirable goal. Each action changes the *state* and the aim is to find the sequence of actions and states that lead from the initial state to a final (goal) state.

Searching through a state space involves the following:

- a set of states;
- operators;
- a start or initial state;
- a test to check for goal state.

A well-defined problem can be described by[1][2][3]:

- *Initial state*;
- Operator or *successor function* - for any state x returns $s(x)$, the set of states reachable from x with one action;
- *State space* - all states reachable from initial state by any sequence of actions;
- *Path* - sequence through state space;
- *Path cost* - function that assigns a cost to a path. Cost of a path is the sum of costs of individual actions along the path;
- *Goal test* - test to determine if at goal state.

2.2 What Is Search?

Search is the systematic examination of states to find a path from the start state to the goal state.

The *search space* consists of the set of possible states, together with operators defining their connectivity.

The solution provided by a search algorithm is a path from the initial state to a state that satisfies the goal test[4][6][7][8][9][11][12][18][20].

C. Grosan and A. Abraham: Intelligent Systems, ISRL 17, pp. 13–52.
springerlink.com

In real life situations search algorithms are usually employed when there is lack of knowledge and the problems cannot be solved in a better way.

Search techniques fall into three groups:

- methods which find *any* start - goal path;
- methods which find the *best* path;
- search methods in the face of adversaries.

The hardship in problem solving is to decide the states and the operator or successor function. Figures 1-3 illustrate some examples depicting the different modeling aspects of the search process.

Example 1: 8-puzzle
In the 8-puzzle example depicted in Figure 2.1 we have[10][26][33]:

States: location of blank and location of the 8 tiles
Operator (*successor*): blank moves left, right, up and down
Goal: match the state given by the Goal state
Path Cost: each step has the cost 1; total cost is considered as being the length of path.

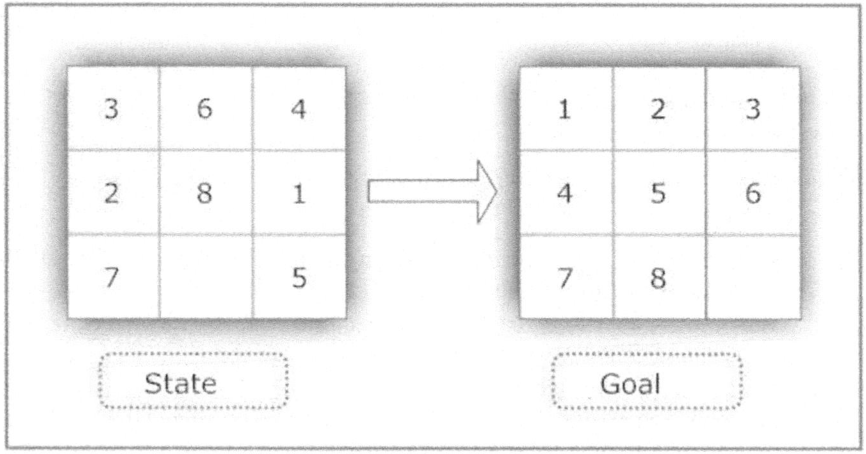

Fig. 2.1 8-puzzle example.

Example 2: N - Queens
The N-Queens problem requires arranging N queens on an N × N (chess) board such as the queens do not attack each other. This problem may be defined as:

States: 0 to N queens arranged on the chess board
Operator (*successor*): place a queen on an empty square

Goal: match a state with N queens on the chess board and no attacks among them (an example of a 5-queen goal state is given in Figure 2.2).
Path Cost: 0.

Fig. 2.2 N-queen (N=5) problem example.

Example 3 - Missionaries and Cannibals Problem

The problem can be stated as follows: three missionaries and three cannibals are on the left bank of a river. They have to cross over to the right bank using a boat that can only carry two at a time. The number of cannibals must never exceed the number of missionaries on any of the banks. The problem is to find a way to get all missionaries and cannibals to the other side, without leaving at any time and place a group of missionaries outnumbered by the cannibals.

For this problem we define:
State: The state consists of:

- the number of missionaries on the left bank,
- the number of cannibals on the left bank,
- the side of the bank the boat is on.

Operator: A move is represented by the number of missionaries and the number of cannibals taken in the boat at one time. Since the boat can carry no more than two people at once, there are 5 possible combinations:

```
(2 Missionaries, 0 Cannibals)
(1 Missionary, 0 Cannibals)
(1 Missionary, 1 Cannibal)
(0 Missionary, 1 Cannibal)
(0 Missionary, 2 Cannibals)
```

Goal: (0, 0, right)
Path cost: number of crossings.

2.3 Tree Based Search

The set of all paths within a state-space can be viewed as a graph of nodes, which are connected by links. If all possible paths are traced out through the graph, and the paths are terminated before they return to nodes already visited (cycles) on that path, a search tree is obtained. Like graphs, trees have nodes, but they are linked by branches. The start node is called the root and nodes at the other ends are leaves. Nodes have generations of descendents. The first generations are children. They have a single parent node, and the list of nodes back to the root is their ancestry. A node and its descendents form a subtree of the node's parent. If a node's subtrees are unexplored or only partially explored, the node is open, otherwise it is closed. If all nodes have the same number of children, this number is the branching factor[27].

2.3.1 Terminology

- *Root node*: represents the node the search starts from;
- *Leaf node*: a terminal node in the search tree having no children;
- *Ancestor/descendant*: node A is an ancestor of node B if either A is B's parent or A is an ancestor of the parent of B. If A is an ancestor of B, B is said to be a descendant of A;
- *Branching factor*: the maximum number of children of a non-leaf node in the search tree;
- *Path*: a path in the search tree represents complete path if it begins with the start node and ends with a goal node. Otherwise it is a partial path.

A node in the tree may be viewed as a data structure containing the following elements:

- a state description;
- a pointer to the parent of the node;
- depth of the node;
- the operator that generated this node;
- cost of the path (sum of operator costs) obtained from the initial (start) state.

It is advisable *not* to produce complete physical trees in memory, but rather explore as little of the virtual tree looking for root-goal paths [1][5].

State space is explored by generating successors of the already explored states. Every state is evaluated in order to see whether this is the goal state. A disadvantage of the tree search is that it can end up repeatedly visiting the same node. A solution to this is to store all the visited nodes but this will require a lot of memory resources. A more general approach is the graph search.

The nodes that the algorithm has generated so far during the search process are kept in a data structure called OPEN or *fringe*. Initially only the start node (the initial state) is in OPEN.

The search starts with the root node. The algorithm picks a node from OPEN for expanding and generates all the children of the node. Expanding a node from OPEN results in a closed node. Some search algorithms keep track of the closed nodes also in a data structure called CLOSED[29].

The search problem will return a solution or a path to a goal node. Finding a path is important in problems like path finding, *n*-puzzle problems, traveling salesman problem and other such problems. There are also problems like the N-queens and cryptarithmetic problem for which the path to the solution is not important. For such problems the search problem needs to return the goal state only.

An Example of search tree for the 8-puzzle problem is depicted in Figure 2.3.

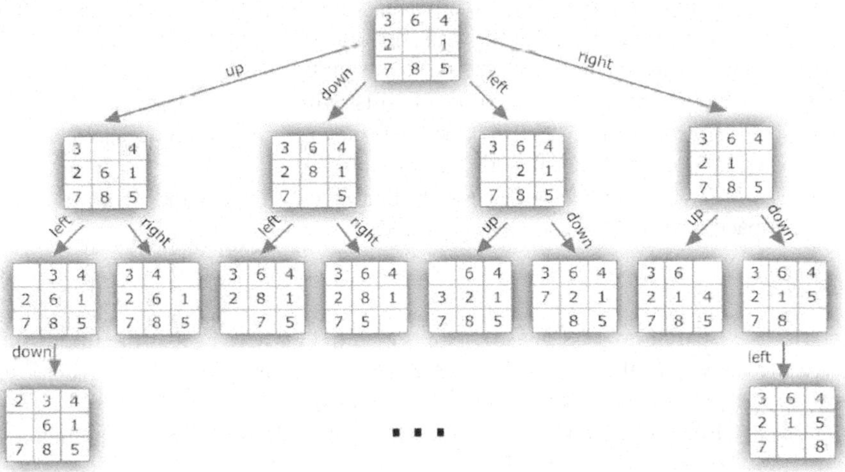

Fig. 2.3 Example of search tree for the 8-puzzle problem.

2.4 Graph Search

If the search space is not a tree, but a graph, the search tree may contain different nodes corresponding to the same state. The state space can be considered a graph G(V, E), where V is the set of nodes and E is a set of vertices, which are directed from a node to another node. Each node contains information including:

- a state description;
- node's parent;
- the operator that generated the node from that parent;
- other information.

Each vertex corresponds to an instance of one of the operators. When the operator is applied to the state associated with the arc's source node, then the resulting state is the state associated with the vertex's destination node. Each vertex has a positive cost associated with it corresponding to the cost of the operator.

Each node has a set of successor nodes corresponding to all of the operators that may be applied at the source node's state. Expanding a node means generating all the successor nodes of it and add them and their associated vertices to the state-space graph.

We have the following correspondence:

- *Initial state*: One or more nodes are designated as start nodes.
- *State space* – Initially, a starting node S is considered and V={S}. Then S is expanded and its generated successors (nodes and vertices) are added to V and E respectively. This process continues until a goal node is found;
- *Path* - each node represents a partial solution path from the start node to the given node. In general, from this node there are many possible paths (and therefore solutions) that have this partial path as a prefix;
- *Path cost*: the sum of the vertices costs on the solution path;
- *Goal test* – test applied to a state to determine if its associated node is a goal node and satisfies all goal conditions;
- *Solution*: a sequence of operators that is associated with a *path* in a state space from a start node to a goal node.

Remarks

(i) Search process constructs a search tree, where root is the initial state and all the leaf nodes are either nodes that have not yet been expanded or nodes that have no successors.

(ii) Because of loops, search tree may be infinite even for small search spaces.

The general search structure is given in Algorithm 2.1. *Problem* describes the start state, operators, goal test and costs. *Strategy* is what differentiates different search algorithms; based on it, several search methods exist. The result of the algorithm is either a valid solution or failure.

Algorithm 2.1

```
General_search (problem, strategy)
Use initial state of the problem to initialize the
search tree
Loop
        If there are no nodes to expand
        Then return failure;
        Based on strategy select a node for extension;
        Apply goal test;
```

```
        If the node is a goal state
        then return the solution
        Else expand the node and add the resulting
        nodes and vertices to the search tree.
End;
```

Remarks

(i) It can be observed that the goal test is not done for each node when it is generated;
(ii) The algorithm does not have any mechanism to detect loops.

2.5 Search Methods Classification

Search methods can be classified into two major categories:

- *uninformed* (blind) search;
- *informed* (heuristic) search.

Some of the well known algorithms, which can be found under the umbrella of uninformed search are:

- breadth-first;
- depth-first;
- iterative deepening depth-first;
- bidirectional;
- branch and bound (or uniform cost search).

In the category of informed search we can find:

- hill climbing;
- beam search
- greedy;
- best first search
- heuristics.

This Chapter treats in detail some of the well known uniformed search methods. The following chapter deals with some of the most important informed search algorithms.

2.6 Uninformed Search Methods

In this section we will talk about blind search or uninformed search that does not use any extra information about the problem domain.

2.6.1 Breadth First Search

In the breadth first search (BFS) algorithm, each state at a given layer is expanded before moving to the next layer of states. This way always the node nearest the root can be cached. This is important if the tree is unbalanced, but is wasteful if all the goal nodes are at similar levels[11][12][13][17].

As observed from Figure 2.4, the root state is expanded to find the states L1, L2, and L3. Since there were no more states at this level, the state L1 is picked and expanded and the states L1.1, L1.2 and L1.3 were produced. There are still two states remaining – L2 and L3 – before expanding the states L1.1-L1.3. So, L2 and L3 are expanded next. If there are no more states at the current level to expand, the first node from the next level is expanded and this is carried on until a solution is found. The process is described in detail in Figure 2.5 for the first 4 steps (a)-(d) and the final configuration obtained at the end of the search process is provided in (e). The colored node is the one that is expanded next. The breath first search algorithm is presented in Algorithm 2.2.

Algorithm 2.2. Breadth first search

```
Step 1. Form a queue Q and set it to the initial state
(for example, the Root).
Step 2. Until the Q is empty or the goal state is found
do:
            Step 2.1 Determine if the first element in
            the Q is the goal.
            Step 2.2 If it is not
            Step 2.2.1 remove the first element in Q.
                  Step 2.2.2 Apply the rule to generate
                  new state(s) (successor states).
                  Step 2.2.3 If the new state is the
                  goal state quit and return this state
                  Step 2.2.4 Otherwise add the new
                  state to the end of the queue.
Step 3. If the goal is reached, success; else failure.
```

The breadth first search algorithm has exponential time and space complexity. The memory requirements are actually one of the biggest problems. Russel and Norvig [2] illustrate an interesting example: consider a complete search tree of depth d varying from 0 to 14 and the branching factor 10. The complexity is $O(10^{d+1})$) nodes. If breadth first search expands 10,000 nodes per second and each node uses 100 bytes of storage, then this will only take 11 seconds for depth 4 but this will increase to 31 hours for depth 8, 35 years for the depth 12 and will take 3,500 years to run in the worst case for depth 14 using 11,100 terabytes of memory.

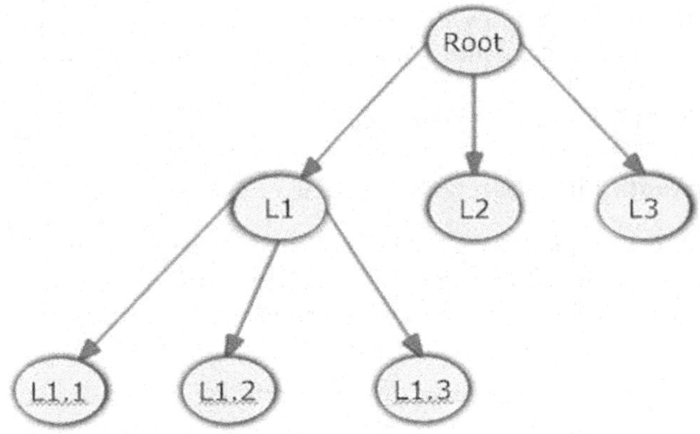

Fig. 2.4 Example of states layers in breadth first search.

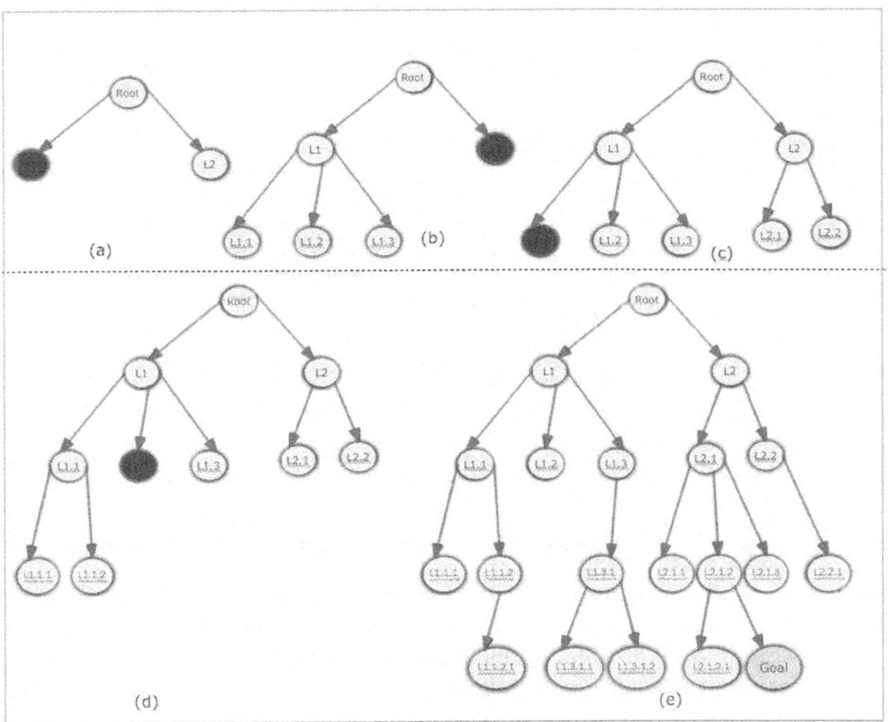

Fig. 2.5 Example of states expansion using breadth first search.

Hence, the conclusion is that the breadth first search algorithm cannot be effectively used unless the search space is quite small. The advantage of the breadth first search is that it finds the path of minimal length to the goal, but it has the disadvantage of requiring the generation and storage of a tree whose size is exponential to the depth of the shallowest goal node.

Example 1: Breadth First Search for 8-puzzle

A simple 8-puzzle example for which the goal state is reached in the third layer of expanded states is presented in Figure 2.6. The goal state is the one in which the blank is on the upper lest corner and the tails are arranged in ascending order.

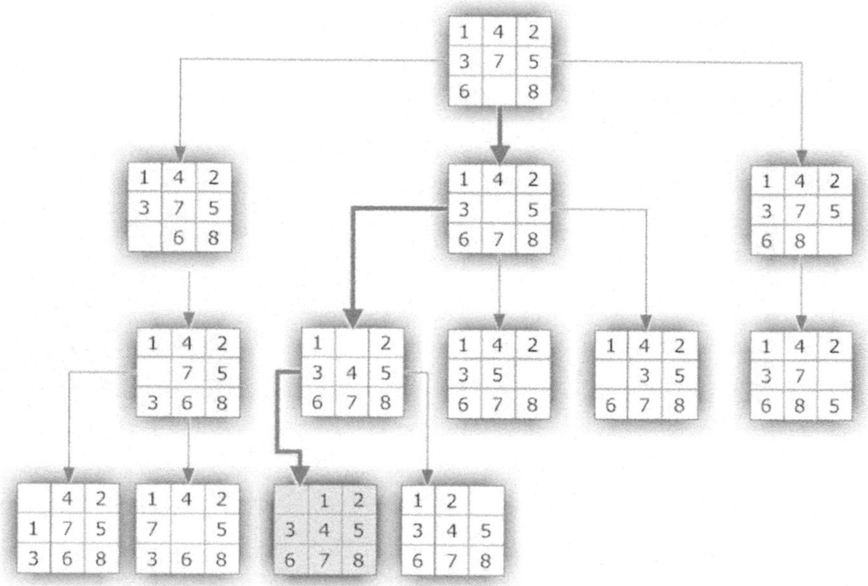

Fig. 2.6 Example of breadth first search for the 8-puzzle problem.

Example 2: Breadth First Search for Missionaries and Cannibals Problem

In order to simplify the explanations, the following notations are used: M for missionaries, C for cannibals and L and R representing the left or right side the boat is in. A graphical illustration of the problem is given in Figure 2.7.

A state can be represented in the following form:

(Left (#M, #C), Boat, Right(#M, #C),

which represents the number of missionaries and cannibals on the left side, the side the boat is, and the number of missionaries and cannibals on the right side respectively. Since the number of missionaries and cannibals should always be 3 on both river banks, we can simplify the notation of the state: (#M, #C, L/R). So,

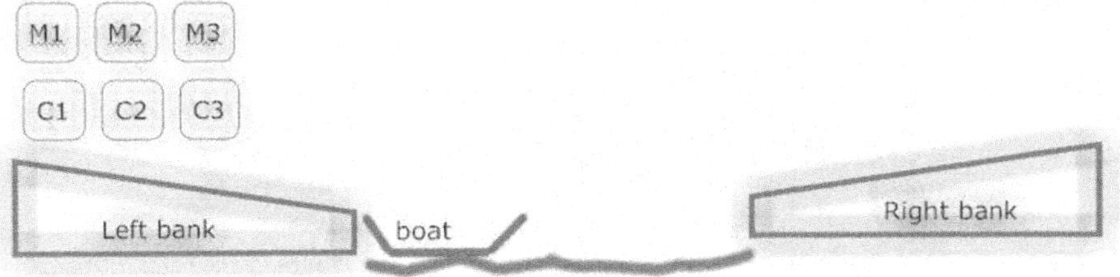

Fig. 2.7 The missionaries and cannibals problem illustration.

the state represents how many people are on the left side of the river and whether the boat is on the left or right side.

There are five possible actions from any state:

- one missionary moves to the right bank;
- two missionaries move to the right bank;
- one cannibal moves to the right bank;
- two cannibals move to the right bank;
- one cannibal and one missionary move to the right bank.

The two important things to note are that each action results in a boat movement and there are at most five actions. Note that, starting from the initial state, 2 of the 5 actions violate the constraints of the problem (the cannibals outnumber the missionaries as in the case of the first two actions).

The search space for this problem consists on 32 states, which are represented in Figure 2.8. The shadowed states correspond to situations in which the problems constraints are violated.

LEFT BANK				RIGHT BANK			
0M 0C L	1M 0C L	2M 0C L	3M 0C L	0M 0C R	1M 0C R	2M 0C R	3M 0C R
0M 1C L	1M 1C L	2M 1C L	3M 1C L	0M 1C R	1M 1C R	2M 1C R	3M 1C R
0M 2C L	1M 2C L	2M 2C L	3M 2C L	0M 2C R	1M 2C R	2M 2C R	3M 2C R
0M 3C L	1M 3C L	2M 3C L	3M 3C L	0M 3C R	1M 3C R	2M 3C R	3M 3C R

Fig. 2.8 The State-space for the missionaries and cannibals problem.

An Example of a solution for this problem is presented in Figure 2.9. It is evident how the situation changes on both sides and also it may be also used to deduce what the boat will be carrying on both directions.

Solution

Left bank	Boat	Right bank
3M 3C	L	0M 0C
3M 1C	R	0M 2C
3M 2C	L	0M 1C
3M 0C	R	0M 3C
3M 1C	L	0M 2C
1M 1C	R	2M 2C
2M 2C	L	1M 1C
0M 2C	R	3M 1C
0M 3C	L	3M 0C
0M 1C	R	3M 2C
0M 2C	L	3M 1C
0M 0C	R	3M 3C

Fig. 2.9 A solution for the missionaries and cannibals problem.

2.6.2 Depth First Search

The depth first search algorithm is almost identical with the breadth first search algorithm with the main difference in Step 2.2.4 where the children is placed in the beginning of the queue compared to the end of the queue in the case of breadth first search (see Algorithm 2.3).

The queue here may be replaced with a stack. Nodes are popped from the front of the queue and new nodes are pushed to the front. The strategy always chooses to expand one of the nodes that is at the deepest level on the search tree. It only expands nodes on the queue that are at the shallower level if the search has reached a dead-end at the deepest level[14][16][19][35].

A path is expanded as much as possible until it reaches a goal node or can be expanded no more prior to expanding other paths.

Algorithm 2.3. Depth first search

```
Step 1. Form a queue Q and set it to the initial state
(for example, the Root).
Step 2. Until the Q is empty or the goal state is found
do:
            Step 2.1 Determine if the first element in
            the Q is the goal.
            Step 2.2 If it is not
                  Step 2.2.1 Remove the first element
                  in Q.
                  Step 2.2.2 Apply the rule to generate
                  new state(s) (successor states).
                  Step 2.2.3 If the new state is the
                  goal state quit and return this state
                  Step 2.2.4 Otherwise add the new
                  state to the beginning of the queue.
Step 3. If the goal is reached, success; else failure.
```

The difference between the way in which breadth first search and depth first search expansion can be observed by comparing Figures 2.4 and 2.10. The search performed by breadth first search in Figure 2.5 can be compared with the search performed for the same data by depth first search in Figure 2.11.

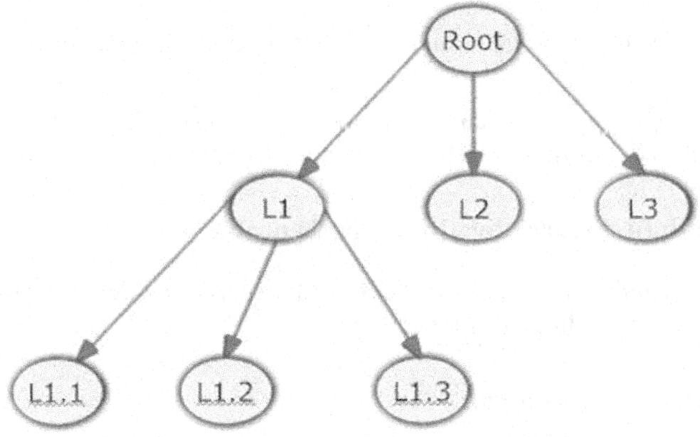

Fig. 2.10 Example of depth first search expansion.

Depth first search algorithm takes exponential time. If *d* is the maximum depth of a node in the search space, the worst case algorithm's time complexity is $O(b^d)$. However the space taken is linear for the depth of the search tree and is given by O(bd).

The time taken by the algorithm is related to the maximum depth of the search tree. If the search tree has infinite depth, the algorithm may not terminate. This can happen in situations where the search space is infinite or if the search space

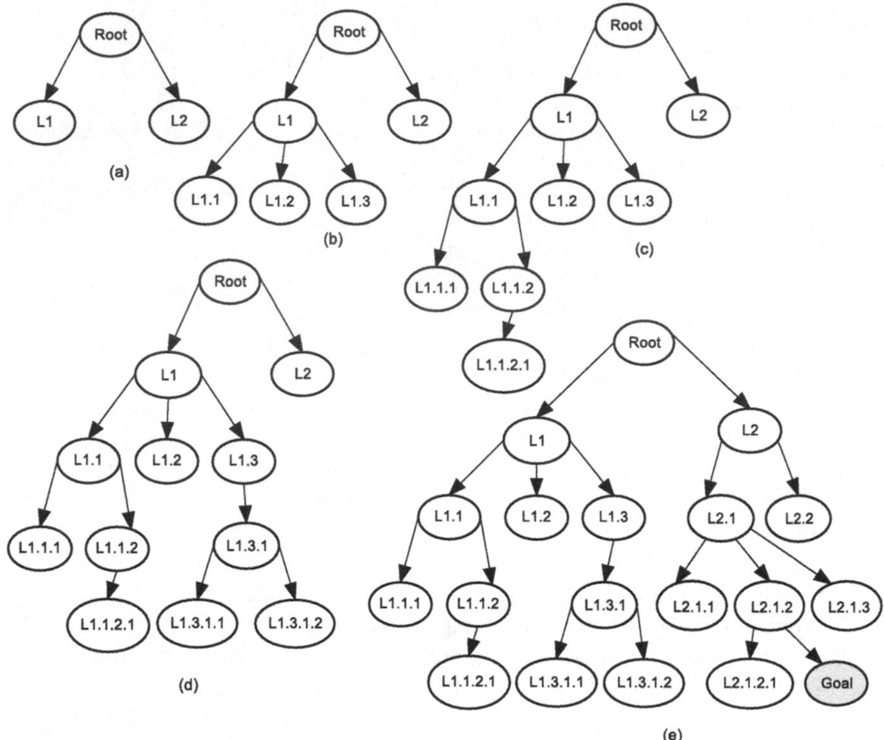

Fig. 2.11 An example of the application of depth first search for reaching a goal state.

contains cycles. The latter case can be handled by checking for cycles in the algorithm. This makes depth first search not to be complete.

Example: Depth first search for the 8- puzzle problem

Figure 2.12 presents the application of depth first search for the 8-puzzle problem (same example as in the breadth first search).

2.6.3 Backtracking Search

Backtracking search is a depth-first search that chooses values for one variable at a time and backtracks when a variable has no legal values left to assign. It uses less memory than depth first search because only one successor is generated at a time but is still not an optimal search technique.

The backtracking search applied for the same 8-puzzle problem given above is presented in Figure 2.13. It is to be noted that Figure 2.13 only presents the way in which a (first) solution is obtained, but the backtracking search algorithm will continue to investigate all other possible situations and will return all the solutions (in case there are more than one).

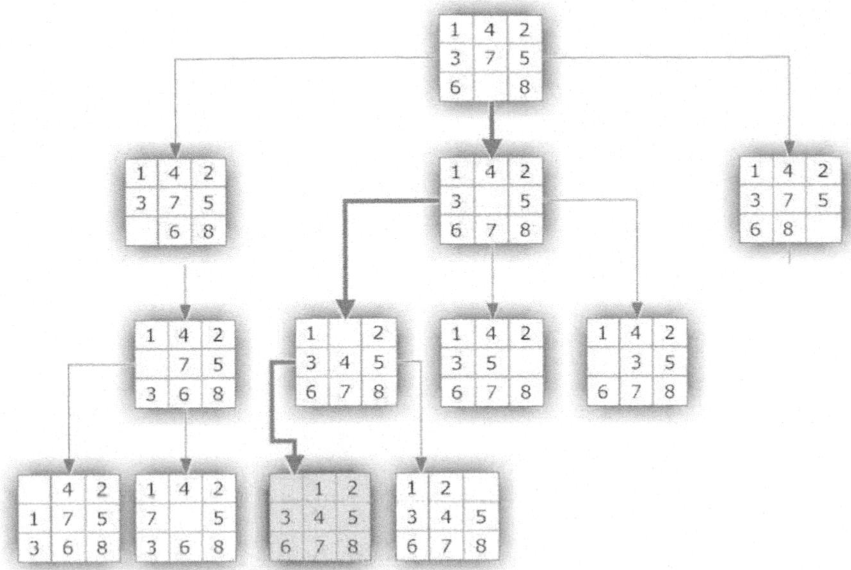

Fig. 2.12 Example of depth first search for the 8-puzzle problem.

The arrows in Figure 2.13 show the order in which states are explored and expanded.

2.6.4 Depth Bounded (Limited) Depth First Search

Depth bounded depth first search (also referred as depth bounded search or depth limited search) is similar with depth first search but paths whose length has reached some limit, *l*, are not extended further. This can be implemented by considering a stack (or a queue but in which nodes are added in the front) but any node whose depth in the tree is greater than *l* is not added.

Figure 2.14 presents the same example as in Figure 2.11. Please refer to Figure 2.14 (a) for *l*= 2 and Figure 2.14 (b) for *l*=4). Depth limited search is similar to standard depth first search but the tree is not explored below some depth-limit *l*. Depth bounded search solves problem of infinitely deep paths with no solutions but will be incomplete if solution is below depth-limit. Depth limit *l* can be selected based on the problem knowledge (e.g., diameter of state-space).

Figure 2.15 presents an example of depth limited search with limit 3 for the 8-puzzle problem.

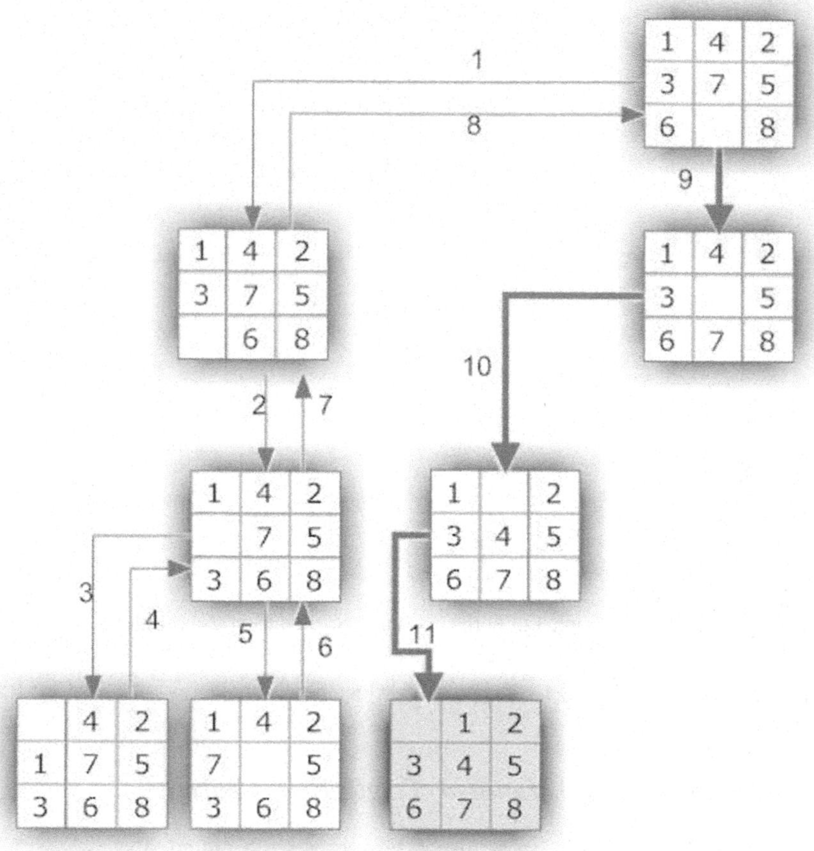

Fig. 2.13 Backtracking search applied for the 8-puzzle problem.

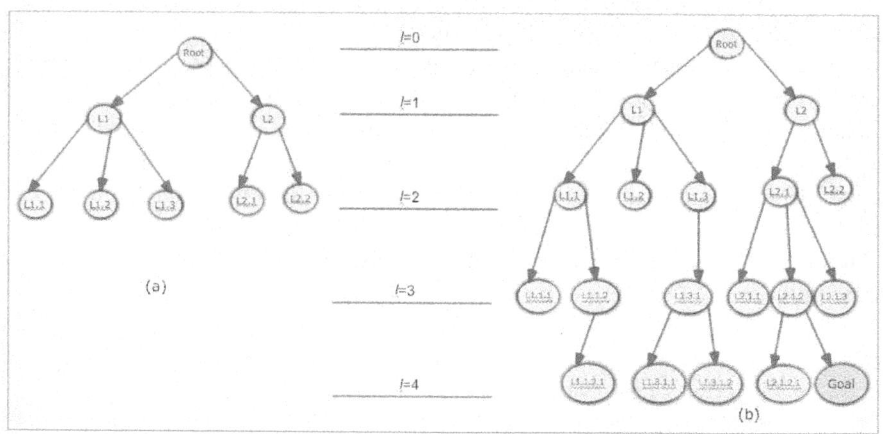

Fig. 2.14 Example of depth bounded search with l=2 – left (a) and l=4 – right (b).

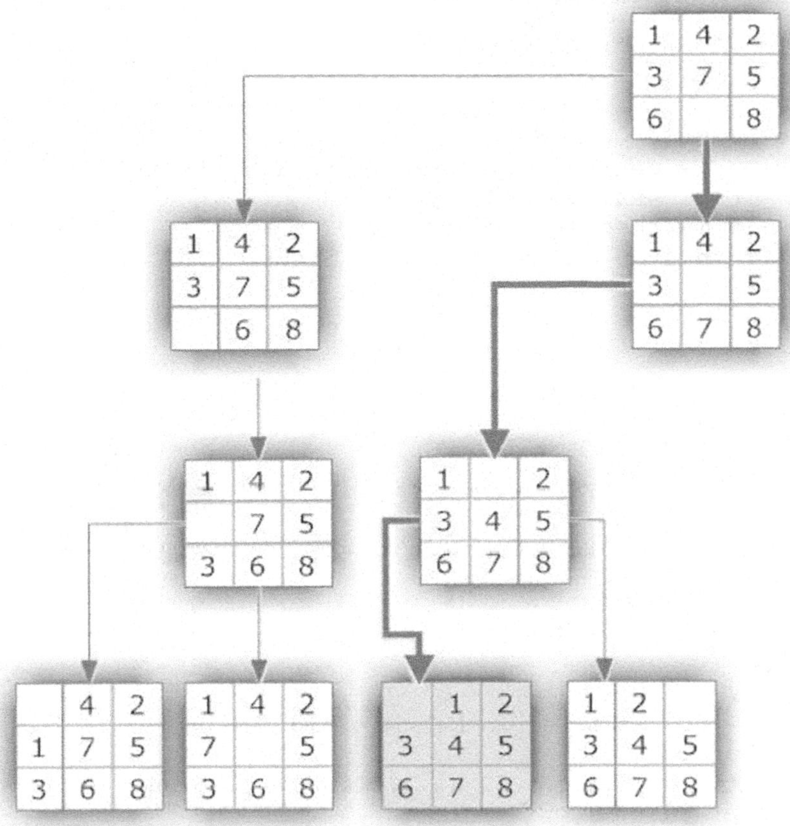

Fig. 2.15 Example of depth limited search for the 8 puzzle problem

2.6.5 Iterative Deepening Depth First Search

Iterative deepening dept bounded depth first search (also referred to as iterative deepening search) consists of repeated depth bounded searches using successive greater depth limits. The Algorithm first attempts a depth bounded search with a depth bound (or limit) 0, then it tries a dept bounded search with a depth limit of 1, then of 2 and so on. Since the search strategy is based on depth bounded search the implementation does not require anything new. The depth bounded searches are repeated until a solution is found[15][28][34][37][38][39].

An example of iterative deepening search with limits from 0 to 3 is depicted in Figures 2.15-2.18.

The iterative deepening search algorithm is simply described in Algorithm 2.4.

Algorithm 2.4. Iterative deepening search

```
Returns a solution or failure;
Input problem;
l = 0
While no solution, do
      Apply depth first search(problem, depth) from in-
      itial state with cutoff l
      If matched the goal
      Then stop and return solution,
      Else increment depth limit l=l+1
End.
```

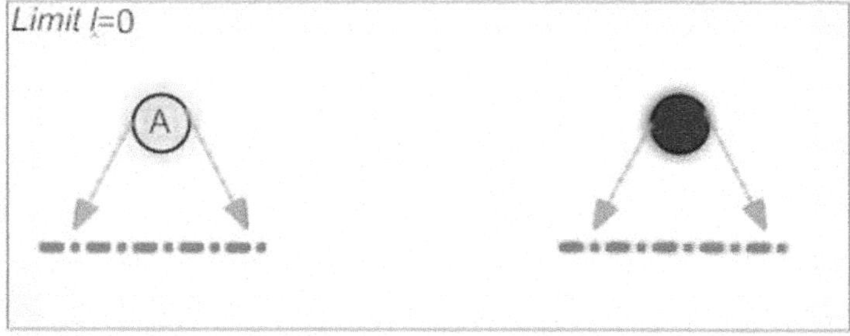

Fig. 2.15 Iterative deepening search for limit *l*=0.

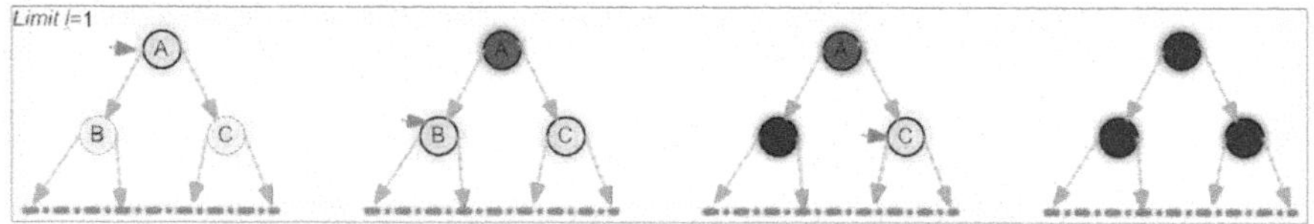

Fig. 2.16 Iterative deepening search for limit *l*=1.

The advantage of the iterative deepening search is that it requires linear memory and it guarantees for goal node of minimal depth. For large depth *d*, the ratio of the number of nodes expanded by iterative deepening search compared to that of depth first search or breadth first search is given by $b/(b-1)$. This implies that for higher values of the branching factor the overhead of repeated expanded states will be smaller. For a branching factor of 10 and deep goals, there will be 11% (10/9) more nodes expanded in iterative deepening search than the breadth first search.

Iterative deepening search combines the advantage of completeness from breadth first search with that of limited space and ability to find longer paths more quickly of the depth first search. This algorithm is generally preferred for large state spaces where the solution depth is unknown. There is a related technique called *iterative broadening*, which works by first constructing a search tree by expanding only one child per node. This algorithm is useful when there are many goal nodes.

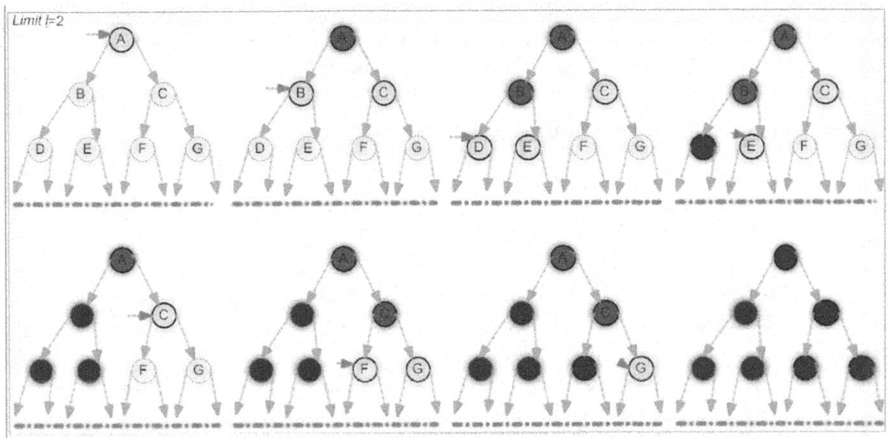

Fig. 2.17 Iterative deepening search for limit l=2.

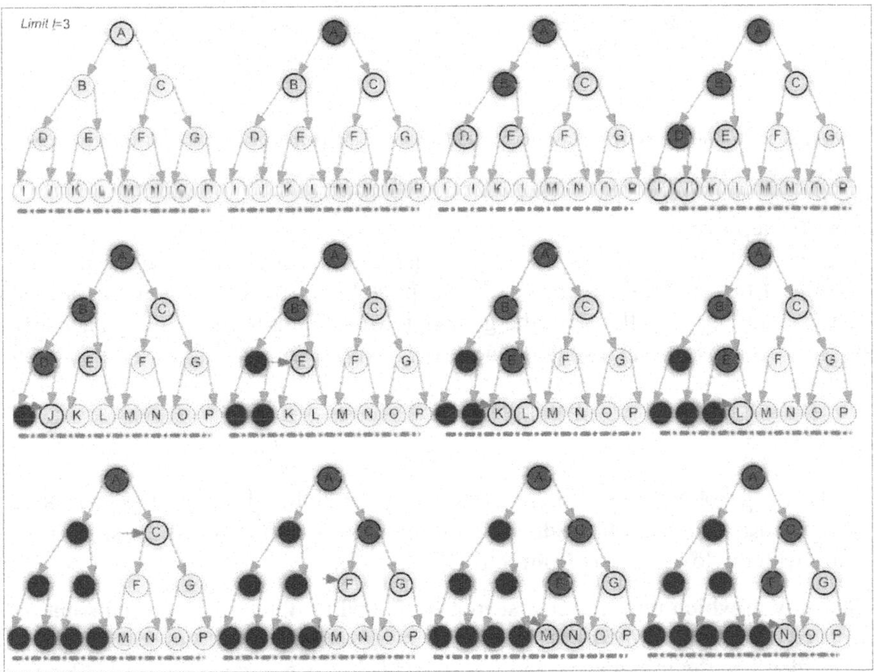

Fig. 2.18 Iterative deepening search for limit l=3.

2.6.6 Branch and Bound (or Uniform Cost Search)

With the branch and bound search, the node with the minimum cost is always expanded. Once a path to the goal is found, it is likely that this path is optimal. In order to guarantee this, it is important to continue generating partial paths until each of them has a cost greater than or equal to the path found to the goal. The branch and bound algorithm is presented in Algorithm 2.5.

Algorithm 2.5. Branch and bound (uniform cost) search

```
Return a solution or failure

Q is a priority queue sorted on the current cost from
the start to the goal
Step 1. Add the initial state (or root) to the queue.
Step 2. Until the goal is reached or the queue is empty
do
      Step 2.1 Remove the first path from the queue;
      Step 2.2. Create new paths by extending the first
      path to all the neighbors of the terminal node.
      Step 2.3. Remove all new paths with loops.
      Step 2.4. Add the remaining new paths, if any, to
      the queue.
      Step 2.5.  Sort the entire queue such as the
least-cost paths are in front.
End
```

Given that every step will cost more than 0, and assuming a finite branching factor, there is a finite number of expansions required before the total path cost is equal to the path cost of the goal state. Hence, the goal is reached within a finite number of steps.

The proof of optimality for the branch and bound search can be done by contradiction. If the solution found is not the optimal one, then there must be a goal state with path cost smaller than the goal state which was found which is actually impossible because branch and bound would have expanded that node first by definition.

Example

Consider the graph given in Figure 2.19 with the initial node S and the goal node G and the cost associated to each edge. Te problem is to find the shortest path (or the path with the lowest cost) from S to G.

The way in which uniform cost search is applied to obtain the optimal solution for this problem is presented in Figure 2.20 and described as follows.

Consider S as the initial state and S is expanded into A and C (Figure 2.20 (a)).

Since the path S – C has the lowest cost until now, C is the next expanded node. C is expanded into B and D.

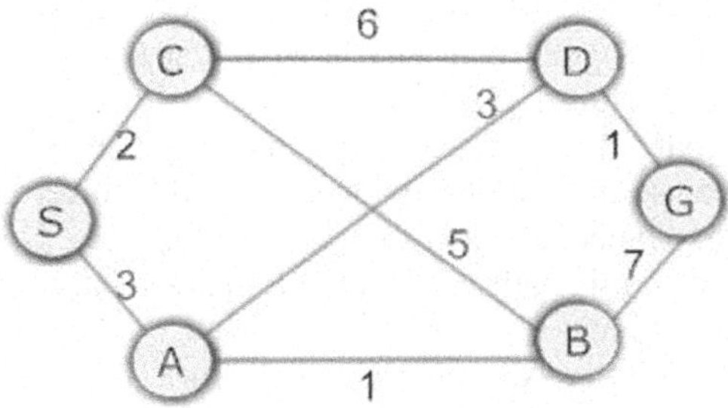

Fig. 2.19 An example of a graph with the cost associated to each edge.

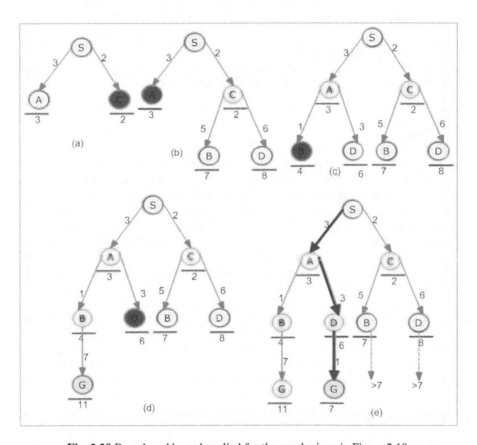

Fig. 2.20 Branch and bound applied for the graph given in Figure 2.19.

At this stage, there are three paths: S – A with cost 3, S – C – B with cost 7 and S – C – D with cost 8 (Figure 2.20 (b)).

Since the path S – A had the lowest cost as of now, A is the next expanded node. A is expanded into B and D. Now there are 4 paths: S – A – B of cost 4, S – A – D of cost 6, S – C – B of cost 7 and S – C – D of cost 8 (Figure 2.20 (c)).

The path S – A – B has the lowest cost, thus B is the next expanded node. B expands to G, which is the goal state and thus, a first solution is obtained. This solution has the cost 11 (Figure 2.20 (d)).

Even though the goal has been reached, it is important to expand the other branches as well to see whether another path exists with a lower cost.

Hence node D is expanded to form the path S – A – D. The goal is reached again and the total cost of this path is 7, which indicates that this as the best solution until now (Figure 2.20 (e)).

There are still two more nodes to be expanded. But it is meaningless to expand any of them because the cost of their paths until now is at least equal or greater than the cost of the best solution obtained until now.

Consequently, the optimal solution found for the graph given in Figure 2.18 is S- A – D – G with a cost of 7.

2.6.7 Bidirectional Search

There are three main known and used search directions:

- forward;
- backward;
- bidirectional.

The forward search starts from the current state and finds the goal, trying all possibilities one by one. The backward search starts from the goal and find current state. This is possible if goal is known. In the bidirectional search, nodes are expanded from the start and goal state simultaneously[21][23]. At each stage it is checked whether the nodes of one have been generated by the other. If so, then the path concatenation is the solution. Instead of searching from the start to the finish, two searches may be performed in parallel: one from start to finish, and one from finish to start. When they meet, a good path should be obtained. The search needs to keep track of the intersection of 2 open sets of nodes.

Suppose that the search problem is such that the arcs are bidirectional. That is, if there is an operator that maps from State A to State B, there is another operator that maps from State B to State A. Many search problems have reversible arcs such as n-puzzle, path finding, path planning etc. However there are other state space search formulations, which do not have this property. If the arcs are reversible then instead of starting from the start state and searching for the goal, one may start from a goal state and try reaching the start state. If there is a single state that satisfies the goal property, the search problems are identical.

The idea behind bidirectional searches is that searching results in a tree that fans out over the map. A big tree is much worse than two small trees, so it's better to have two small search trees.

Sometimes it might be hard to perform backward search from the goal because of the following situations:

- specify the predecessors of the goal;
- deal with situations where there are multiple goals.

The *retargeting* approach abandons simultaneous searches in the forward and backward directions. Instead, it performs a forward search for a short time, chooses the best forward candidate, and then performs a backward search not to the starting point, but to that candidate. After a while, it chooses a best backward candidate and performs a forward search from the best forward candidate to the best backward candidate. This process continues until the two candidates are the same point.

Example

Consider the graph given in Figure 2.19. Figure 2.21 shows an example on how bidirectional search may be applied. Depth first search is used as search algorithm. Two depth first searches are performed in parallel, one starting from S and one starting from G (see Figure 2.21 (a)). In level 2 of expansions from the initial node and first level of expanded nodes from the goal node one of the paths meets the other one (see Figure 2.21 (b)).

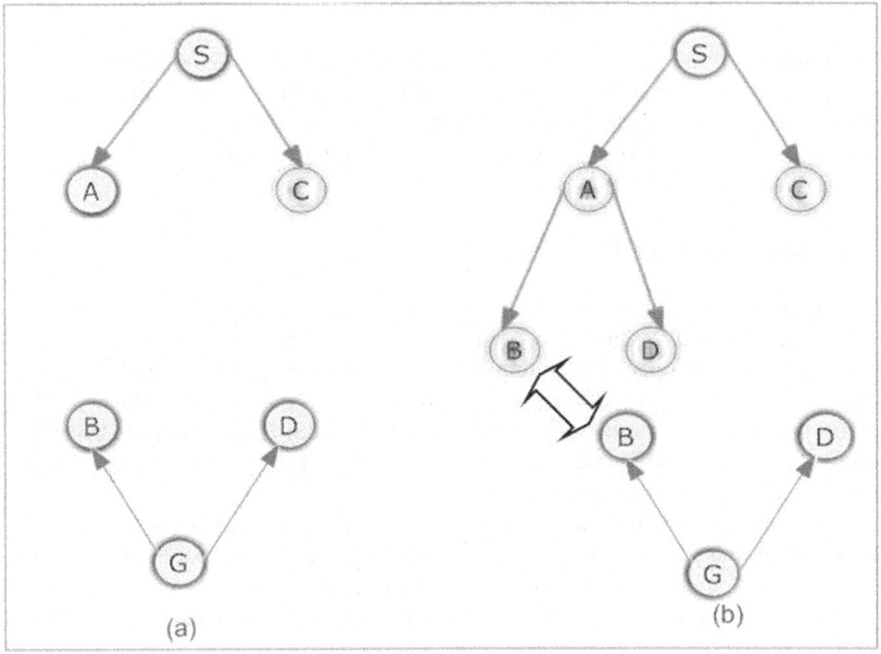

Fig. 2.21 Example of bidirectional search (depth first search is the search technique used) for the graph depicted in Figure 2.19.

2.7 Performance Evaluation of the Uninformed Search Strategies

The different search strategies can be compared in terms of:

(i) *completeness*: guarantees finding a solution whenever one exists;
(ii) *time complexity*: it measures how long it takes to find a solution;
(iii) *space complexity*: it measures how much space is used by the algorithm. This measure is usually done in terms of the maximum size of the nodes list expanded during the search.
(iv) *optimality* (quality of solution): it reflects what is the guarantee of a solution once find to be optimal; whether it is or not the one with minimum cost.

Time and space complexity are measured in terms of:

- b: maximum branching factor of the search tree;
- d: depth of the least-cost solution;
- m: maximum depth of the state space (which may be also ∞);
- l: depth limit for the depth limited search.

A comparison of all the six main uninformed search techniques described in the previous Sections (breadth first search, depth first search, depth limited search, iterative deepening search, uniform search and bidirectional search) in terms of completeness, time complexity, space complexity and optimality is illustrated in Table 2.1.

2.7.1 Remarks and Discussions

Depth first search possesses the benefit of lower space complexity. Breadth first search is guaranteed to find an optimal path. A combination of these two search techniques is the iterative deepening search. Iterative deepening search involves trying all possible depths in turn and stopping once a goal state is reached. The benefits are that it is complete and optimal like Breadth First Search, but has the modest memory requirements of depth first search. Like depth-first search, its memory requirements are $O(bd)$[36][40]

Like breadth-first search, it is complete when the branching factor is finite and optimal when the path cost is a non-decreasing function of the depth of the node. If the path cost limit is increased instead of the search depth the phenomenon is called *iterative lengthening*. However, this has more overheads and does not inherit all advantages of iterative deepening search.

The depth first search is appropriate when:

- the space is restricted;
- the solutions tend to occur at the same depth in the tree;
- there is a known way to order nodes in the list of neighbors so that solutions will be found relatively quickly.

Table 2.1 Comparison of some of the most important uninformed search techniques in terms of completeness, time complexity, space complexity and optimality.

Search method	Completeness	Time complexity	Space complexity	Optimality
Breadth first search	It is complete if *b* is finite	$1+b+b^2+b^3+\ldots+b^d+b(b^d-1)=O(b^d+1)$	$O(b^d+1)$ – keeps every node in memory	Optimal if all operators have the same cost. Otherwise it finds a solution with the shortest path length.
Depth first search	No: fails in infinite-depth spaces, spaces with loops	$O(b^m)$ Not a good situation when *m* is much higher than *d*. If solutions are dense, may be much faster than breadth first search.	$O(bm)$ – linear space	No
Depth limited search	Complete if solution is at depth < *l*	$O(b^l)$	$O(bl)$	No
Iterative deepening search	Yes	$(d+1)b^0 + d\,b^1 + (d-1)b^2 + \ldots + b^d = O(b^d)$	$O(bd)$	Optimal if the cost is a constant
Uniform search	Yes	$O(b^{floor(Cost/\varepsilon)})$ where Cost is the cost of the optimal solution, ε is a positive constant and it is assumed that every step costs at lest ε. floor(x) is the largest integer not greater than *x* floor(Cost/ε) ~ depth of solution if all costs are approximately equal	$O(b^{floor(Cost/\varepsilon)})$	Yes
Bidirectional search	It is complete if *b* is finite	$O(b^{d/2})$	$O(b^{d/2})$	Yes

The depth first search is inappropriate when:

- some paths have infinite length;
- the graph contains cycles;
- some solutions are very deep, while others are very shallow.

Complexity is a motivation for the bidirectional search: it is obvious that $b^{d/2} + b^{d/2} << b^d$

For example, by starting the search from both direction (intial state and goal) using the breadth first search and considering d=6 and b=10, only 22,200 nodes will be generated by the bidirectional search compared to 11,110,000 nodes generated by a standard application of the breadth first search from a single direction. There are various trade-offs among these algorithms; best algorithm will depend on the nature of the search problem

2.7.2 Repeated States

Repeated states can be the source of great inefficiency: identical sub-trees will be explored many times.

In many search problems, one can adopt some less computationally intense strategies. Such strategies do not stop duplicate states from being generated, but are able to reduce many of such cases. Some of such strategies are as follows:

1. Not to return to the state the algorithm just came from (prevents cycles of length one). This simple strategy avoids many node re-expansions in n-puzzle like problems.
2. Check that paths with cycles in them are not constructed. This strategy only needs to check the nodes in the current path so is much more efficient than the full checking algorithm. This strategy can be employed successfully with depth first search and not require additional storage.
3. Do not generate any state that was ever created before.
4. Avoid infinite-depth trees (for finite-state problems) but do not avoid visiting the same states again in other branches
5. Maintain Close-List beside Open-List (fringe). If current node is on the closed-list, it is discarded, not expanded

The user should decide which strategy to be employed by considering the frequency of loops in the state space. Failure to detect repeated states can turn a linear problem into an exponential one. However, for dealing with problems with many repeated states (but small state-space), graph-search can be much more efficient than tree-search.

Summary

This Chapter outlined the basic search algorithm and the various variations of this algorithm. A search space consists of states and operators and it can be easily seen as a graph. Corresponding to a search algorithm, we get a search tree which contains the generated and the explored nodes. A search tree represents a particular exploration of search space. The search tree may be unbounded. This may happen if the state space is infinite. This can also happen if there are loops in the search space.

Search techniques are used in artificial intelligence to find a sequence of steps that will get us from some initial state to some goal state (or multiple goal states). Search can be forward, from the initial state or backwards, from the goal state and sometimes can be from both directions (bidirectional). Whichever direction is chosen, various search algorithms can be employed to do the search. The appropriate direction of search and the appropriate algorithm depend on the nature of the problem to be solved, and in particular the properties of the search space.

There are different search strategies:

- Blind search strategies or uninformed search, which include:
  - Breadth first search
  - Depth first search
  - Depth limited search
  - Iterative deepening search
  - Uniform cost search
  - Bidirectional search
- Informed Search
- Constraint Satisfaction Search
- Adversary Search.

This chapter mainly focused on the uninformed search techniques. Breadth first search technique was introduced first. The next one is the depth first search followed by a short description of backtracking search as a particular case of it and the by depth bounded (limited) search as a variant of depth first search. Iterative deepening search is the fifth method presented followed by uniform cost search and ending with the bidirectional search. A comparison of all these techniques in terms of completeness, optimality, time and space complexity was also illustrated.

Some important ideas, which can be derived from this Chapter, are summarized below[22][24][25][30][31][32]:

- Breadth first search algorithm is optimal if all operators have the same cost. Otherwise, breadth first search finds a solution with the shortest path length. The algorithm has exponential time and space complexity and for a search tree of depth 15 and branch factor 10 it takes thousands of years to find the solution.
- Depth first search is exponential in time but linear in search space. The time taken by the algorithm is related to the maximum depth of the search tree. Note that if the search tree has infinite depth, the algorithm may not terminate. This can also happen if the search space is infinite or contains cycles.
- Depth first and breadth first search both have some advantages. Which is best depends on properties of the problem you are solving. For tree search at least, depth first search tends to require less memory. If there are lots of solutions, but all at a comparable depth in the tree, then a solution can be reached just by exploring a very small part of the tree. On the other hand, that may not be the best solution. Depth first search may get stuck exploring long (and sometimes infinite) paths, when there is a solution path of only one or two steps. This can be prevented by setting the depth limit.
- Depth first is good when there are many possible solutions and we are only looking for one solution. It is less suitable when there is only one solution or we are looking for the shortest one.

- Breadth first search may use more memory, but will never get stuck and will always find the shortest path first or at least the path that involves the least number of steps. It may be more appropriate when exploring very large search spaces where there is an expected solution which takes a relatively small number of steps, or when we are interested in finding all the solutions.
- Iterative deepening search uses only linear space and not much more time than other uninformed search algorithms.
- Uniform cost search is complete and optimal but exponential in time and search space.
- Forward search builds a tree from the initial state until the goal set is reached or the termination condition is satisfied. Backward search builds a tree from the goal state until the initial state is reached. Bidirectional search performs forward and backward search simultaneously until the trees meet. So, a breadth first search can be forward, backward, or bidirectional, for example.

Problem formulation and representation is the key in solving many real world problems. Implementation as expanding directed graph of states and transitions is appropriate for problems where no solution is known and many combinations must be tried. Problem space is of exponential size in the number of world states (NP-hard problems). The failures occur due to lack of space and/or time.

References

[1] http://www-g.eng.cam.ac.uk/mmg/teaching/artificialintelligence/nonflash/problemframenf.htm (accessed on July 30, 2009)
[2] Russel, S., Norvig, P.: Artificial Intelligence: A modern approach. Prentice Hall, Englewood Cliffs (1995)
[3] Amarel, S.: On representations of problems of reasoning about actions. In: Michie, D. (ed.) Machine Intelligence 3, vol. 3, pp. 131–171 (1968)
[4] Bellman, R.E., Dreyfus, S.E.: Applied Dynamic Programming. Princeton University Press, Princeton (1962)
[5] Deo, N., Pang, C.: Shortest path algorithms: Taxonomy and annotation. Technical Report CS-80-057, Computer Science Department, Washington State University (1982)
[6] Dijkstra, E.W.: A note on two problems in connexion with graphs. Numerische Mathematik 1, 269–271 (1959)
[7] Doran, J., Michie, D.: Experiments with the graph traverser program. Proceedings of the Royal Society of London 294, Series A, 235–259 (1966)
[8] Floyd, R.W.: Algorithm 96: Ancestor. Communications of the Association for Computing Machinery 5, 344–345 (1962)
[9] Floyd, R.W.: Algorithm 97: Shortest path. Communications of the Association for Computing Machinery 5, 345 (1962)
[10] Johnson, W.W., Story, W.E.: Notes on the 15 – puzzle. American Journal of Mathematics 2, 397–404 (1879)

[11] Knut, D.E.: The Art of Computer Programming, 1: Fundamental Algorithms, 2: Seminumerical Algorithms, 3: Sorting and Searching. Addison-Wesley, Reading (1997) (1998)
[12] Karger, D.R., Koller, D., Phillips, S.J.: Finding the hidden path: time bounds for all-pairs shortest paths. SIAM Journal on Computing 22(6), 1199–1217 (1993)
[13] Knoblock, C.A.: Learning abstraction hierarchies for problem solving. In: Proceedings of the Eighth National Conference on Artificial Intelligence (AAAI 1990), vol. 2, pp. 923–928. MIT Press, Boston (1990)
[14] Korf, R.E.: Depth-first iterative-deepening: an optimal admissible tree search. Artificial Intelligence 27(1), 97–109 (1985)
[15] Korf, R.E.: Iterative-deepening An optimal admissible tree search. In: Proceedings of the Ninth International Joint Conference on Artificial Intelligence (IJCAI 1985), pp. 1034–1036. Morgan Kaufmann, Los Angeles (1985)
[16] Korf, R.E.: Optimal path finding algorithms. In: Kanal, L.N., et al. (eds.) Search in Artificial Intelligence, Berlin, vol. ch. 7, pp. 223–267 (1988)
[17] Korf, R.E.: Linear-space best-first search. Artificial Intelligence 62(1), 41–78 (1993)
[18] Netto, E.: Lehrbuch der Combinatorik. B. G. Teubner, Leipzig (1901)
[19] Nilsson, N.J.: Problem-Solving Methods in Artificial Intelligence. McGraw-Hill, New York (1971)
[20] Nilsson, N.J.: Principles of Artificial Intelligence. Morgan Kaufmann, San Mateo (1980)
[21] Pohl, I.: Bi-directional and heuristic search in path problems. Technical Report 104, SLAC (Stanford Linear Accelerator Center, Stanford, California (1969)
[22] Pohl, I.: First results on the effect of error in heuristic search. In: Meltzer, B., Michie, D. (eds.) Machine Intelligence 5, pp. 219–236 (1970)
[23] Pohl, I.: Bi-directional search. In: Meltzer, B., Michie, D. (eds.) Machine Intelligence 6, pp. 127–140. Edinburgh University Press, Edinburgh (1971)
[24] Pohl, I.: The avoidance of (relative) catastrophe, heuristic competence, genuine dynamic weighting and computational issues in heuristic problem solving. In: Proceedings of the Third International Joint Conference on Artificial Intelligence (IJCAI 1973), Stanford, CA, pp. 20–23 (1973)
[25] Pohl, I.: Practical and theoretical considerations in heuristic search algorithms. In: Elcock, E.W., Michie, D. (eds.) Machine Intelligence 8, pp. 55–72 (1977)
[26] Ratner, D., Warmuth, M.: Finding a short solution for the n xn extension of the 15-puzzle is intractable. In: Proceedings of the Fifth National Conference on Artificial Intelligence (AAAI 1986), vol. 1, pp. 168–172. Morgan Kaufmann, Philadelphia (1986)
[27] Russell, S.: Tree-structured bias. In: Proceedings of the Seventh National Conference on Artificial Intelligence (AAAI 1988), vol. 2, pp. 641–645. Morgan Kaufmann, St. Paul (1988)
[28] Russell, S.: Efficient memory-bounded search methods. In: Proceedings of the 10th European Conference on Artificial Intelligence (ECAI 1992), pp. 1–5. Wiley, Vienna (1992)
[29] Schofield, P.D.A.: Complete solution of the eight puzzle. In: Dale, E., Michie, D. (eds.) Machine Intelligence 2, pp. 125–133 (1967)
[30] Simon, H.A., Newell, A.: Heuristic problem solving: The next advance in operations research. Operations Research 6, 1–10 (1958)
[31] Simon, H.A., Newell, A.: Computer simulation of human thinking and problem solving. Datamation, 35–37 (1961)

[32] Slate, D.J., Atkin, L.R.: CHESS 4.5—The Northwestern University chess program. In: Frey, P.W. (ed.) Chess Skill in Man and Machine, pp. 82–118. Springer, Berlin (1977)
[33] Tait, P.G.: Note on the theory of the "15- puzzle". Proceedings of the Royal Society of Edinburgh 10, 664–665 (1880)
[34] Korf, R.E.: Depth-first iterative-deepening: An optimal admissible tree search. Art. Intell. 27, 97–109 (1985)
[35] Powley, C., Ferguson, C., Korf, R.E.: Depth-first heuristic search on a SIMD machine. Art. Intell. 60, 199–242 (1993)
[36] Rao, V.N., Kumar, V., Korf, R.E.: Depth-first vs. best-first search. In: 9th Nat. Conf. on Art. Int. AAAI 1991, Anaheim, CA, pp. 434–440 (1991)
[37] Reinefeld, A., Marsland, T.A.: Enhanced iterative-deepening search. Univ. Paderborn, FB Mathematik-Informatik, Tech. Rep. 120 (March 1993)
[38] Reinefeld, A., Marsland, T.: Enhanced iterative deepening search. IEEE Trans. on Pattern Analysis and MachineIntelligence 16, 701–710 (1994)
[39] Russell, S.: Efficient memory-bounded search methods. In: European AI-Conference, ECAI 1992, Vienna, pp. 1–5 (1992)
[40] Korf, R.: Linear-space best-first search. Artificial Intelligence 62, 41–78 (1993)

Verification Questions

1. Briefly discuss the advantages and disadvantages of depth and breadth first search. What sort of problem is each appropriate for?

2. How does the use of a *closed* node list reduce the amount of search required in graph search?

3. What are the main advantages and what are the disadvantages of the breadth first search technique?

4. What are the main advantages and what are the disadvantages of the depth first search technique?

5. How does the iterative deepening search behave while compared to depth first search?

6. What is the advantage of depth limited search compared to depth first search and what are the inconveniences of this technique?

7. What is the time complexity of breadth first search and how you calculate it?

8. What is the time and space complexity of depth first search and how you obtain them?

9. What is the time and space complexity of iterative deepening search and how you calculate them?

10. What is the complexity of bidirectional search? Explain.

11. Give an example of a state space in which iterative deepening search performs much worse than depth first search.

12. Suppose that we are searching a tree with a branching factor b and depth d. There is a single goal state located at depth g ($0 \le g \le d$).

a) In the worst case, how many nodes are explored by breadth first search?
b) In the best case, how many nodes are explored by breadth first search?
c) In the worst case, how many nodes are explored by depth first search?
d) In the best case, how many nodes are explored by depth first search?

Exercises

2.1 Consider the n-queens problem.

a) Define the search elements: states space, initial state, operators, and goal state.

b) Solve the problem for n=5 and n= 8 using breadth first search and depth first search and compare the results.

2.2 Given the search tree bellow (Figure 1), state the order in which the nodes will be searched for breadth first search, depth first search, depth limited search (for *l*=1, 2, 3) and iterative deepening search.

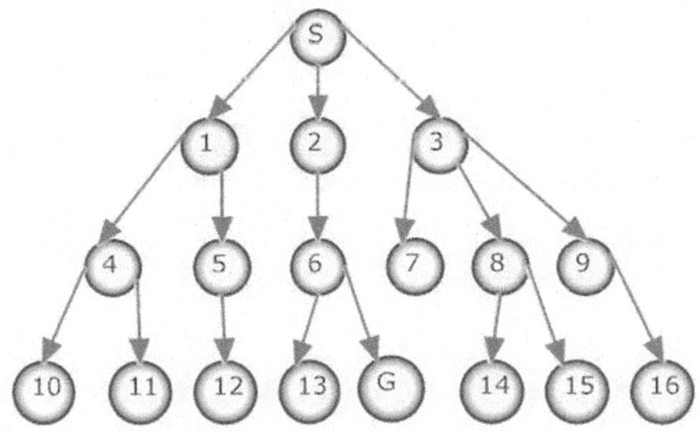

Fig. 1 Search tree example fir problem 2.2

2.3. Let us consider a *Sudoku* game. This game is a logic puzzle whose idea is to take a partially-filled n × n square and fill in the missing cells with numbers which are assigned uniquely for, each row, each column, and each of the n squares of size $\sqrt{n} \times \sqrt{n}$ (as shown in Figure 2 (a) and (b) for a puzzle of size 4×4 and one of size 9×9 respectively).

A well-formed Sudoku puzzle is one where there exists a unique solution to these constraints.

a) Apply any of the uninformed search methods you like to find a solution for the 4×4 puzzle given in Figure 2 (a). At each step in the game, you can fill in the empty positions with a number from 1 to 4. A solution to the puzzle is a fully filled in game board in which each number from 1 to 4 appears exactly once in each row, column and block.

b) What is the number of search states for the problem considered at a)?

c) What can you tell about time and space complexity of this problem (the one at a))?

d) How will the time and space complexity increase if you remove the numbers from the cells (3, 4) and (4, 2) of the initial configuration?

e) Consider the 9×9 Sudoku given in Figure 2 (b). Try to solve it using the same search algorithm as for the 4×4 one and compare the complexity.

f) Pick one of the uninformed search algorithms which will reach the solution in the shortest time (if this is not the one you already used).

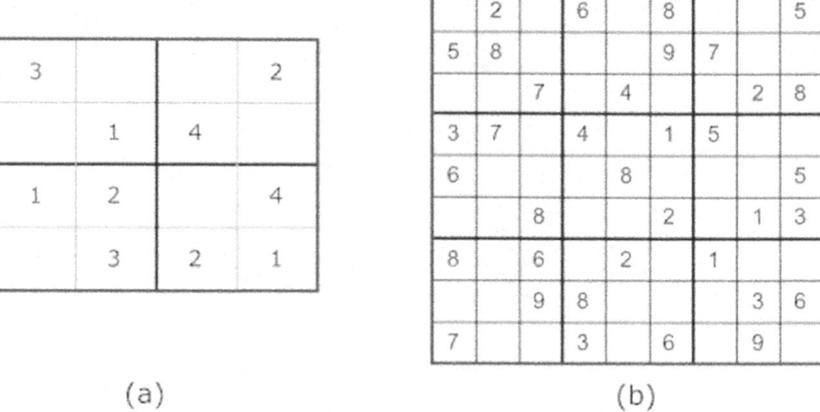

Fig. 2 Example for the problem 2.3.

2.4. Develop a general uninformed search algorithm, which can efficiently solve any of the cryptarithmetic problems given in Figure 3.

```
 SEND+      TAKE+       EAT+    NEVER-
 MORE=         A       THAT=    DRIVE=
------      KAKE=     ------    ------
MONEY      ------      APPLE      RIDE
            KATE
```

Fig. 3. Examples for the problem 2.4.

The following are known:

- Each letter represents only one digit throughout the problem;
- When letters are replaced by their digits, the resultant arithmetical operation must be correct;
- The numerical base, unless specifically stated, is 10;
- Numbers must not begin with a zero;
- There must be only one solution to the problem.

2.5 Consider the geometric shapes given in Figure 4. Find a way to arrange them into the 5 × 4 rectangular area given on the left side such as the area will be entirely filled, all the shapes will be used and none of the shapes overlap. Use breadth first search, iterative deepening search and backtracking for this problem and compare the results (in terms of completeness, optimality, time and space complexity).

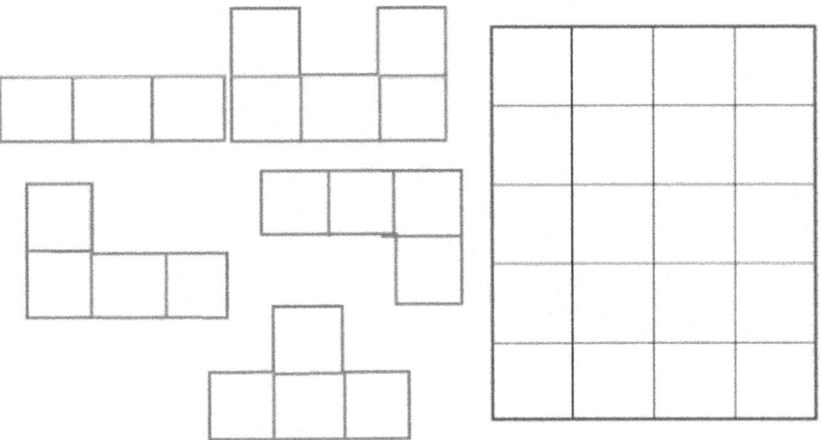

Fig. 4 Example for the problem 2.5.

2.6 Sort a list of objects (for instance, a list of integers) using as operator swapping of two objects in the list. Use bidirectional search technique for this.

2.7 Consider the graph in Figure 5 which represents some of the European cities and distances between them (Figure 6). The problem requires finding the shortest path from Madrid to Bucharest.

a) What is the state space for this problem and what is its size?
b) Draw the search tree resulting from breadth first search. How many nodes are expanded?
c) Draw the search tree resulting from depth first search. How many nodes are expanded?
c) Draw the search tree resulting from uniform cost search. How many nodes are expanded?
d) Use backtracking;
e) Use iterative deepening search;
g) Explain each of the techniques used above gives the best result. Explain which of the techniques might get stuck into loops and will not reach a solution in reasonable time.

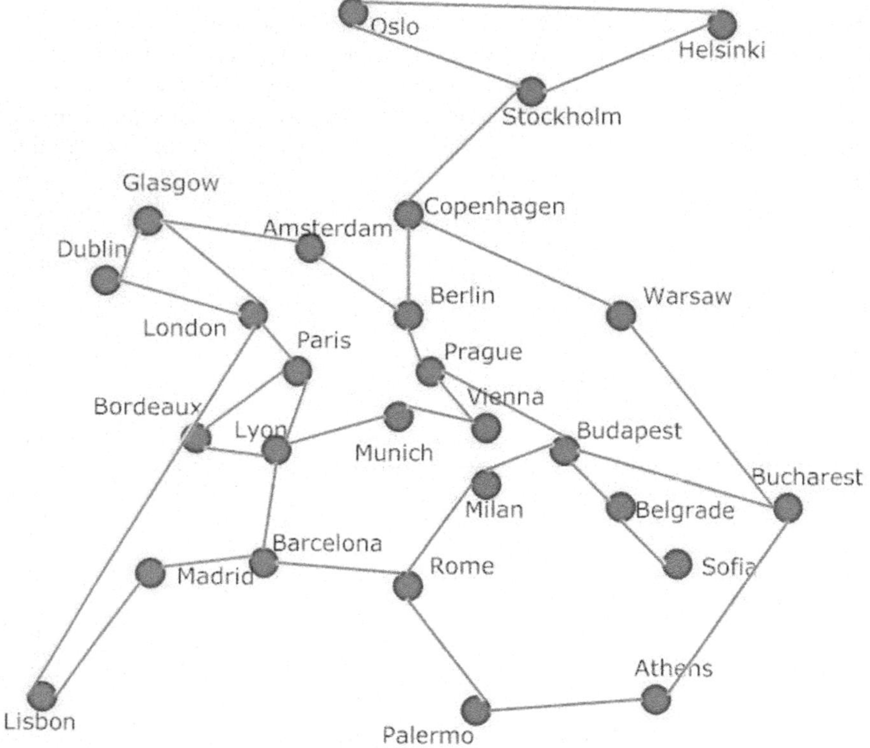

Fig. 5 Example for the problem 2.7

Olso - Helsinki:	970	Rome: Milan:	681	Madrid - Barcelona:	628
Helsinki - Stockholm:	400	Milan - Budapest:	789	Madrid - Lisbon:	638
Oslo - Stockholm:	570	Vienna - Budapest:	217	Lisbon - London:	2210
Stockholm - Copenhagen:	522	Vienna - Munich:	458	Barcelona - Lyon:	644
Copenhagen - Warsaw:	668	Prague - Vienna:	312	Paris - London:	414
Warsaw - Bucharest:	946	Prague - Berlin:	354	London - Dublin:	463
Bucharest - Athens:	1300	Berlin - Copenhagen:	743	London - Glasgow:	667
Budapest - Bucharest:	900	Berlin - Amsterdam:	648	Glasgow - Amsterdam:	711
Budapest - Belgrade:	316	Munich - Lyon:	753	Budapest - Prague:	443
Belgrade - Sofia:	330	Lyon - Paris:	481	Barcelona - Rome:	1471
Rome - Palermo:	1043	Lyon - Bordeaux:	542	Paris - Bordeaux:	579
Palermo - Athens:	907			Glasgow - Dublin:	306

Fig. 6 Edge cost for the graph in Figure 5 (for the problem 2.7).

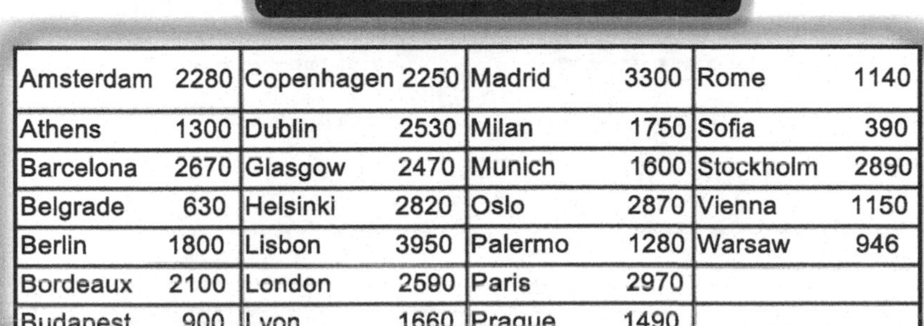

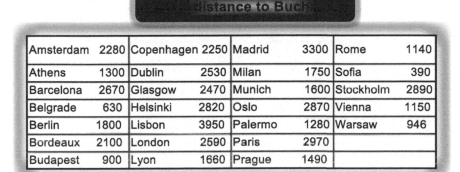

Amsterdam	2280	Copenhagen	2250	Madrid	3300	Rome	1140
Athens	1300	Dublin	2530	Milan	1750	Sofia	390
Barcelona	2670	Glasgow	2470	Munich	1600	Stockholm	2890
Belgrade	630	Helsinki	2820	Oslo	2870	Vienna	1150
Berlin	1800	Lisbon	3950	Palermo	1280	Warsaw	946
Bordeaux	2100	London	2590	Paris	2970		
Budapest	900	Lyon	1660	Prague	1490		

Fig. 7 Direct distances to Bucharest from any node (city) of the graph depicted in Figure 5 (for problem 2.7).

2.8 Implement the missionaries and cannibals problem described in this Chapter using any of the uninformed search techniques you prefer.

2.9 On the bank of a river are 1 adult, 2 children and a small boat. The people have to cross the river. The boat can only carry:

- 2 children or
- a single child or
- a single adult.

a) What is the search space for this problem?
b) Draw the search tree using any of the uninformed search techniques.

c) How will the complexity be improved if reduce de branching factor by consider that logically both children are same? Explain.

2.10 Consider the 3 – puzzle and 8 – puzzle given in Figure 8 a) and b).

a) Draw the tree searched obtained by breadth first search for the 3 - puzzle.
b) Draw the tree searched obtained by depth first search for the 3 – puzzle.
c) What is the branching factor for the 3- puzzle problem?
d) What is the solution depth for the 3-puzzle?
e) How the branching factor and depth modify for the 8 – puzzle problem (Figure 8 b) ?

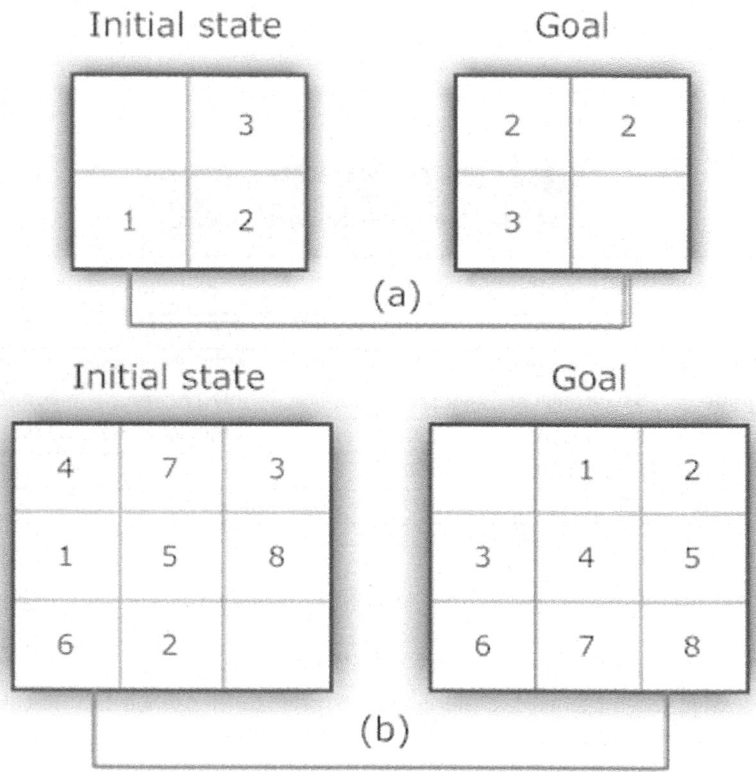

Fig. 8 Puzzle example for problem 2.10.

2.11 Consider the breadth first search algorithm. Given a branching factor b of size 10, the depth d, knowing that 1,000 nodes can be checked and expanded per second and that a node requires 100 bytes of storage, calculate the time and memory requirement for a depth d varying from 0 to 20.

2.12 Use any of the uninformed search techniques for solving the following problem such as the solution is obtained in optimal time. Motivate your choice.

A farmer, a goat, a wolf and a bag of cabbage are on the left side of a river. They all need to get to the other side of the river. The following constrains hold:

- *Only 2 can cross at a time (here cabbage is also considered as an item);*
- *The wolf and the goat cannot be left together on the same side;*
- *The cabbage and the goat cannot be left together on the same side.*

2.13 Find a way (using an uninformed search technique) to get down from the top of the pyramid depicted in Figure 9 with a maximum score cumulated. Each cube stands on top of 4 other cubes. Each cube has a score associated (this you can assign by yourself). Getting down from one cube can be only made by stepping on one of the 4 cubes the cube stands on (so, at each step there are 4 possibilities).

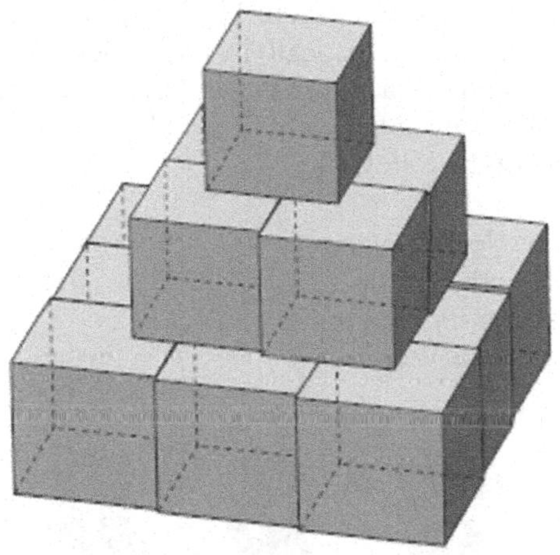

Fig. 9 Example for the problem 2.13.

2.14. Find a way the get down from the top of the pyramid such as the sum of cumulated points (each cell has a number of points associated) is maximal. From a cell only three movements are allowed: down, left and right (this means each state can expand into three new states). Employ an uninformed search technique which will provide the optimal solution. Which algorithm is this?

							6							
						8	4	6						
					2	7	1	8	5					
				2	8	5	5	9	4	7				
			3	5	3	5	8	3	3	2	6			
		4	5	6	2	4	3	9	2	1	5	7		
	3	1	2	4	2	3	4	5	8	4	2	1	4	
2	2	3	3	5	2	3	4	9	6	7	8	4	3	2

Fig. 10 Example for the problem 2.14.

2.15 Consider the labyrinth given in Figure 11. The dark cells represent walls, the white cells represent alleys and the cell in the middle marked with two "*" represents the starting point. Each light cell has a number associated which represents the cost of using it (in the path). The only possible directions to follow when at a point are any of the following:

- North-West, North, North-East or
- North-East, East, South-East or
- South-West, West, North-West or
- South-East, South, South-West;

	8		3		9		5	
	9		1		7		8	
	5	2	5		8	4	5	
	1		2	5	6	8	7	5
1	3			*1*	5	5	9	
	2	3	1	5	6		2	
		2	1	4	4		8	5
5	4	5		3	5	7	5	
		5				6		

Fig. 11 Example for the problem 2.15.

which means at one point can only go into 3 directions following any of the four groups.

a) Use an uninformed search technique which returns all the possible ways to get out of the labyrinth.
b) Use an uninformed search technique which finds the cheapest way to get out of the labyrinth.

2.16. In an office building with 5 floors and 8 office rooms which a number of people occupying each room (as shown in Figure 12) at each floor and internet cable should be installed. The cable cannot pass through all rooms and it has a given length. Find a way to distribute this cable and to use as less length of it ass possible such as the number of people which will benefit from internet is maximum.

The cable starts in the upper left corner and has to ends in the lower right corner and can only be transferred from the current room in the up, down, left and right neighboring rooms. Employ an uninformed search technique for solving it.

3	4	5	1	0	1	2	1
1	2	4	1	1	2	1	0
2	1	3	0	1	2	1	1
1	0	4	2	4	3	3	3
0	1	1	0	1	2	1	2

Fig. 12 Example for the problem 2.16.

2.17 In an airport, 9 watches showing the time in different cities in the world are placed on a big wall. The time shown by the watches is the one given in Figure 13 a).

But the time shown is not the real one and it should be modified. There are 2 buttons which allow modifications as follows:

- Button 1: If moved up, will increase the time with one hour for all the watches in a selected row. If moved down, it will decrease the time with one hour for all the watches on that row.
- Button 2: If moved up, will increase the time with one hour for all the watches in a selected column. If moved down, it will decrease the time with one hour for all the watches on that column. We have to notices that no matter which button is used, it cannot act only on a single watch but on 3 at a time.

Use an uninformed search algorithm that is able to set the correct time (the one given in Figure 13 b)) by using the buttons as less as possible.

Fig. 13 Example for the problem 2.17.

Chapter 3
Informed (Heuristic) Search

3.1 Introduction

In the previous Chapter, we have presented several blind search or uninformed search techniques. Uninformed search methods systematically explore the search space until the goal is reached. As evident, uninformed search methods pursue options that many times lead away from the goal. Even for some small problems the search can take unacceptable amounts of time and/or space. The blind search techniques lack knowledge about the problem to solve and this makes them inefficient in many cases. Using problem specific knowledge can significantly improve the search speed.

Informed search (also called *directed search* and *heuristic search*), tries to reduce the amount of search that must be done by making intelligent choices for the nodes that are selected for expansion. The nodes, which are likely to lead to a good solution, are placed towards the front. This implies the existence of some way of evaluating the likelihood that a given node is on the solution path. In general this is done by using a heuristic function. Informed search strategies use problem-specific knowledge to find solution faster. The concept of heuristic function is an important component of the informed search techniques.

The uninformed search techniques keep a priority ordered queue. By always taking nodes from the front of the queue, the path selected to be extended is always the shortest (or cheapest if there is a cost associated) so far. In informed search, the priority ordered queue is still preserved. The ordering in this case is determined by an *evaluation function*, which for each node on the fringe returns a number that signifies the promise or the potential of that node. One of the most important kinds of knowledge to use when constructing an evaluation function is an estimate of the cost of the cheapest path from the (current) state to a goal state. Functions that calculate such estimates are called *heuristic functions*. We should note that in the AI field the word 'heuristic' is not only in the context of 'heuristic functions' but also used for any techniques that might improve the average case performance but does not necessarily improve worst-case performance.

A heuristic function is used to evaluate the promise of a *state*. We choose, which node to expand next using the heuristic value of its state. Heuristic functions do not evaluate *operators*, i.e. if several operators can be used to expand a

C. Grosan and A. Abraham: Intelligent Systems, ISRL 17, pp. 53–81.
springerlink.com

node, heuristic functions do not say which operator is the most promising one [11][13][14][15][16][17].

Heuristic functions are problem-specific. We usually design (or learn) different functions for different problem domains. There are a variety of search techniques that rely on the estimate provided by a heuristic function. In all cases - the quality (accuracy) of the heuristic is important for the real-life application of the technique.

The following Section presents the most important aspects of heuristics. General presentation of the heuristics is then followed with the description of some of the most important informed search techniques namely:

- Best-first search
- Greedy best-first search
- A* search
- IDA* search

3.2 Heuristics

Heuristics (Greek *heuriskein* = find, discover) can be defined as "the study of the methods and rules of discovery and invention". Some snapshots about heuristics from Judea Pearl's book – Heuristics: Intelligent Search Strategies for Computer Problem Solving [1] are quoted below:

> *"...popularly known as rules of thumb, educated guesses, intuitive judgments or simply common sense. In more precise terms, heuristics stand for strategies using readily accessible though loosely applicable information to control problem-solving processes in human beings and machine[s]."*
>
> *"The study of heuristics draws its inspiration from the ever-amazing observation of how much people can accomplish with that simplistic, unreliable information source known as intuition..."*
>
> *"Heuristics are criteria, methods, or principles for deciding which among several alternative courses of action promises to be the most effective in order to achieve some goal. They represent compromises between two requirements: the need to make criteria simple and, at the same time, the desire to see them discriminate correctly between good and bad choices.*
>
> *A heuristic may be a rule of thumb that is used to guide one's actions."*

The principal advantage of using a heuristic function is the reduction of the state space. For example, the full tree for Tic-Tac-Toe has 9! leaves. If we consider symmetries, the tree becomes six times smaller.

Consider some examples of heuristic functions for the 8-tiles puzzle. Please consider the puzzle given in Figure 3.1 (the initial state (or a current state) on the left and the goal state in the right).

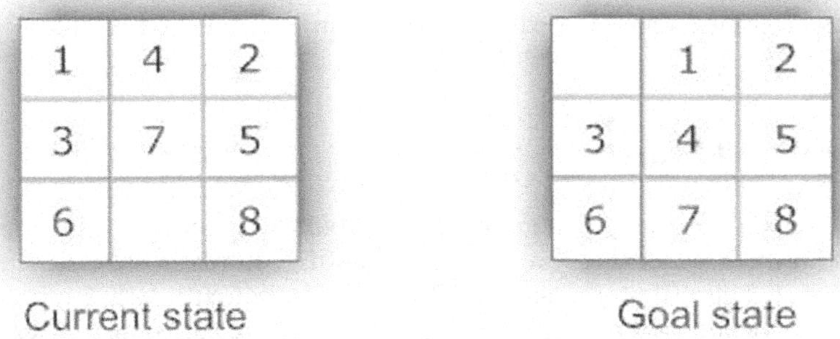

Fig. 3.1 Current (initial) and goal state for an 8-puzzle example.

We can define a first heuristic $h_1(n)$ (where n refers to n-th node) as follows:

$h_1(n)$ = the number of misplaced tiles in the current state relative to the goal state.

For the example given in Figure 3.1, $h_1(n)= 4$ because tiles 1, 4, 7 and 6 are out of place. Obviously, lower values for h_1 are preferred. Another heuristic function example – h_2 – which can be defined for the puzzle problem is Manhattan distance. This heuristic sums the Manhattan distance for each tile in the puzzle (Manhattan distance of a tile represents the number of squares from desired location)

For the 8-puzzle given in Figure 3.1 we have:

$h_2(n) = 1+0+0+1+0+0+1+0 = 3.$

Both functions estimate the number of moves we'll have to make to modify the current state into a goal state. In fact, both h1 and h2 underestimate the costs of the cheapest paths in this state space, and this turns out to be a significant property. It is obvious for any heuristic h that h(n) = 0 if n is a goal.

Definition 1
A heuristic function h(n) is called admissible if for all nodes one has $h(n) < k(n)$ where k(n) is the actual distance to the goal from n.

An admissible heuristic never overestimates the cost to reach the goal, i.e., it is optimistic.

Remark: Both the heuristics given above are admissible.

Definition 2
*Let $h_1(n)$ and $h_2(n)$ be two heuristic functions, both admissible. If $h_2(n) \geq h_1(n)$ for all n then $h_2(n)$ **dominates** $h_1(n)$ and is better for search.*

If we have k non-overestimating heuristics for a problem $h_1(n), h_2(n), \ldots, h_k(n)$, then $\max_{i \leq k} h_i(n)$ is a more powerful non-overestimating heuristic [22][23][24][25] [28].

3.3 Best First Search

Best first search uses an evaluation function and always chooses the next node to be that with the best score. However, it is exhaustive, in that it should eventually try all possible paths. It uses a queue as in breadth/depth first search, but instead of taking the first node off the agenda (and generating its successors) it will take the *best* node (or will arrange ascending the queue and then will take the first node). The successors of the best node will be evaluated (a score will be assigned to them) and added to the list. A cost function f(n) is applied to each node. The nodes are put in OPEN in the order of their f values. Nodes with smaller f(n) values are expanded earlier.

The standard best first search algorithm is outlined in Algorithm 3.1.

Algorithm 3.1 Best first search

```
Step 1. Let Q be a priority queue containing the
        initial state (starting state).
Step 2. Until Q is empty or failure
      Step 2.1 if queue is empty return failure
      Step 2.2 Remove the first node from the queue
               (take it from the OPEN list and move it
               into the CLOSED list)
      Step 2.3 If the first node is the goal
                 then return the path to it from the
           initial state
                 Else generate all successors of the
               node and put them into the queue ac-
               cording to their score (f(n) value)
               (best ones in the front).
Step 3. If a solution is found return it, else return
failure.
```

There are different ways of defining the function f. This leads to different search algorithms. There are two basic categories of approaches:

- one which tries to expand the node closest to the goal;
- another one, which tries to expand the node on the least-cost solution path.

3.4 Greedy Search

In greedy search, the idea is to expand the node with the smallest estimated cost to reach the goal (or the node which appears to be closest to the goal). In an informal way, an algorithm follows the Greedy search if it makes a series of choices, and each choice is locally optimized, or, in other words, when viewed in isolation, that step is performed optimally. Similar to depth first search, Greedy search tends to follow a single path to the goal. The heuristic function is:

$$f(n) = h(n)$$

where h(n) estimates the remaining distance to a goal. Greedy algorithms often perform very well. They tend to find good solutions quickly, although not always the optimal ones. The heuristics usually perform well on typical problems (ones that arise in practice). So, with a well-chosen heuristic function, greedy search might perform well on the problems we place in front of it. The time and space demands might be quite reasonable on these problems.

Example 1: 8-puzzle

Let us start with a simple example to illustrate how Greedy search works. Consider the 8-puzzle example for which we will take two heuristic functions: number of misplaces tiles and Manhattan distance. For the first heuristic function, the path to the solution obtained by Greedy search is depicted in Figure 3.2. We can observe that there are three possibilities to move from the initial state:

- blank moves left;
- blank moves right and
- blank moves up.

When blank moves left, the newly obtained state will have 4 misplaced tiles (1, 4, 6 and 7). It is same situation for the state obtained where the blank moves right, with the tiles 1, 4, 7, and 8 misplaced. When the blank is moved up, the new obtained state will only have 2 misplaced tiles: 1 and 4. The Greedy algorithm will decide to follow this path since this seems to be the closest one to the final state.

From this new state, blank can be again moved left, right and up. By moving the blank up, the number of misplaced tiles will be 1 (only the tile 1 is misplaced). By moving the blank either left or right, the number of misplaced tiles will be 3 (tiles 1, 4, 3 and 1, 4, 5 respectively will be misplaced). So, the first state will be chosen to expand further. From this new state, there are two possible new states which can be obtained by moving the blank left or right. It can be observed that by moving the blank left the goal state it reached (for this state the heuristic function value is 0). This is actually the shortest path to the solution in this case and Greedy search is able to find it.

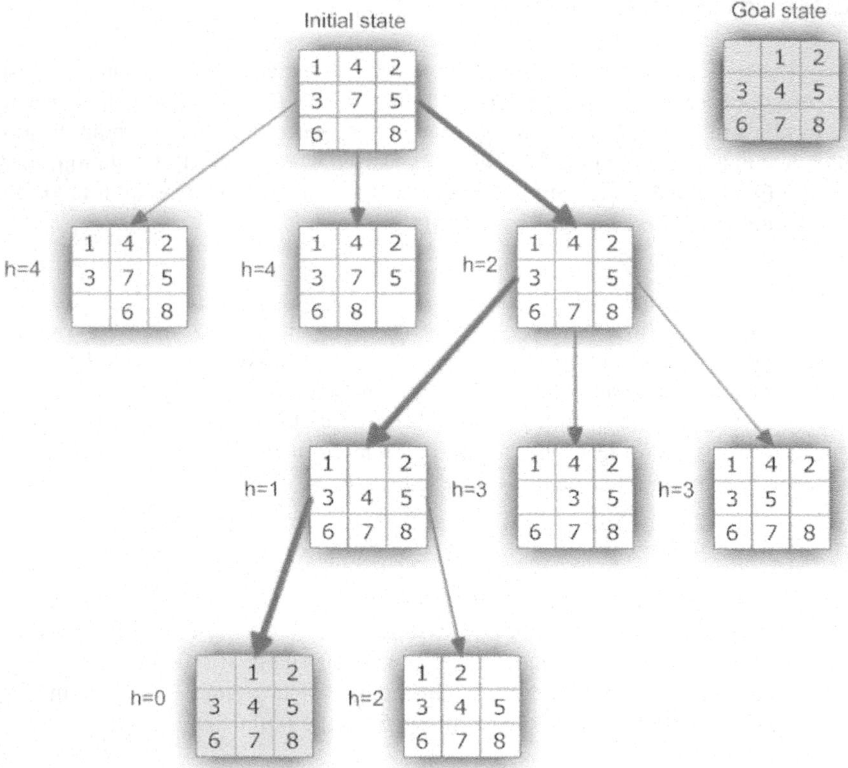

Fig. 3.2 Greedy search applied for the 8-puzzle problem using the heuristic function as the number of misplaced tiles.

Let us now consider using Manhattan distance as the heuristic function. The path to the solution in this case is depicted in Figure 3.3. The Manhattan distance represents the number of tiles from the desired location of each tail. We compare at each step the current state with the goal state to calculate this.

Once the initial state is expanded three new states are obtained (by moving the blank left, right and up). By applying the Manhattan distance, the value of the heuristic function for each new state is given by:

- *blank moves left*: the heuristic function value is 4 (1+0+0+1+0+1+1+0) because:
    - number of states from the location of tile 1 to its desired location is 1;
    - number of states from the location of tile 4 to its desired location is 1;
    - number of states from the location of tile 7 to its desired location is 1;

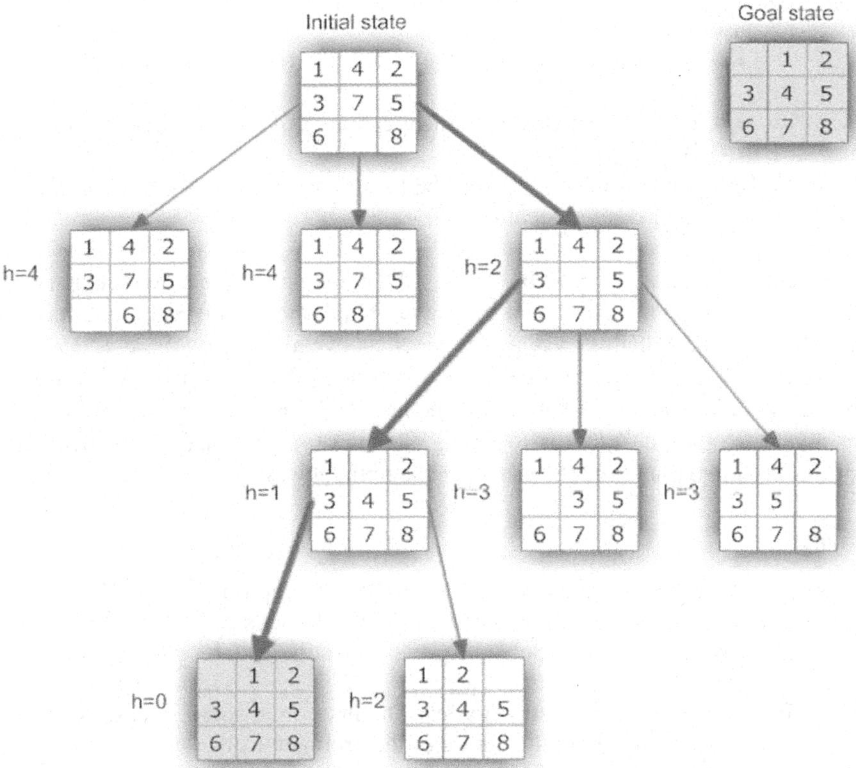

Fig. 3.3 Greedy search applied for the 8-puzzle example using the heuristic function as Manhattan distance.

  - number of states from the location of tile 6 to its desired location is 1;
  - number of states from the location of tiles 2, 3, 5 and 7 to their desired location is 0;
- *blank moves right*: the heuristic function value is 4 (1+0+0+1+0+0+1+1) because:
  - number of states from the location of tile 1 to its desired location is 1;
  - number of states from the location of tile 4 to its desired location is 1;
  - number of states from the location of tile 7 to its desired location is 1;
  - number of states from the location of tile 8 to its desired location is 1;
  - number of states from the location of tiles 2, 3, 5 and 6 to their desired location is 0;

- *blank moves up*: the heuristic function value is 2 (1+0+0+1+0+0+0+0) because:
  - number of states from the location of tile 1 to its desired location is 1;
  - number of states from the location of tile 4 to its desired location is 1;
  - number of states from the location of tiles 2, 3, 5, 6, 7 and 8 to their desired location is 0;

Consequently, the state obtained by moving the blank up is chosen to be expanded next. Here, 3 situations arise: blank moves up, blank moves left and blank moves right. The heuristic functions values (obtained as explained above) are 1, 3 and 3 respectively. The first node will be expanded further. There are two possibilities: blank moves left or blank moves right. The heuristic functions values are 0 and 2 respectively, so the goal state is reached. We can observe that for both heuristic functions used, the goal state is reached in the optimal way.

Remarks

(i) Greedy algorithm does not always produce the optimal results. The question which arises is when this strategy which looks at each step individually and ignores the global aspects can still lead to globally optimal solutions.

(ii) In fact, when a greedy strategy leads to an optimal solution, it says something interesting about the nature of the problem itself.

(iii) In several cases, even if Greedy approach does not give the optimal solution, in many cases it leads to provably good solution (not too far from the optimum).

In what follows, we will consider two very simple examples for which Greedy search does not provide the best solution.

Example 3: Shortest path

Consider the graph given in Figure 3.4. The nodes in the graph represent some European cities. Some of the nodes are connected and the distances between any two connected cities are known. The problem is related to finding the shortest path from Barcelona to Bucharest. The direct distances from each city (node in the graph) to Bucharest are also given.

Let us consider Greedy search for this problem with the heuristic h giving the straight line distance from the current state to Bucharest. The initial state is Barcelona and the goal state is Bucharest. The solution obtained by the Greedy algorithm is presented in Figure 3.5.

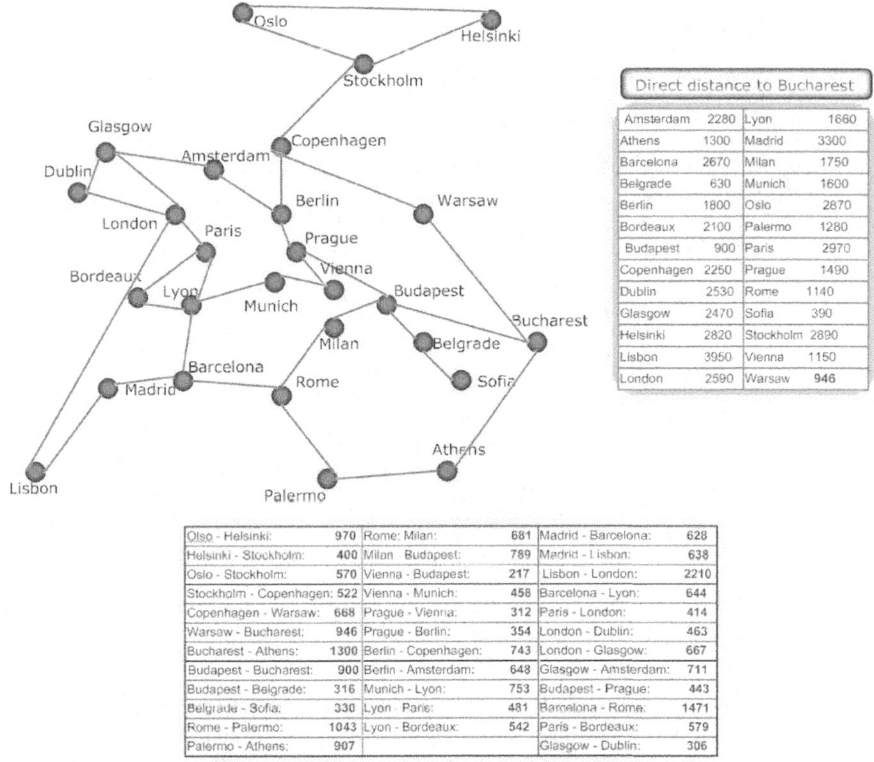

Direct distance to Bucharest

Amsterdam	2280	Lyon	1660
Athens	1300	Madrid	3300
Barcelona	2670	Milan	1750
Belgrade	630	Munich	1600
Berlin	1800	Oslo	2870
Bordeaux	2100	Palermo	1280
Budapest	900	Paris	2970
Copenhagen	2250	Prague	1490
Dublin	2530	Rome	1140
Glasgow	2470	Sofia	390
Helsinki	2820	Stockholm	2890
Lisbon	3950	Vienna	1150
London	2590	Warsaw	946

Olso - Helsinki:	970	Rome: Milan:	681	Madrid - Barcelona:	628
Helsinki - Stockholm:	400	Milan - Budapest:	789	Madrid - Lisbon:	638
Oslo - Stockholm:	570	Vienna - Budapest:	217	Lisbon - London:	2210
Stockholm - Copenhagen:	522	Vienna - Munich:	458	Barcelona - Lyon:	644
Copenhagen - Warsaw:	668	Prague - Vienna:	312	Paris - London:	414
Warsaw - Bucharest:	946	Prague - Berlin:	354	London - Dublin:	463
Bucharest - Athens:	1300	Berlin - Copenhagen:	743	London - Glasgow:	667
Budapest - Bucharest:	900	Berlin - Amsterdam:	648	Glasgow - Amsterdam:	711
Budapest - Belgrade:	316	Munich - Lyon:	753	Budapest - Prague:	443
Belgrade - Sofia:	330	Lyon - Paris:	481	Barcelona - Rome:	1471
Rome - Palermo:	1043	Lyon - Bordeaux:	542	Paris - Bordeaux:	579
Palermo - Athens:	907			Glasgow - Dublin:	306

Fig. 3.4 A graph example containing some European cities with distances between them (corresponding to the arcs between the given pair of connected cities) and with the direct distance to Bucharest from each city.

Starting from Barcelona, we can expand this node into two new states: Rome and Lyon. The straight line distance from Lyon to Bucharest is 1660 while the straight line distance from Rome to Bucharest is 1140. This means that the node Rome will be expanded next.

As of now we have the path Barcelona – Rome of cost 1471.

Rome is to be further expanded and the two new states obtained are Milan and Palermo. The straight line distance from Palermo to Bucharest is 1280 and the straight line distance from Milan to Bucharest is 1750. Greedy search will chose to expand the node Palermo.

The path at this moment is Barcelona – Rome – Palermo of cost 1471+1043.

The result of Palermo node expansion is the unique node, Athens, which will be the next state.

The path until now is Barcelona – Rome – Palermo – Athens of cost 1471+1043+907.

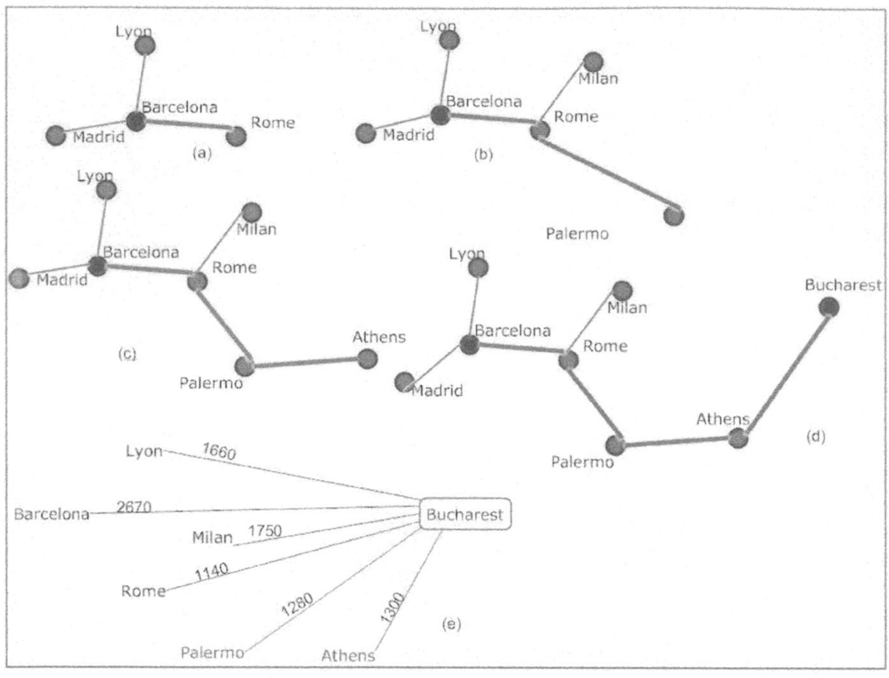

Fig. 3.5 (a)-(d): Solution obtained by Greedy search using straight line distance heuristic function for Example 3. (e): The direct distances to the goal state from the expanded nodes.

Athens is directly linked to the goal state so the path is complete and the final solution is:

Barcelona – Rome – Palermo – Athens – Bucharest of cost 1471+1043+907+1300 = 4,721.

But this is not the shortest path from Barcelona to Bucharest in the given graph. The shortest path has the cost 3,510 (as obtained by A* algorithm in the following section

Example 4: Coin Changing

Let $C = \{c_1, c_2, \ldots, c_k\}$ be a finite set of distinct coin denominations. We make the following assumptions:

each c_i, $i=1,.., k$ is an integer and $c_1 > c_2 > \ldots > c_k$;
each denomination is available in unlimited quantity.

The problem is to make change for the sum N, using a minimum total number of coins.

Remark
There is always a solution if c_k=1.

The Greedy search method has been widely applied for solving these type of problems. The basic heuristic function use is: repeatedly choose the largest coin less than or equal to the remaining sum, until the desired sum is obtained.

For the coin set {25, 10, 5, 1} and sum 30, the greedy method always finds the optimal solution. But if we have the sum 30 and the coins {25, 10, 1} then Greedy will return:

$1 \times 25 + 0 \times 10 + 5 \times 1$ (6 coins)

While the optimum is:

$0 \times 25 + 3 \times 10 + 0 \times 1$ (3 coins).

Again, Greedy search will not get the optimal solution for the case C= {12, 5, 1} and N = 15. Greedy search obtains $1 \times 12 + 0 \times 5 + 3 \times 1$, while the optimum is $0 \times 12 + 3 \times 3 + 0 \times 1$ (a total of 3 coins instead of 4 obtained by greedy).

Some facts about the Greedy search technique, which can be derived from the above examples:

- tend to find good solutions quickly, although not always optimal ones ;
- they can get into loops, so they are not complete;
- they are not admissible; sometimes heuristics may underestimate;
- if there are too many nodes, the search may be exponential;
- worst case time complexity is same as for depth first search;
- worst case space complexity is same as breadth first search;
- a good heuristic can give significant improvement;
- Greedy search is used for small problems to have quick answers.

3.5 A* Search

The A* algorithm combines the uniform cost search and the Greedy search in the sense that it uses a priority (or cost) ordered queue (like uniform cost) and it uses an evaluation function (like Greedy) to determinate the ordering [5]. The evaluation function f is given by:

$$F(n) = h(n) + g(n)$$

where:

- h(n) is a heuristic function that estimates the cost from n to goal and
- g(n) is the cost so far to reach n.

Therefore, f(n) estimates total cost of path through n to goal.

The queue will be then sort based on estimates of full path costs (not just the cost so far, and not just the estimated remaining cost, but the two together).

It can be proven that if h(n) is admissible, then A* search will find an optimal solution.

A* is optimally efficient, i.e. there is no other optimal algorithm guaranteed to expand fewer nodes than A*. But it is not the answer to all path search problems as it still requires exponential time and space in general

Theorem 1
If h(n) is admissible then A* using tree search is optimal.

Proof
Suppose the goal is G and some suboptimal goal G' has been generated and is in the fringe. Let n be an unexpanded node in the fringe such that n is on a shortest path to an optimal goal G.

Since G' is a (suboptimal) goal, then h(G') = 0. This implies f(G')=g(G').
Since G is a goal, then h(G') = 0. This implies f(G)=g(G).
Since G' is suboptimal then g(G') > g(G).
This implies f(G') > f(G).
Since h is admissible $h(n) \leq h^*(n)$
Thus $g(n) + h(n) \leq g(n) + h^*(n)$
Then $f(n) \leq f(G)$.
Hence f(G') > f(n) (and from the above) A* will never select G' for expansion.

Definition 3
A heuristic h is *consistent* (or *monotone*) if, for every node n and every successor n' of n generated by any operator (action) a, the estimated cost of reaching the goal from n is no greater than the step cost of getting to p plus the estimated cost of reaching the goal from p (see Figure 3.6). In other words:

$h(n) \leq c(n, n')+h(n')$ and

$h(g)=0$

where:

- *h* is the consistent heuristic function;
- n refers to any node in the graph;
- n' is any child of n;
- G is any goal node.

If h is consistent, we have:

$$f(n') = g(n') + h(n') = g(n) + c(n, n') + h(n') \geq g(n) + h(n) = f(n)$$

This shows that f(n) is non-decreasing along any path.

Theorem 2
If h(n) is consistent, A* using graph search is optimal.

There are two useful properties of a consistent heuristic:

1) Any consistent heuristic is also admissible.
2) There are some specific benefits for graph search: in situations where we encounter a state we have already seen before, if we are using a consistent heuristic we can be sure that the second time we encounter the state it will be via a path which is at least as costly as the path we have already found to this state, and therefore that the search will remain optimal if we just throw away the second state.

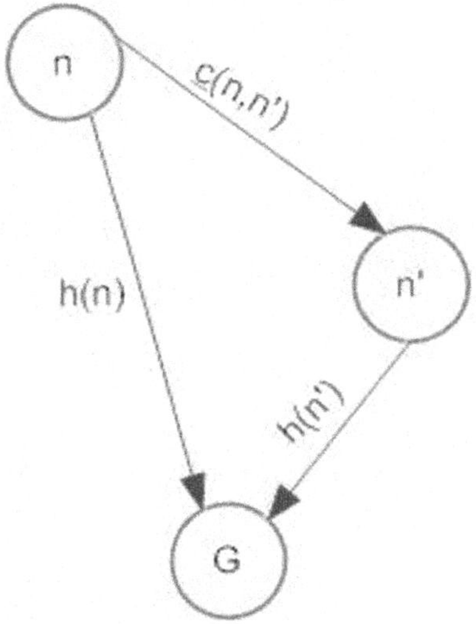

Fig. 3.6 Consistent heuristic.

The A* algorithm is outlined in Algorithm 3.2.

```
Algorithm 3.2. A* search
Step 1. Let Q be a queue of partial paths (initially
root to root, length 0);
  Step 2. While Q is not empty or failure
      Step 2.1 if the first path P reaches the goal
      node then return success
      Step 2.2.remove path P from the queue;
      Step 2.3 extend P in all possible ways and add
      the new paths to the queue
      Step 2.4 sort the queue by the sum of two values:
      the real cost of P until now (g) and an estimate
      of the remaining distance (h);
      Step 2.5 prune the queue by leaving only the
      shortest path for each node reached so far;
Step 3. Return the shortest path (if success) or
failure.
```

Example 1: Shortest path

Let us consider the same example as we used for Greedy search given in Figure 3.4. The goal is the same: finding the shortest path from Barcelona to Bucharest. In what follows we describe in detail how A* works for this example. Barcelona is the starting point, so Barcelona will expand into 3 nodes: Madrid, Lyon and Roma (see Figure 3.7 (a)).

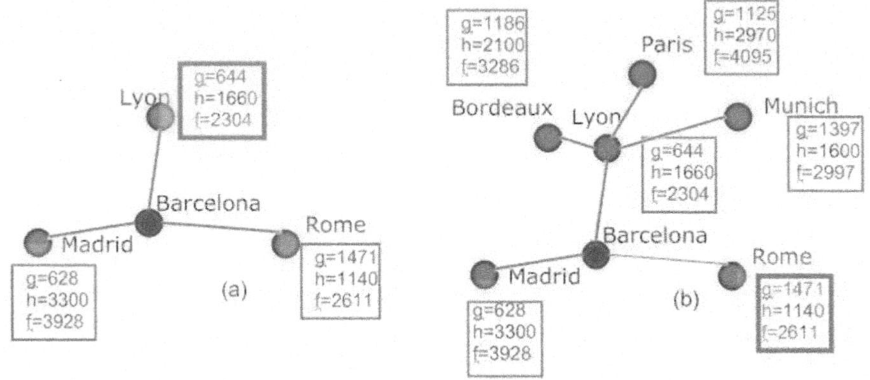

Fig. 3.7 First and second steps obtained by applying A* for Example 3.4.

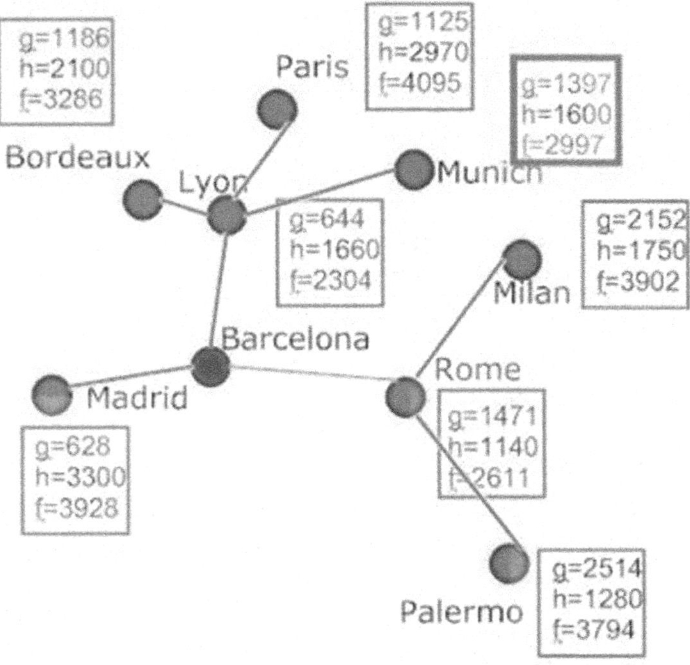

Fig. 3.8 Third step in the application of A* for Example 3.4.

For each node n we have to calculate 3 entities: h(n) – which is the direct distance to Bucharest from the node n, g(n), which represents the cost of the path so far (i.e. the sum of all costs in the path from Barcelona to the current node) and f(n) which is g(n)+h(n).

For Madrid, we obtain:

- g = 628 (distance from Barcelona to Madrid)
- h=3,300 (straight line distance from Madrid to Bucharest)
- f= 3,300+628=3,928

Similarly, we obtain g=644, h=1,660 and f=2,304 for Lyon and g=1,471, h=1,140 and f=2,611 for Rome respectively.

Since Lyon is the node with the lowest f value among all possible to expand nodes, Lyon will be expanded next (Figure 3.7 (b)).

Lyon expands and 3 new nodes are obtained:

- Munich (with g=1,397 (obtained from summing the distances Barcelona – Lyon (644) and Lyon – Munich (753), h=1,600 (straight line distance from Munich to Bucharest and f=2,997);
- Paris (with g=644+481)=1,125, h=2,970 and f= 4,095) and
- Bordeaux (with g=644+542=1,186, h=2,100 and f= 3,286)

The lowest f value in the whole tree is the one of Rome's node which is expanded next (see Figure 3.8).

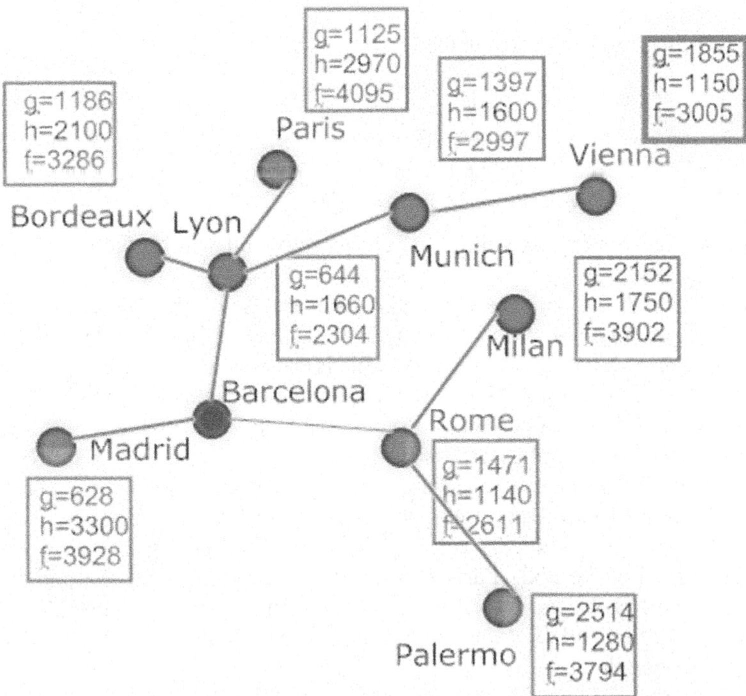

Fig. 3.9 Forth step in the application of A* for Example 3.4

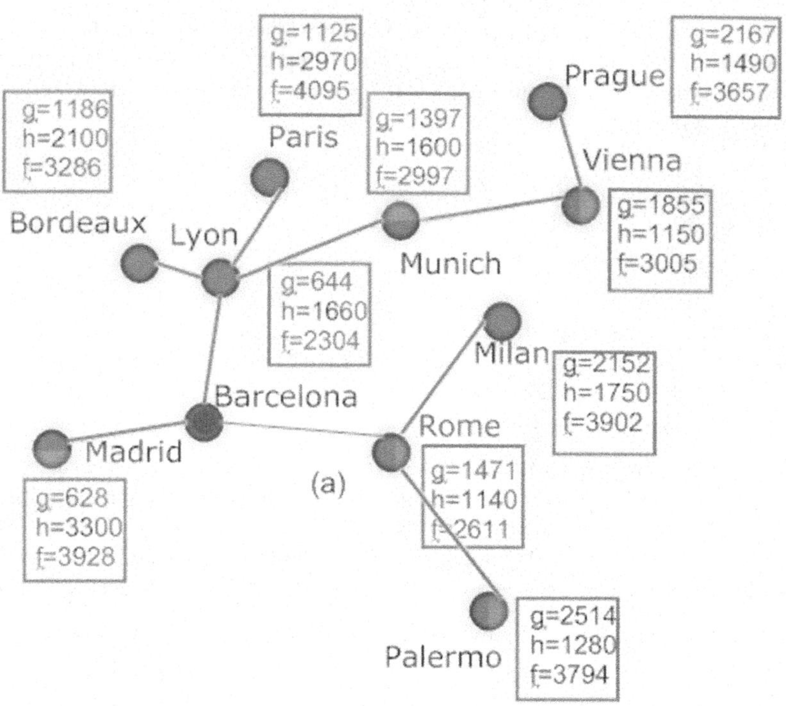

Fig. 3.10 Fifth step in the application of A* for Example 3.4

Rome expands into 2 new nodes:

- Milan (whose g is 2,152 (Barcelona – Rome (1,471) + Rome – Milan (681), h=1,140 and f=2,611) and
- Palermo (for which g = 2,514, h=1,280, f=3,794)

Munich is having the lowest value for f at this step so it will be the next expanded node (Figure 3.9).

Munich has only one successor: Vienna. For this node the value of g is 1,855 obtained by summing Barcelona – Lyon (644) + Lyon – Munich (753) + Munich – Vienna (458). h is 1,150 and f is 3,005. This is the lowest value of f in the whole tree obtained until now, thus Vienna will expand next (Figure 3.10).

By expanding Vienna, a new node Prague (with g=2,167, h=1,490 and f= 3,657) is obtained. The lowest value of f is now the one of Bordeaux's node. But Bordeaux can only expand to Paris and the new values which will be obtained for the node Paris in the path Barcelona – Lyon – Bordeaux – Paris will be: g= 644+542+579 = 1,765, h = 2,970, f = 4,735, which is higher that the information already contained in the node Paris. Thus, Bordeaux is not expanded and the next one with lowest f is Prague which is expanded at this step (see Figure 3.11). Two new nodes – Berlin and Budapest – are added to the current tree for which we have the following data:

Berlin: g = 1,125, h = 2,970, f = 4,095 and
Budapest: g = 2,610, h = 900 and f = 3,510.

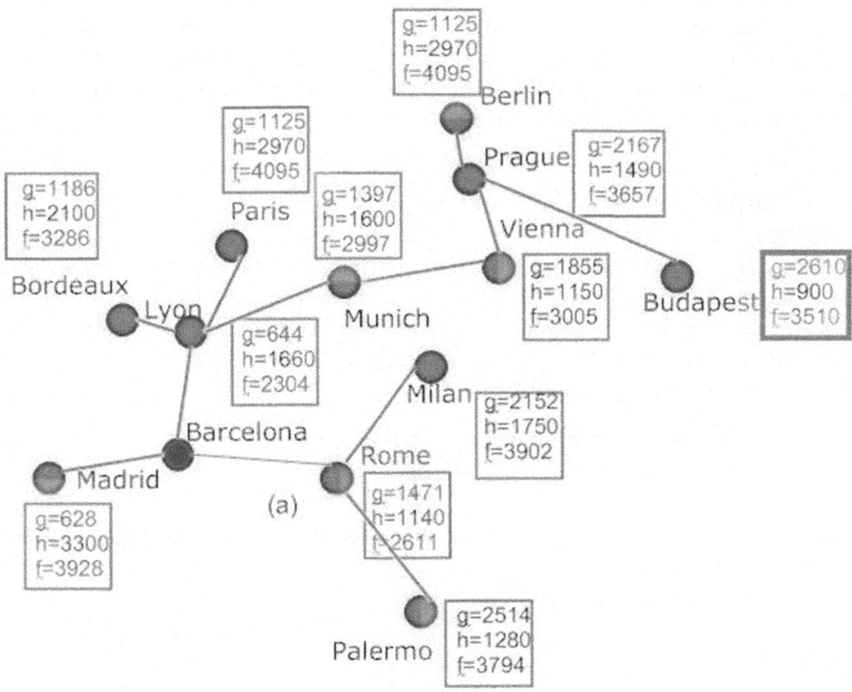

Fig. 3.11 Sixth step in the application of A* for Example 3.4

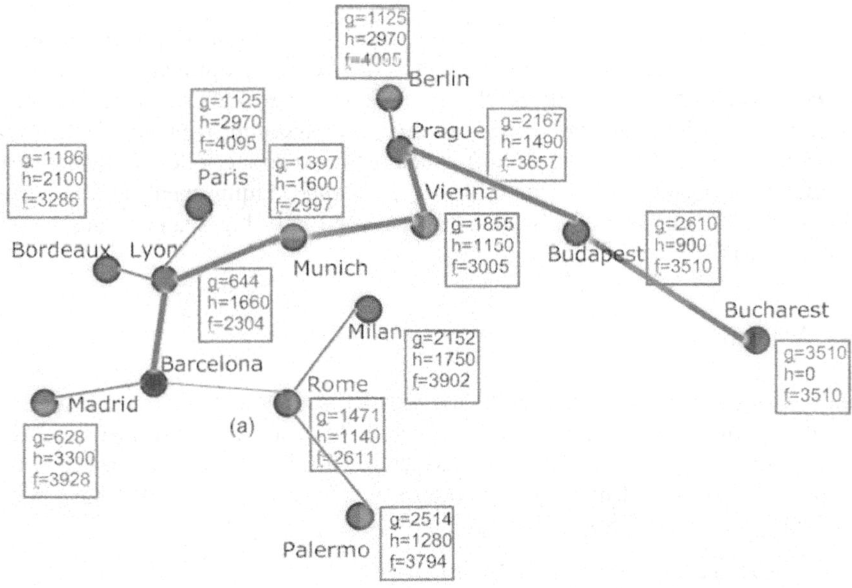

Fig. 3.12 The optimal path between Barcelona and Bucharest obtained by A* search.

As evident from the new obtained tree, Budapest node is having the smallest *f* and is further expanded. It is directly connected to the goal node – Bucharest. So the obtained path from Barcelona to Bucharest is depicted in Figure 3.12. This is actually the shortest path (its cost is 3510) and is evident that A* is able to find the optimal path while compared to the Greedy search.

3.6 Comparisons and Remarks

As in the case of uninformed search techniques, it would be interesting to compare the Greedy and A* search in terms on completeness, optimality, time complexity and space complexity [10]. Table 3.1 summarizes the behavior of these two algorithms with respect to the four attributes mentioned above.

We use the following notations:

- b: maximum branching factor of the search tree;
- m: maximum depth of the state space

Table 3.1 Comparison of Greedy search and A* search in terms of completeness, optimality, time and space complexity.

	Greedy search	**A* search**
Complete	No	Yes
Time Complexity	$O(b^m)$	$O(b^m)$
Space Complexity	$O(b^m)$	$O(b^m)$
Optimal	No	Yes

As evident from Table 3.1, both Greedy search and A* search have the same space and time complexity, but compared to Greedy, A* is optimal and complete. The worst-case time complexity is $O(b^m)$, the same as uniform cost search (and for similar reasons), but a good heuristic can give dramatic improvement. Since both approaches have to hold on to all unfinished paths (keep all nodes in memory) in case they later wish to explore them further, their space requirements are therefore again similar to uniform cost search: $O(b^m)$. For really big search spaces, both algorithms will run out of memory.

3.7 A* Variants

If all the costs are positive and the heuristic is admissible then A* terminates and finds the shortest path. Like breadth first search, A* can use a lot of memory. The memory usage is one of A*'s biggest issues. In the worst case, it must also remember an exponential number of nodes. Several variants of A* [9] have been developed to cope with this, including:

- iterative deepening A* (IDA*),
- memory-bounded A* (MA*),

- o simplified memory bounded A* (SMA*),
- o recursive best-first search (RBFS) and
- o Dynamic A* (D*).

3.7.1 *Iterative Deepening A* (IDA*)*

If the idea of iterative deepening search is combined with A*, an algorithm called IDA* (for iterative deepening A*) is obtained [7]. The space is searched depth first for successively larger bounds on the heuristic function f(n).

Like iterative deepening, it is complete and optimal but it has linear space requirements while compared to A*. It does repeated depth-first searches but the searches are not limited by a simple depth bound. Instead, a path in one of these depth first searches is discontinued if its f value exceeds some cut-off value. In the first search, the cut-off is the heuristic value of the start node. In subsequent searches, the cut-off is the lowest f(n) for nodes n that were visited but not expanded in the previous search.

With an admissible heuristic estimate function h, IDA* is guaranteed to find an optimal (shortest) solution path [1819]. Moreover, IDA* obeys the same asymptotic branching factor as A* [2], if the number of newly expanded nodes grows exponentially with the search depth [6][18][19]20][21]. The growth rate, (heuristic branching factor), depends on the average number of applicable operators per node and the discrimination power of the heuristic estimate h.

The IDA* procedure (pseudo code) is described bellow:

```
Function depth_first_search(n, limit)
    If f(n) > limit
    Then Return f(n)
    If h(n) = 0
    Then successful
      Return  lowest  value  of  depth_first_search(ni,
    limit) for all successors ni of n
end

Procedure IDA*(n)
    limit=h(n)
    repeat
       limit=depth_first_search(n, limit)
    until successful

end
```

3.7.2 *Simplified Memory Bounded A* (SMA*)*

Simplified memory-bounded A* places a size limit on the queue. SMA* makes full use of memory to avoid expanding previously expanded nodes. It discards the least-promising nodes from the queue, if it needs to, in order to keep the queue within the size limit. However, it keeps enough information to allow these discarded paths to be quickly re-generated should they ever be needed [8][12][26][27].

It works as follows: if memory is full and we need to generate an extra node then:

- Remove the highest f-value leaf from the queue;
- Remember the f-value of the best 'forgotten' child in each parent node.

Simplified Memory-Bounded A*

```
if (the initial state is a goal state), then return it
Step 1. Add the root node to the queue.
Step 2. While queue is not empty or failure do
        Step 2.1 If the queue is empty return failure
        Step 2.2.Take the top priority node from the
                 queue as current node
        Step 2.3 If this is a goal state then return it
        Step 2.4 Take a successor of the current node
        Step 2.5 If the successor is not a goal state
                 and it reached the depth-limit
                 then set f(successor) to INFINITE
                 else f(successor) = MAX(f(current),
                 f(successor))
        Step 2.6 If the successor is the last one
                 then update the ancestors' f-cost to be
                 the minimum of their successors's f-cost
        Step 2.7 If no more memory for the successor
                 then delete the shallowest highest-f-
                 cost node and remember the best forgot-
                 ten f-cost
        Step 2.8 Insert the child into the queue
end
```

Example 1

Consider the tree given in Figure 3.13 with 6 nodes (node 1 is the initial state and nodes 5 and 6 are both goal states) and with the known values of f function for each node. The problem is to find the shortest path to a goal using SMA* with a memory of maximum 3 nodes.

We first expand node 1 and we obtain node 2, whose f value is 12. There is still memory for one node and we expand node 1 to obtain node 3. Now the memory is full and none of the children is the goal (see Figure 3.14 (a) and (b). We proceed as follows:

- update f of the node 1 with minimun f of the children (which is 12 for the node 2)
- expand node 2

- drop the higher f-cost leaf (node 3)
- Memorize the f of the node 3 (13).

We are now in the situation depicted in Figure 3.14 (c). The node 4 is further expanded. This is not a goal node so we will do the following:

- memorize node 3 into the node 1
- memory is full
- node 4 is not a goal node, mark it to infinite

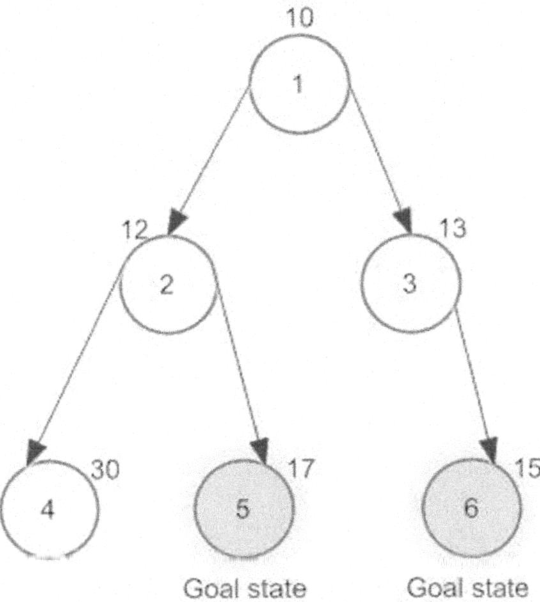

Fig. 3.13 The graph used for exemplifying SMA*.

Now we are in the situation (d) of Figure 3.14. Here we will do the following:

- drop node 4 and add node 5
- node 2 memorizes node 5
- update f of node 2 with the minimum f of the children
- update f for the node 1

Node 5 is a goal node. But since we still have in the memory a value of 13 for an unexpanded node which is lower than the f of the obtained goal, we continue to expand the memorized node.

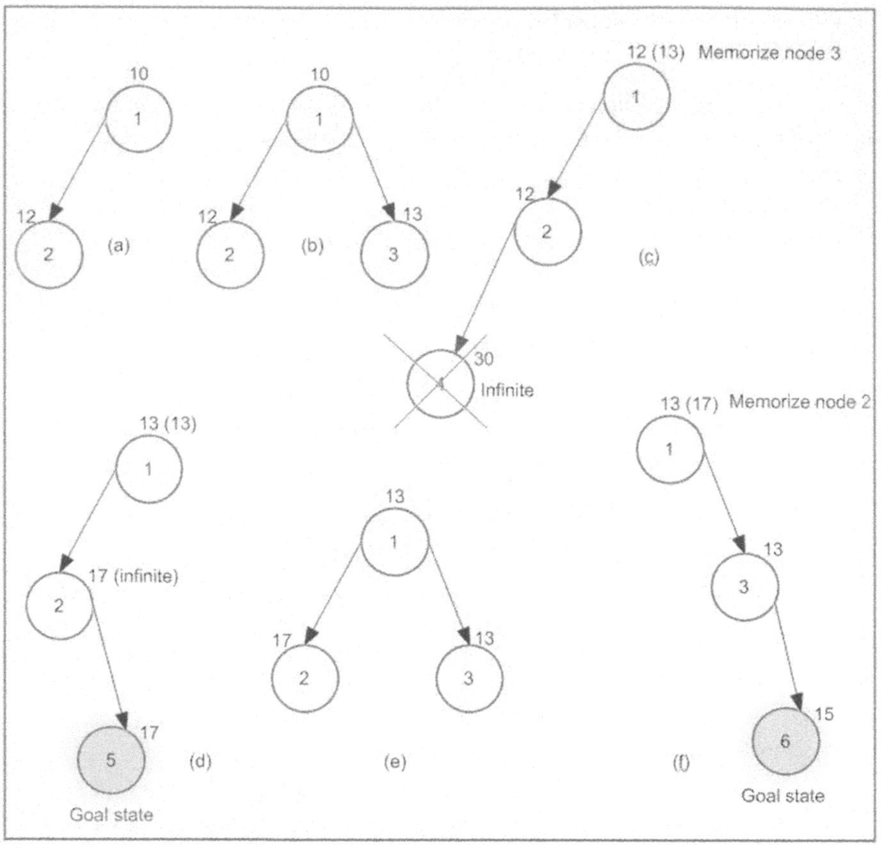

Fig. 3.14 Example of SMA* behavior for Example 1

Thus, node 2 is again expanded (situation depicted in Figure 3.14 (e)). The steps given below are followed:

- drop node 2 and add node 6
- memorize node 6 into the node 1
- node 6 is a goal node and it is lowest f-cost node
- terminate.

Some facts about SMA*:

- utilizes all memory available;
- it has the ability to avoid repeated states;
- is complete if enough memory is available: to store the shallowest solution path;

- Is optimal if enough memory is available: to store the shallowest solution path. Otherwise it returns the best solution that can be reached given the available memory.
- It is optimally efficient if enough memory is available for the entire search tree.

3.7.3 Recursive Best-First Search (RBFS)

Recursive best-first search (RBFS) is a linear-space algorithm that expands nodes in best-first order even with a non-monotonic cost function and generates fewer nodes than iterative deepening with a monotonic cost function. RBFS is similar to depth first search but keeps track of the f-value of the best alternative path available from any ancestor of the current node. If current node exceeds this limit, recursion unwinds back to the alternative path, replacing the f-value of each node along the path with the best f-value of its children. RBFS also remembers the f-value of the best leaf in the forgotten subtree.

The RBSF procedure is presented below.

```
RBFS (node: n ,limit l)
    if f( n) > l return f(n)
    if n is a goal, then exit
    if n has no children, return infinity
    else for each child n_i of n, set f_i  = f(n_i)
    sort n_i and f_i (ascending based on f_i value)
    if there is only one child then f_2 = infinity
    while f_1 ≤ l and f_1 < infinity
          f_1 = RBFS (n_1, min(l, f_2))
          insert n_1 and f_1 in sorted list
    return f_1
```

3.7.4 D* Algorithm

D* is a provably optimal and efficient path planning algorithm and have been proposed for sensor-equipped robots [3]. D*, resembles A* [2], but it is *dynamic* in the sense that arc cost parameters can change during the problem solving process. Like A*, D* maintains an OPEN list of states. The OPEN list is used to propagate information about changes to the arc cost function and to calculate path costs to states in the space. Every state n has an associated *tag* t(n) which is "new" if n has never been on the OPEN list, "open" if n is currently on the OPEN list, and "closed" if n is no longer in the OPEN list.

For each state n, D* maintains an estimate of the sum of the arc costs given by the *path cost* function h(n). This estimate is equivalent to the optimal (minimal) cost from state n to G. For each state n on the OPEN list, the *key* function k(n) is defined to be equal to the minimum of h(n) before modification and all values assumed by since was placed on the list. The key function classifies a state n on the list into one of two types: a *raise* state if k(n) < h(n) and a *lower* state if k(n)=h(n). D* uses raise states on the OPEN list to propagate information about path cost increases and lower states to propagate information about path cost reductions. The propagation takes place through the repeated removal of states from the list. Each time a state is removed from the list, it is *expanded* to pass cost changes to its neighbors. These neighbors are in turn placed on the list to continue the process.

States on the OPEN list are sorted by their key function value. The parameter k_{min} is defined to be min(k(n)) for all n in the OPEN list. The parameter k_{min} represents an important threshold in D*: path costs less than or equal to k_{min} are optimal, and those greater than k_{min} may not be optimal. The parameter k_{old} is defined to be equal to k_{min} prior to most recent removal of a state from the OPEN list. If no states have been removed, k_{old} is undefined [3].

The D* algorithm consists primarily of two functions: Process_state and Modify_cost. Process_state is used to compute optimal path costs to the goal and Modify_cost is used to change the arc cost function and enter affected states on the OPEN list. Initially, t is set to new for all states, h is set to zero, and G is placed on the OPEN list. Process_state is repeated until the state n is removed from the OPEN or a value of -1 is returned. The second function Modify_cost is then used to correct the arc cost function and place affected states on the OPEN list.

D* can handle any path cost optimization problem where the cost parameters change during the search for the solution. D* is most efficient when these changes are detected near the current starting point in the search space. More applications of D* can be found in [4]. D* is intended for use when you don't have complete information. If you don't have all the information, A* can make mistakes; D*'s contribution is that it can correct those mistakes without taking much time.

3.7.5 Beam Search

In the main A* loop, the OPEN set stores all the nodes that may need to be searched to find a path. The Beam Search is a variation of A* that places a limit on the size of the OPEN set. If the set becomes too large, the node with the worst chances of giving a good path is dropped. One drawback is that you have to keep your set sorted to do this, which limits the kinds of data structures you'd choose.

Beam search may be also uses as any heuristic search f(n) = g(n) + h(n). However, it is parameterized by a positive integer k (like depth limited search is parameterized by a positive integer *l*). Once the successors of a node are computed, it only places onto the agenda the best k of those children (those k children have the lowest f(n) values).

Beam search is not complete and not optimal.

Summary

Informed search makes use of problem-specific knowledge to guide progress of search and this can lead to a significant improvement in the performance of search. This chapter presented informed search; concepts of heuristic functions and some of the well known search approaches using them such as Greedy search and A* (including some of its variants).

In practice we often wish the goal with the minimum cost path, which can be accomplished by exhaustive search for small problems but it is practically impossible for other problems. Heuristic estimates of the path cost from a node to the goal can be efficient in reducing the search space. Heuristics can help speed up exhaustive, blind search, such as depth first search and breadth first search. Coming up with a good heuristic is a challenging task: the better the heuristic function, the better the resulting search method will be.

Two main heuristic search algorithms were presented in this chapter: Greedy search and A* search. Greedy search minimizes the estimated cost to the goal, $f(n)$, and it usually decreases the search time but is neither complete nor optimal. If h(n) is an admissible heuristic function, A* search is complete and optimal. However for most of the problems, the number of nodes within the search space is exponential in the length of the solution.

Memory space is the main drawback of A* search (rather than time complexity) because it keeps all the generated nodes in memory. It usually runs out of space long before it runs out of time. Several variants of the A* search have been proposed to overcome some of the A* drawbacks. This chapter presents some of them such as: iterative deepening A* (IDA*), memory-bounded A* (MA*), simplified memory bounded A* (SMA*), recursive best-first search (RBFS) and Dynamic A* (D*).

Admissible heuristics are optimistic: they think the cost of solving the problem is less than it actually is. The depths of the solutions found can be different with different search algorithms and/or heuristics. Quality of a heuristic may be measured by the effective branching factor. Well designed heuristic would have a value of the branching factor close to 1.

References

1. Pearl, J.: Heuristics: Intelligent Search Strategies for Computer Problem Solving. Addison-Wesley, Reading (1984)
2. Nilsson, N.J.: Principles of Artificial Intelligence. Tioga Publishing Company (1980)
3. Stentz, A.: Optimal and Efficient Path Planning for Partially-Known Environments. In: Proceedings IEEE International Conference on Robotics and Automation, pp. 3310–3317 (1994)
4. Stentz, A.: Optimal and Efficient Path Planning for Unknown and Dynamic Environments, Carnegie Mellon Robotics Institute Technical Report CMU-RI-TR-93-20 (August 1993)
5. Dechter, R., Pearl, J.: Generalized best-first search strategies and the optimality of A*. Journal of the ACM 32(3), 505–536 (1985)
6. Mahanti, A., Ghosh, S., Nau, D.S., Pal, A.K., Kanal, L.: Performance of IDA* on trees and graphs. In: 10th Nat. Conf. on Art. Int., AAAI 1992, San Jose, CA, pp. 539–544 (1992)

7. Reinefeld, A.: Complete solution of the Eight-Puzzle and the benefit of node-ordering in IDA*. In: Procs. Int. Joint Conf. on AI, Chambéry, Savoi, France, pp. 248–253 (1993)
8. Chakrabarti, P., Ghosh, S., Acharya, A., DeSarkar, S.: Heuristic search in restricted memory. Artificial Intelligence 47, 197–221 (1989)
9. Ikeda, T., Imai, H.: Enhanced A* algorithms for multiple alignments: Optimal alignments for several sequences and k-opt approximate alignments for large cases. Theoretical ComputerScience 210, 341–374 (1999)
10. Gaschnig, J.: Performance measurement and analysis of certain search algorithms. Technical Report CMU-CS-79-124, Computer Science Department, Carnegie Mellon University (1979)
11. Hansson, O., Mayer, A., Heuristic, A.: search as evidential reasoning. In: Proceedings of the Fifth Workshop on Uncertainty in Artificial Intelligence. Morgan Kaufmann, Windsor (1989)
12. Kaindl, H., Khorsand, A.: Memory-bounded bidirectional search. In: Proceedings of the Twelfth National Conference on Artificial Intelligence (AAAI 1994), Seattle, Washington, pp. 1359–1364. AAAI Press, Menlo Park (1994)
13. Kanal, L.N., Kumar, V.: Search in Artificial Intelligence. Springer, Berlin (1988)
14. Kumar, V., Kanal, L.N.: The CDP: A unifying formulation for heuristic search, dynamic programming, and branch-and-bound. In: Kanal, L.N., Kumar, V. (eds.) Search in Artificial Intelligence, ch. 1, pp. 1–27. Springer, Berlin (1988)
15. Mostow, J., Prieditis, A.E.: Discovering admissible heuristics by abstracting and optimizing: a transformational approach. In: Proceedings of the Eleventh International Joint Conference on Artificial Intelligence (IJCAI 1989), vol. 1, pp. 701–707. Morgan Kaufmann, Detroit (1989)
16. Newell, A., Ernst, G.: The search for generality. In: Kalenich, W. (ed.) Information Processing Proceedings of IFIP Congress 1965, vol. 1, pp. 17–24 (1965)
17. Nilsson, N.J.: Problem-Solving Methods in Artificial Intelligence. McGraw-Hill, New York (1971)
18. Korf, R.E.: Depth-first iterative-deepening: an optimal admissible tree search. Artificial Intelligence 27(1), 97–109 (1985)
19. Korf, R.E.: Iterative-deepening A*: An optimal admissible tree search. In: Proceedings of the Ninth International Joint Conference on Artificial Intelligence (IJCAI 1985), pp. 1036–1043. Morgan Kaufmann, Los Angeles (1985)
20. Korf, R.E.: Optimal path finding algorithms. In: Kanal, L.N., Kumar, V. (eds.) Search in Artificial Intelligence, ch. 7, pp. 223–267. Springer, Berlin (1988)
21. Korf, R.E.: Linear-space best-first search. Artificial Intelligence 62(1), 41–78 (1993)
22. Pohl, I.: Bi-directional and heuristic search in path problems. Technical Report 104, SLAC (Stanford Linear Accelerator Center), Stanford, California (1969)
23. Pohl, I.: First results on the effect of error in heuristic search. In: Meltzer, B., Michie, D. (eds.) Machine Intelligence 5, pp. 219–236 (1970)
24. Pohl, I.: The avoidance of (relative) catastrophe, heuristic competence, genuine dynamic weighting and computational issues in heuristic problem solving. In: Proceedings of the Third International Joint Conference on Artificial Intelligence (IJCAI 1973), Stanford, California, pp. 20–23 (1973)
25. Pohl, I.: Practical and theoretical considerations in heuristic search algorithms. In: Elcock, E.W., Michie, D. (eds.) Machine Intelligence 8, pp. 55–72 (1977)

26. Russell, S.J.: Efficient memory-bounded search methods. In: Proceedings of the 10th European Conference on Artificial Intelligence (ECAI 1992), Vienna, Austria, pp. 1–5 (1992)
27. Zhou, R., Hansen, E.A.: Memory-Bounded A* Graph Search. In: Proceedings of 15th International FLAIRS Conference, Pensecola, Florida (2002)
28. Russell, S.J., Norvig, P.: Artificial Intelligence: A Modern Approach, 2nd edn. Pearson Education, London (2003)

Verification Questions

1. Are there any similarities between Greedy search and uninformed search techniques?
2. How can we compare the complexity (time and space) of Greedy search with the ones of depth first search and breadth first search?
3. How can we compare breadth first search and A* search?
4. How can we compare uniform cost search and A* search?
5. What are the variants of A*?
6. Which of the A* variants perform the best?
7. Which type of applications is D* meant for?
8. How SMA* improves the performance of A*?
9. Which of the heuristic search methods are optimal and complete?
10. When a heuristic function does dominate another one? Give an example.
11. What are the strengths and weaknesses of Greedy search?
12. What are the strengths and weaknesses of A* search?
13. How is IDA* memory complexity and time complexity?
14. What are the proprieties of heuristics?
15. Out of a set of k non-overestimating heuristics which is the most powerful one?

Exercises

3.1. Frobenius Problem.

Let $N = \sum_i x_i c_i$ denote the sum of money that can be represented with coins $c_1, c_2, ..., c_n$.

If $c_1 = 1$, then obviously any quantity of money N can be represented.

Suppose coins are 2, 5 and 10. Then N=1 and N=3 cannot be represented. All other N can be represented. Given coins of denomination $c_1, c_2, ..., c_n$, so that no two have a common factor, find the largest integer N that cannot be changed using these coins.

3.2. A* search uses an evaluation function f:

$$f(n)=g(n) + h(n)$$

where g(n) is the cost of the path from the start node to node n and h(n) is an estimate of the cost of the cheapest path from node n to a goal node.

1. Define g and h so that the search will be the one performed by breadth-first search.
2. Define g and h so that the search will be the one performed by uniform cost search.

3.3. Represent a state space and apply an admissible heuristic to get a solution path, but give a value for k for which beam search would fail to find that solution path.

3.4. Represent a state space and show how an admissible heuristic will find the solutions (consider two solution paths), but give a value for k for which beam search on your state space would only find the more costly of the two solution paths.

3.5. Consider the 8 puzzle presented in the beginning of the chapter.

a) Define two heuristics h_1 and h_2 different from the ones used in this chapter.
b) Solve the puzzle using your heuristics with both Greedy and A* search algorithms and see if you getting a better solution (less number of moves).
c) Take another puzzle (can be 8-puzzle or bigger) and apply Greedy and A* with your heuristics for it. Compare the results with the ones obtained by using the heuristics presented in the chapter (number of misplaced tiles and Manhattan distance). Evaluate the methods by using effective branching factor.
d) Analyze whether any of the heuristics proposed by you is providing better results than using Manhattan distance. If so, explain why.

3.6. Consider an 8-puzzle example as in the case of problem 3.6 (the one you picked, different from the one presented in this chapter).

a) Take the heuristic proposed by you for which you obtained the best results.
b) Implement at least 3 variants of A* with this heuristic function and compare the results.

3.7. Dating game

The dating game consists of three males (M), three females (F), and an empty chair.

The initial configuration is in Figure 3.16 (a).

The game has two legal moves with associated costs:

a) A person (male or female) may move into an adjacent empty chair. This has a cost of 1.

b) A person can jump over one or two other persons into the empty chair. This has a cost equal to the number of persons jumped over.

The goal is to pair each male with some female. There are several ways to get a final configuration. But in this case consider as goal state the state depicted in Figure 3.15 (b).

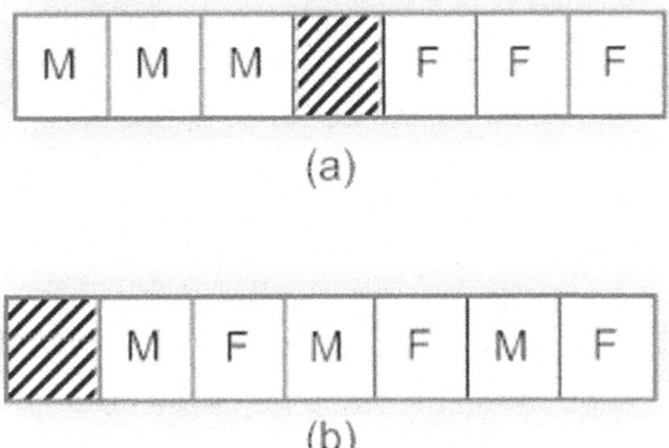

Fig. 3.15 The configuration for the problem 3.7.

Requirements

1) Defining initial state, operators, goal-test, and the path-cost function.

2) Find and define an appropriate heuristic. Specify whether it is admissible or not.

3) Implement greedy and A* search algorithms and compare them.

4) Compare the best from the above algorithms which one of the uninformed search techniques (say depth first search).

3. 8. Suppose we have an admissible heuristic function h for a state space and all action costs are positive.

State which of the following is true and justify your answer.

a) If n is a goal state, then h(n) = 0.
b) If h(n) = 0, then n is a goal state.
c) If n is a 'dead-end' (i.e. it is a non-goal state from which a goal state cannot be reached), then h(n) = 1.
d) If h(n) = 1, then n is a dead-end.

3.9. Consider the problem of coloring a map using 3 colors such as no neighboring countries on the map have the same color.

a) Define a heuristic function for this problem.
b) Apply A* search and IDA* search using this heuristic.
c) Compare and explain the results.

Chapter 4
Iterative Search

4.1 Introduction

This chapter continues with the presentation of other informed search strategies (which are heuristics). They appear to be very useful for certain kind of problems even tough for certain categories of problems the quality of solution(s) provided may be unsatisfactory.

These strategies try to improve space and time complexity but are sacrificing completeness and optimality. In many optimization problems, path to the solution is irrelevant; the goal state itself is a solution. Then the state space is a set of complete configurations and the task of the algorithm is to find the optimal configuration or the configuration which satisfies the constraints. So, these kinds of approaches are suitable when the solutions are states not paths.

Iterative refinement algorithms keep just a single (current) state and try to improve it and usually do not need to keep track of an agenda. Only the current state is kept track of. When the current state is expanded, one of its successors is selected and made the new current state. All other successor states are discarded. They are not placed on an agenda and there is no intention of visiting these unexplored states later. In effect, one path is pursued relentlessly, to the exclusion of all other paths. The idea is to start with a state configuration that violates some of the constraints for being a solution, and make gradual modifications to eliminate the violations. One way to visualize iterative improvement algorithms is to imagine every possible state laid out on a landscape with the height of each state corresponding to its goodness. Optimal solutions will appear as the highest points. Iterative improvement works by moving around on the landscape seeking out the peaks by looking only at the local vicinity. Obviously, the lack of systematic search means that local search is neither complete nor optimal in general. Its success in practice depends crucially on the function used to pick the most promising successor. The very basic iterative search algorithm is given in Algorithm 4.1. It is assumed that the search starts from an initial state, which will be the first current state for the local search algorithm.

C. Grosan and A. Abraham: Intelligent Systems, ISRL 17, pp. 83–109.
springerlink.com

```
Algorithm 4.1 Iterative search
Step 1. Current_state = initial state
Step 2. while current state does not satisfy goal test
        and time limit is not exceeded
        Step 2.1 generate the successors of the
        current_state
        Step 2.3 set as new current_state the successor
        with highest promise (i.e. its heuristic
        value is lowest)
Step 3. if current_state satisfies goal test
        then return path of actions that led to the
        current_state
        else return failure
end
```

In the search algorithms that we have looked at so far, what was important was finding a solution path, i.e. a sequence of actions that transforms the initial state to the goal state. However, sometimes we are not interested in this path of actions. Sometimes we are interested only in finding the goal state itself. Local search is often used for such problems. The method used by local search can be seen as a variation of problem solving by search and we can observe the following analogy:

- *Start (initial) state*: is a complete configuration in the case of local search compared with a single node on a path (which is a solution);
- *Operators*: Changes applied to the current state to (heuristically) improve quality
- *Evaluation function:* Instead of a goal state or goal test, an evaluation function is used. The problem to solve may not have sometimes an exact goal (for instance we are looking for the minimum or maximum value (which is unknown) of a function to optimize);
- *Search:* Consists on improving the quality of current state until some number of iterations of algorithm has been reached or some other termination condition has fulfilled. Typically does not keep track of repeated states.

Iterative search algorithms discussed in this Chapter are:

- Hill-climbing search (ascent/descent search);
- Simulated annealing search;
- Local beam search.

4.2 Hill Climbing

In the hill climbing algorithm (also known as gradient ascent or gradient descent) the idea is to keep improving current state, and stop when we can't improve any

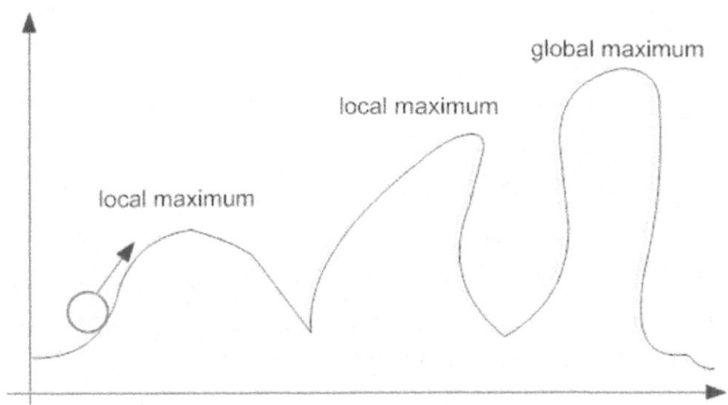

Fig. 4.1 Hill climbing.

more. Hill -climbing is analogous to the way one might set out to climb a mountain in the absence of a map: always move in the direction of increasing altitude (see Figure 4.1). This is especially useful if we don't know ahead of time what the final outcome will be, and want to find the highest ground. Like real-world hill climbing, it suffers from the problems of false peaks: one can reach a non-goal node from which there is no way to go but down.

The hill-climbing algorithm is illustrated in Algorithm 4.2.

```
Algorithm 4.2. Hill-climbing
Step 1. Set current_state to take initial state (starting
        state).
Step 2.   loop
      Step 2.1 Generate successors of current_state;
      Step 2.2 Get the successor with the highest value;
      Step 2.3 if value(successor) < value(current_state)
               then Return current_state
               else currenst_state = successor
   end
```

Depending on the initial state, hill-climbing may get stuck into local optima. Once it reaches a hill the algorithm will stop since any new successor will be down the hill. Figure 4.2 presents two different initial states situation; the one on the left will lead the search process to a local optimum (maximum in this case) and the one on the right will reach the global maximum.

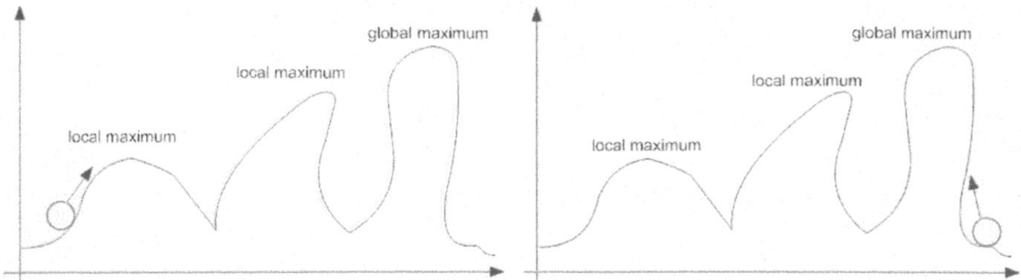

Fig. 4.2 Example of different starting states for the hill-climbing search leading to a local maximum (left) and global maximum respectively (right).

There are some potential dangerous situations for hill-climbing:

- *plateau:* successor states have same values, no way to choose;
- *foothill:* local maximum, can get stuck on minor peak;
- *ridge:* foothill where N-step look ahead might help.

Example 1

We illustrate a very simple practical example for which hill-climbing gets stuck in a local optimum. Consider the 5 geometric figures of sizes 1, 2, 3, 4 and 5 as given in Figure 4.3.

The goal state and the initial state are also given. Just one piece (the top most piece) can be moved at one time and only 2 additional stacks can be used to arrange the pieces.

For simplifying the explanations, let us denote each figure with a number corresponding to its size (piece of size 1 will be 1, piece of size 2 will be 2 and so on).

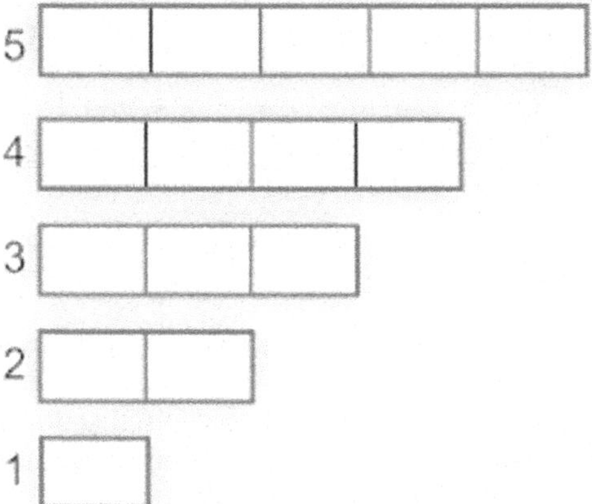

Fig. 4.3 Geometric figures used in Example 1 for hill-climbing.

Let us analyze the behavior of hill climbing using two heuristics. First heuristic will lead to a local solution and is as follows:

- count +1 for every figure that sits on the correct figure. The goal state has the value +5;
- count -1 for every figure that sits on an incorrect figure.

In the initial state (which is the current state) figures 1, 2 and 3 are correctly situated and figures 4 and 5 are wrong situated. This gives the value +3-2=1 for this state.

Since only one piece can be moved at a time, a single successor can be obtained at this step (see figure 4.4 move 1).

The value of this new successor is +4-1 = 3 (the figures 1, 2, 3 and 5 are sitting correctly while 4 is wrong).

The value of this successor is better than the value of the initial state, so the current state will be replaced by the successor.

There are now two possible moves which will conduct to 2 different successors (see figure 4.4 move 2a and move 2b).

The value of the first successor (move 2a) is +3 -2 = 1 (2, 3 and 5 are sitting on correct pieces while 1 and 4 not) and the value of the second successor (move 2b) is again +3-2=1, same like for the first successor.

Both successors have lower values than their parent (the current state) which has the value 3. This leads us to the conclusion that move 1 is the optimum. As evident, this is just a local optimum, not the global solution. So, hill-climbing fails to find the solution in this case.

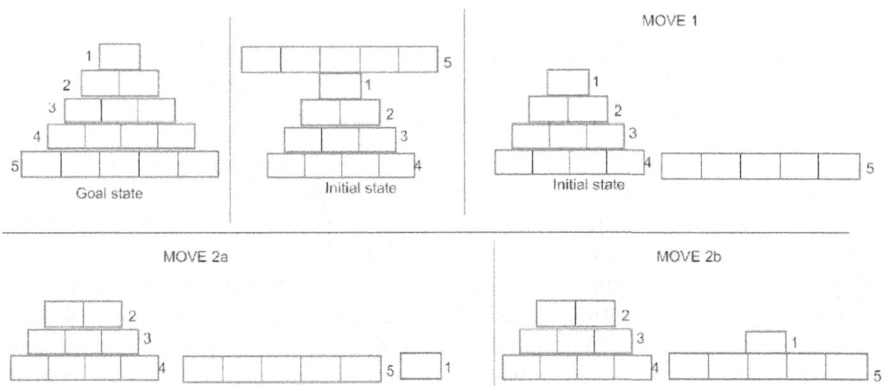

Fig. 4.4 Hill-climbing behavior using heuristic h_1 for Example 1.

Let us now use another heuristic, h2, which is going to be a global heuristic. This heuristic is defined as follows:

- count +n for every piece that sits on a correct stack of n pieces. The goal state has the value +10.
- count -n for every block that sits on an incorrect stack of n.

This heuristic applies a higher penalty for geometric figures placed in a wrong way.

Initial state has the value (-1)+(-2)+(-3)+(-4) for the figures 3 sitting on top of one wrong piece, figure 2 sitting on top of 2 wrong pieces, figure 1 sitting on top of 3 wrong pieces and figure 5 sitting on top of 4 wrong pieces.

The successor obtained by move 1 has the value -6 (same as for the initial state but piece 5 sits now correctly, so we only have (-1)+(-2)+(-3)).

The move 2 generates two successors with values (-1)+(-2) for move 2a (piece 3 sits on top of one wrong piece and piece 2 sits on top of 2 wrong pieces) and (-1)+(-2)+(-1) from the penalty of the pieces 3, 2 and 1 in the move 2b.

As evident, the local optimum is avoided using this second heuristic.

Example 2: Missionaries and cannibals

Let us consider again the missionaries and cannibals problem. 3 missionaries and 3 cannibals are on the left bank of a river. They have to cross the river to the right bank. A small boat is available but this boat can only carry 2 people at a time. Also, the missionaries should not be outnumbered by the cannibals at any time and on any of the river sides.

A straight forward heuristics which comes to anyone's mind first is that of evaluation the number of people on the left bank: on the initial state there are 6 people and the goal state should contain 0 people.

At the first step, there are 3 possible moves (which also respects the problem constraints):

- 2 cannibals cross the river;
- one cannibal and one missionary cross the river;
- one cannibal crosses the river.

By using any of the first 2 moves, the situation on the left bank will get closer to the goal state, so hill-climbing will prefer one of these moves.

However, by doing so, hill-climbing will reach a false optimum (a local optimum). At least one person who already crossed the river has to bring the boat back and this will increase again the number of people on the left side. Hill-climbing will not accept such a solution and will report as final solution the one with 4 people remaining on the left bank and 2 and the right bank. Hill climbing approach stops whenever it can find no move that improves the situation. In this case, it will stop after the first move.

A heuristic which might guide hill-climbing to a better solution for this problem (which we will see it is again a local optimum) is to consider a goal state for each of the river sides and to compare the successors of a state to the goal state of the side they are in at that moment.

For instance, the current state and goal state on the both sides is as follows (1 or 0 on the third position refers to whether the boat is on that side or not):

Left side		Right side
(3, 3, 1)	current state	(0, 0, 0)
(0, 0, 0)	goal state	(3, 3, 1)

We first start from the left side and obtain 3 successors. We compare these successors with respect to the goal situation and we expect to reach on the right side. Again, one of the moves where 2 people reach the right bank is preferred to the one in which just one person crosses the river. Thus, we can get in any of the situations:

Left side	Right side
(2, 2, 0)	(1, 1, 1) or

Left side	Right side
(3, 1, 0)	(0, 2, 1)

Any of these situations is better than the initial state we had on the right side. Suppose the second one is chosen. Now, one or two persons have to return with the boat. There are two possible moves (without violating problem's restrictions), which will lead to the following two situations:

Left side	Right side
(3, 2, 1)	(0, 1, 0) or

Left side	Right side
(3, 3, 1)	(0, 0, 0)

We now compare the new states with the state of the left river side. Always the comparison and evaluation is made with respect to the goal and current state of the side the boat is. First state is better than the previous state on the left side (when the boat was there). So, this will be preferred and the search will continue getting the situations below:

Left side	Right side
(3, 2, 1)	(0, 1, 0)

Left side	Right side
(3, 0, 0)	(0, 3, 1)

Left side	Right side
(3, 1, 1)	(0, 2, 0)

Left side	Right side
(1, 1, 0)	(2, 2, 1)

Left side	Right side
(2, 2, 1)	(1, 1, 0)

At this stage, 2 people have to go back with the boat (otherwise the missionaries will be outnumbered) and the goodness of the new state is same as the one of the current state ((3, 1, 1) and (2, 2, 1)). So, we are again in a local optimum, but this time the solution is closer to the goal state (the state (2, 2, 1) on the right side and (1, 1, 0) on the left side will be reported as final solution).

Example 3: 8-puzzle

We consider a simple example for which hill-climbing works. The result is depicted in Figure 4.5. Heuristic used is the number of misplaced tails, which has to be 0.

For this example hill climbing – like other heuristic search based methods – is able to obtain the global optimum and in a shortest number of steps (which is actually not taken into account by the hill-climbing).

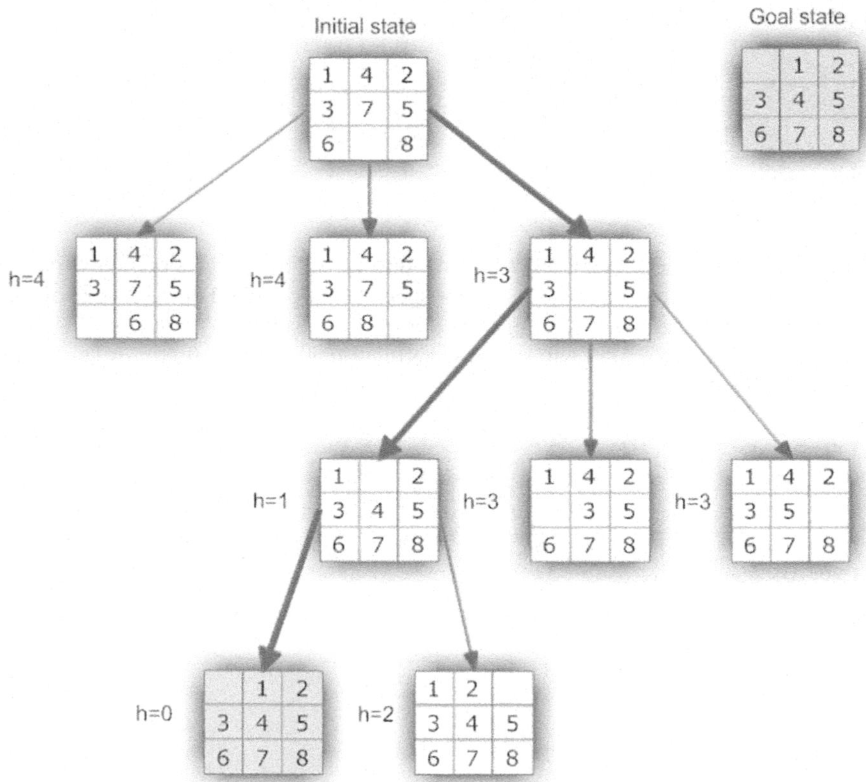

Fig. 4.5 Hill-Climbing for the 8-puzzle problem.

Example 4: Traveling Salesman problem (TSP)

Traveling salesman problem (TSP) is one of the most intensively studied problems.

It can be stated as follows: Given a set of n cities and the cost of travel between each pair of them, the problem is to find the cheapest way of visiting all of the cities and returning to the starting city. For this problem, Hill-climbing starts with a configuration, which is a permutation of the n cities representing the order they are visited. In order to get to a better configuration (or to the optimal one) several operators may be used. The simplest one consists on swapping (or changing) the order two cities are visited. This can be generalized to more changes: 2, 3 or k. This is usually referred to as k-opt, k representing the number of changing or swaps.

Please consider the graph given in Figure 4.6 with nodes A, B, C, D, E and F and with the given cost for each edge. Let us suppose that the starting city is A. Hill-climbing will start with a random state and then will try to optimize it.

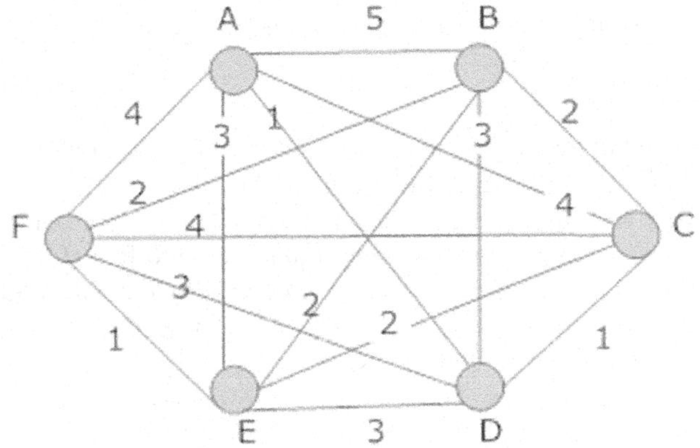

Fig. 4.6 The graph for the TSP example.

Let us consider that the starting configuration is A B C D E F (and suppose from F we will return to A). The solution obtained by hill-climbing when just a consecutive pair of cities is swapped at a time has the cost 15 and is depicted in Figure 4.7.

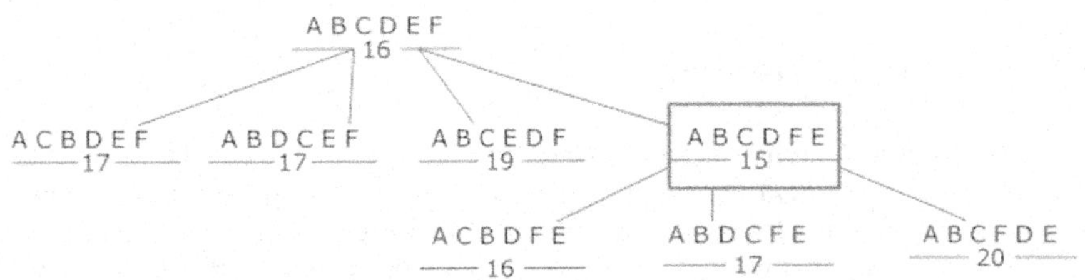

Fig. 4.7 The (local) solution obtained by hill-climbing for TSP.

This solution is just a local optimum. If we consider that any pair of cities can be swapped at a time, then hill-climbing is able to obtain a better solution (see Figure 4.8, with a cost of 10).

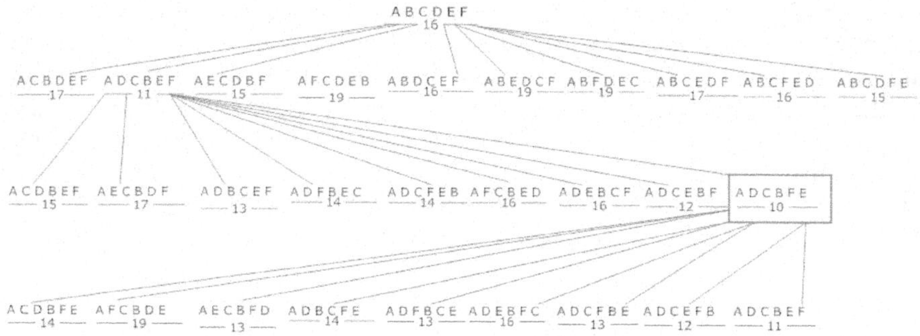

Fig. 4.8 The global solution obtained by hill-climbing for TSP.

There are some advantages as well as some disadvantages of using hill-climbing.

Advantages:

1. easy to implement; the algorithm is very simple and easy to reproduce;
   - requires no memory (since there is no backtracking);
   - since it is very simple, can be easily used to get an approximate solution when the exact one is hard or almost impossible to find (for instance, for very large TSP instances (or other similar NP Complete problems), an approximate solution may be satisfactory when the exact one is not known and difficult to find).

Disadvantages:

- the evaluation function may be sometimes difficult to design;
- if the number of moves is enormous, the algorithm may be inefficient;
- by contrary, if the number of moves is less, the algorithm can get stuck easily;
- it's often cheaper to evaluate an incremental change of a previously evaluated object than to evaluate from scratch;
- inner-loop optimization often possible.

4.3 Simulated Annealing

The hill-climbing search gets stuck, whenever the best of the current's state successors is not better than the parent. The search process stops and many a times only a local solution is found. While compared to hill-climbing, simulated annealing allows moves in the wrong direction on a probabilistic basis; the probability of a backward move decrease as the search continues. The idea is as follows: early in the search, when far from the goal, heuristic may not be good heuristic and it should improve as it gets closer to the goal. The base of simulated annealing is an analogy with thermodynamics, mainly with the way that liquids freeze and crystallize, or metals cool and anneal. At high temperatures, the molecules of a liquid move freely with respect to one another. If the liquid is cooled slowly, thermal mobility is lost. The atoms are able to line themselves up and form a pure crystal that is completely ordered over a distance up to billions of times the size of an individual atom in all directions.

This crystal is the state of minimum energy for this system. It is interesting that, for slowly cooled systems, nature is able to find this minimum energy state. In fact, if a liquid metal is cooled quickly or "quenched", it does not reach this state but rather ends up in a polycrystalline or amorphous state having somewhat higher energy. So the essence of the process is slow cooling, allowing ample time for redistribution of the atoms as they lose mobility. This is the technical definition of annealing, and it is essential for ensuring that a low energy state will be achieved [12].

For a better understanding of Simulated Annealing (SA) technique, we will start with the description of Monte Carlo simulation. A Monte Carlo simulation represents a large number of random trials. Information is obtained by tabulating the results of these trials.

For example, we want to determine the probability of a coin to flip on heads if the coin initially starts with the heads-face showing. In the Monte Carlo simulation, we flip the coin a large number of times, each time with heads showing before you flip, and recording the number of times the coin lands on heads. The probability of getting heads if we start from heads will be given by the number of times it lands on heads divided by the total number of times the coin is flipped.

Monte Carlo simulations use random moves to explore the search space to find out some information about the space. In a simple Monte Carlo simulation, all random moves are accepted such that a different region of search space is sampled at each step. In 1953, Nicholas Metropolis [4] proposed a new sampling procedure, which incorporates a temperature of the system. In contrast with the simple Monte Carlo simulation, a new point in the search space is sampled by making a slight change to the current point. In 1983, Kirkpatrick [6] proposed a method of using a Metropolis Monte Carlo simulation to find the lowest energy (most stable) orientation of a system. Their method is based upon the procedure used to make the strongest possible glass. This procedure heats the glass to a high temperature so that the glass is a liquid and the atoms can move relatively freely. The temperature of the glass is slowly lowered so that at each temperature the atoms can move enough to begin adopting the most stable orientation. If the glass is cooled slowly enough, the atoms are able to "relax" into the most stable orientation. This slow cooling process is known as annealing, and so their method is known as Simulated Annealing.

A Simulated Annealing search starts with a Metropolis Monte Carlo simulation at a high temperature. This means that a relatively large percentage of the random steps that result in an increase in the energy will be accepted. After a sufficient number of Monte Carlo steps, or attempts, the temperature is decreased. The

Metropolis Monte Carlo simulation is then continued. This process is repeated until the final temperature is reached [5]. A Simulated Annealing search starts with an initial solution to the problem, which is also the best solution so far, and the temperature set at the initial, high temperature T_i. This solution becomes the current state. The number of Monte Carlo attempts is set to 0. It is incremented by 1 and is tested to see if it has reached the maximum number of attempts at this temperature. If so, the current temperature is checked. If it is equal to the final temperature, T_f, the simulation is finished and both the final solution and the best solution found during the simulation are stored. If the current temperature is above the final temperature, it is reduced using a cooling schedule. The number of Monte Carlo attempts is reset to 1 [5].

If the number of attempts at this temperature has not been reached, or the temperature has been decreased, the current solution is modified to generate a new solution. This constitutes the Monte Carlo step. If the energy of the new solution is lower than that of the current solution, it is checked to see if it is the best solution found to date. If it is, it is stored separately. Whether or not it is the best, it becomes the new current solution for the next Monte Carlo step. If the energy of the new solution is higher than the parent's by an amount dE, the Boltzmann probability ($e^{-\frac{dE}{kT}}$, where k is Boltzmann's constant and T is the current temperature) is calculated. If this probability is greater than a random number between 0 and 1, this new solution is accepted and becomes the parent solution for the next iteration, and the current solution. Conversely, if the Boltzmann probability is less than the random number, the new solution is rejected and the current/parent solution stays the same and is used in the next iteration [5].

The algorithm employs a random search, which not only accepts changes that decrease the evaluation function f, but also some changes that increase it. The latter are accepted with a probability $p = e^{-\frac{\Delta f}{T}}$, where Δf is the increase in objective function, and f and T are control parameters.

The simulated annealing algorithm is outlined in Algorithm 4.3.

```
Algorithm 4.3. Simulated Annealing
Step 1. Pick a temperature from a sequence (schedule)
of temperatures.
            (Start with hotter temperatures, then
      gradually cool it.)
Step 2. Set t (time)=0
Step 3. While temperature is not 0 do
      Step 3.1. t=t+1;
      Step 3.2 If schedule(t) = 0 (temperature is 0)
               Then return the current solution and
      exit the algorithm.
      Step 3.3 Apply a random operator from set of
               operators to current state
```

```
        Step 3.4 if new state is better than old state
                 then decrease and use this new state
                 if new state is worse than old state
                 then use worse state with some probability
                 else stick with previous better state
end.
```

Remark

As evident, there is an annealing schedule; this means that there is an initial temperature and rules for lowering it as the search progresses. When Δf is negative then large T's increase the probability to take a bad move. Large Δf's reduce probability to take worse move. There are several ways to set the probability of accepting a worsening move, for instance:

- Probability is 0.1;
- Probability decreases with time;
- Probability decreases with time, and also as ΔE increases.

The Metropolis algorithm can be used in other systems than thermodynamic systems by identifying the following elements:

- system configurations;
- a random changes generator (in the configuration);
- an objective function E (analog of energy) whose minimization is the goal of the algorithm;
- a control parameter T (analog of temperature);
- an annealing schedule which tells how the temperature it is lowered from high to low values, e.g., after how many random changes in configuration is each downward step in T taken, and how large is that step. The high and low values can be determined using some physical insight and/or trial-and-error experiments [12].

The way in which physical annealing can be mapped to simulated annealing is presented below (as given in [16]):

Thermodynamic Simulation	Combinatorial Optimization
System States	Feasible Solutions
Energy	Cost
Change of State	Neighbouring Solutions
Temperature	Control Parameter
Frozen State	Heuristic Solution

Using these mappings any combinatorial optimisation problem can be converted into an annealing algorithm [6][14]. There are a few natural questions that may arise while trying to implement simulated annealing for a particular problem. These questions are related to the settings of the parameters involved, but mostly referring to the temperature. For instance, it is important to know how to set the initial temperature, how high it should be (of course, there is no generic value, it depends very much on the problem and many a times can be fixed only by some experiments). Another question is related to the final temperature, whether we should let the algorithm run until it reaches temperature 0 (this may take really long time in certain situations) or another low value will be enough (but this "low" should also be set somehow). It is then naturally to set a temperature decrement; again, there is no general rule, but many algorithms use a high decrement in the beginning and a lower one once the search process advances. Finally, it is important to know how much iterations are to be performed with a temperature, before decreasing it.

Initial temperature. The initial temperature must be hot enough to allow a move to almost any neighbouring state. If this is not high enough then the ending solution will be the same or very similar to the initial state and will play the role of a hill-climbing algorithm.

On the other hand, if the temperature starts at a too high value then the search can move to any neighbour and thus transform the search into a random search. The search will be random until the temperature is cool enough to start acting as a simulated annealing algorithm. The problem is finding the correct starting temperature. There is no known method for finding a suitable starting temperature for a whole range of problems. But this has been suggested in different ways: if the maximum distance (cost function difference) between one neighbour and another is known then this information can be used to calculate a starting temperature.

Some of the proposed methods are similar to how physical annealing works in that the material is heated until it is liquid and then cooling begins. The method suggested in [21] starts with a very high temperature and cools it rapidly until about 60% of worst solutions are being accepted. After this, the temperature will be cooled more slowly.

Dowsland suggested [16] to rapidly heat the system until a certain proportion of worse solutions are accepted and then slow cooling can start. In the beginning a higher number of worse solutions are accepted and in the final cooling stages, this number is dramatically decreased. We can then say that simulated annealing does most of its work during the middle stages of the cooling schedule. Connolly [15] even suggests annealing at a constant temperature. As a conclusion, we cannot set the temperature once for all. This can be done experimentally for a certain class of problems only, not in general, and it is better to be particularly set for each problem (or even for each different instances of the same problem).

Final temperature. The usual final temperature (as state in the algorithm also) is that of value 0. But this can make the algorithm run for a very long time. In practice, it is not necessary to let the temperature reach zero because as it approaches zero the chances of accepting a worse move are almost the same as the temperature being equal to zero.

Therefore, the stopping criteria can either be a suitably low temperature or when the system is frozen at the current temperature (i.e. no better or worse moves are being accepted).

Decrement step. Once the initial and final temperature is set, a step to reach from start to end should be defined. The temperature should be decremented with this step at each time until it will reach the final temperature (if this is the stopping criterion of the algorithm). The way in which the temperature is decremented is critical. Enough iterations are to be allowed at each temperature so that the system stabilises at that temperature. But this can be sometimes exponential to the problem size. We can then either allow a large number of iterations at a few temperatures and a small number of iterations at many temperatures or a balance between the two. The temperature can be decremented linear ($T=T+\alpha$) or geometric ($T=T\cdot\alpha$; In this case α should be between 0 and 1. Experimentally, it was shown that values of α between 0.8 and 0.99 are satisfactory, with preference given to higher values. The higher the value of α, the longer it will take to decrement the temperature to the stopping criterion [13]).

Number of iterations at each temperature. There are several ways to set the number of iterations which are to be performed at a temperature. Of course, none of them is proven to be optimal, but any of them can be considered as an alternative. Some of the possibilities are:

- A constant number of iterations at each temperature;
- A dynamic change of the number of iterations as the algorithm progresses: less iterations at high temperature and large number of iterations at low temperature (so that the local optimum can be fully explored). At higher temperatures, the number of iterations can be less.
- One iteration at each temperature but decrease the temperature very slowly (suggested by [19]) so that you can play around with the parameters, if you are interested.

It has been proved that by carefully controlling the rate of cooling the temperature, SA can find the global optimum[7]. However, this requires infinite time. Several simulated annealing variants have been developed with annealing schedule inversely linear in time (fast simulated annealing), exponential function of time (very fast simulated re-annealing)[9]. There are a couple of advantages and disadvantages of simulated annealing method:

Advantages:

- it is a general technique and can deal with highly nonlinear models, chaotic and noisy data and many constraints;
- its advantages over other local search methods (such as hill climbing for instance) is the ability to approach global optimum[8];
- simple to implement and adapt the code to various problems (even thought it is a bit more complicated than hill-climbing for instance).

Disadvantages:

- there is a clear tradeoff between the quality of the solutions and the time required to compute them;
- it is not easy to design the annealing schedule;
- the precision of the numbers used in implementation is of SA can have a significant effect upon the quality of the outcome.

4.4 Tabu Search

Tabu search (TS) was proposed by Glover [25] (see also [24], [26]-[36]) as an iterative procedure designed for the solution of optimization problems. "Taboo" refers to a strong social prohibition relating to any area of human activity or social custom declared as sacred and forbidden. Breaking of the taboo is usually considered objectionable or abhorrent by society. The word comes from Tongan language and appears in many Polynesian cultures [23]. The most important association with traditional usage, however, stems from the fact that tabus as normally conceived are transmitted by means of a social memory which is subject to modification over time. This creates the fundamental link to the meaning of "tabu" in tabu search. The forbidden elements of tabu search receive their status by reliance on an evolving memory, which allows this status to shift according to time and circumstance. As Glover noted in [28], the origins of TS are in the late 1970s, in combinatorial procedures applied to nonlinear covering problems. There are two important elements in the TS procedure:

(i) tabu moves: determined by a non-Markovian function that uses information from the search process, taking into account last t generations;
(ii) tabu conditions: can be linear inequalities or logical relationships expressed directly in terms of current trial solution. Their role is in choosing the tabu moves (elements which violate the tabu conditions).

If we consider the single optimization problem and the following notations:

(i) S the search space
(ii) N the neighborhood
(iii) T - the set of tabu moves
(iv) f - the evaluation function
(v) x* - optimal solution found so far
(vi) k - number of iterations

then the TS algorithm can be simply described as illustrated in Algorithm 4.4 [28]:

```
Algorithm 4.4. tabu search
Step1. Select x∈S and set x*=x.
          Set k=0.
          Set T=∅.
Step2. Set k=k+1
      Generate a subset of solutions in the neighbor-
hood N-T of x.
Step 3. Choose the best solution s from this neighbor-
hood and set x=s.
Step 4. if f(x)<f(x*)
         then x*=x.
Step 5. if termination criteria
        then stop
        else
           Update T.
           Go to step 2.
end
```

Remarks

(i) Termination criteria used in the algorithm is a fixed number of generations. Once this number is reached, the search process will stop. But there are several other stopping criteria, which may be taken into account:

a. a number of iterations when no improvements occur.;
b. if the solution is known, then the algorithm may stop if the solution has been approximated well enough;
c. there are no more solutions to check in the neighborhood which are not in the tabu list.

(ii) In cases where exclusion from T can be expressed as a requirement to satisfy a set of inequality constraints and the set S can be similarly characterized, the solution s obtained by defining f in this manner represents the outcome obtained by solving an auxiliary optimization problem [28].

(iii) If S-T is large and processed by itemization rather than auxiliary solution, the function f may be based on a strategy for sampling this region, shrinking this set for identifying the minimum f(s).

The usage of a tabu list may prevent cycles of size at most |T|, but, one the other hand, keeping this list in memory may be extremely impractical. Some variants have been proposed to overcome some of the drawbacks. The usage of a moves list instead of a tabu list is one of them.

For each solution i in the search space S, a set M(i) of moves which can be applied to i in order to obtain a new solution j is kept (for simplification, we will use the notation: $j=i\oplus m$ as in [22]). Then neighborhood of i can be defined as $N(i)=\{j / \exists m\in M(i) \text{ with } j=i\oplus m\}$.

So instead of keeping a list T of the last |T| solutions visited, it can simply kept track of the last |T| moves or of the last |T| reverse moves associated with the moves actually performed. For efficiency purposes, it may be convenient to use several lists Tr at a time. Then some constituents tr (of i or of m) will be given a tabu status to indicate that these constituents are currently not allowed to be involved in a move. Generally the tabu status of a move is a function of the tabu status of its constituents, which may change at each iteration. A move m (applied to a solution *i*) will be a tabu move if all conditions are satisfied.

By replacing the solutions by moves in the tabu list, solutions, which may be unvisited so far may be given a tabu status. Since the list of visited moves can grow very much, there should be a way to restrict it. An improvement, which may help is the one of keeping only the most recent visited states in the memory. A recency function may be used to restrict the size of the list in some way; it keeps the most recently visited states in the list - discarding the others.

The easiest (and most usual) implementation of this function is to simply keep the list at a fixed size and use a queue working as FIFO (First-In, First-Out) to maintain the list.

The list-size parameter may be dynamic and change as the algorithm runs. One method of doing this is to keep the list small when states are not being repeated very often but, when repeated states keep being generated the list is made larger so that the search is forced to explore new areas. Another concept is further used: aspiration level. A tabu move m may appear attractive because it gives a solution better than the best found so far. m can be then accepted in spite of its status if it has an *aspiration level* which is better than a threshold given value. The objective function can be modified by introducing two more terms – intensification and diversification – which will penalize certain solutions:

- By introducing the intensification term in the objective function priority will be given to the solutions which have common features with the current solution and solutions far from the current one will be penalized.
- By introducing the diversification term in the priority function diversification of the exploration will try to spread over different regions of the search space. This term will penalize (at some stage) solutions which are close to the current one.

Example: 0/1 Knapsack problem

The 0/1 knapsack problem is a widely studied NP-Complete problem and can be stated as follows: a set of n items is given and for each itcm i its utility u_i and its weight w_i is known. A knapsack of capacity C is also given.

The problem is to find a subset of the items set which can be taken into the knapsack such as the following constraints hold:

- the knapsack capacity is not overloaded;
- the total utility of the items taken is maximum.

We illustrate how to apply tabu search for this problem. Consider a tabu list of moves and also take into account the aspirations conditions (a move from the tabu list can be re-considered if no better solution can be obtained in 3 consecutive trials). At each step, consider a neighborhood of the current solution consisting of 6 neighbors.

Consider the following data:
Knapsack capacity: C= 50;
Number of items n= 7 (each item will be identified by its number)
The weight and utility for each of the items are provided below:
Item 1: weight 10, utility 2
Item 2: weight 12, utility 1
Item 3: weight 15, utility 3
Item 4: weight 27, utility 4
Item 5: weight 30, utility 1
Item 6: weight 20, utility 3
Item 7: weight 7, utility 1

For simplicity, a state will be denoted by a binary array of size 7 (the number of items). 0 or 1 on a position has the meaning that that item is not considered (if 0) or considered (if 1) to be taken into the backpack.

For instance, the solution 0 1 1 0 1 0 0 corresponds to the items 2, 3, and 5 to be considered. We now revise the basic elements requited by tabu search for this problem:

- *Initial solution*: we will just consider a random starting state.
- *Possible moves*: we consider the moves which allow deleting as well as adding an item to the current configuration (just one move at a time). We will thus have the moves *add*$_i$ and *delete*$_i$ for i=1 to 7.
- *Evaluation function*: in order to evaluate a solution we will first check whether the problem constraints are satisfied. In this case, the knapsack capacity should not be overloaded. We will penalize any overload of the knapsack capacity with an amount of 50. Thus, we have a first criterion f_1 which is the difference (in absolute value) between the total weight of the selected items and the knapsack capacity + the penalty (50) in case the capacity is overloaded (and this criterion is to be minimized). But we have to take into account the utility of the selected objects which should be as high as possible (so this has to be maximized). Let us denote this by f_2. Naturally, one will tell that the overall evaluation function f will be f_1+f_2. This is not possible in our case because f1 is to be minimized and f2 is to be maximized. We can then consider $f=f_1+(-f_2)$ and we wish to minimize it.

The reader may observe how tabu search will work for a couple of iterations.

Iteration 1
Initial state: 1 0 1 0 0 1 1
The value of f for the initial state is 52-9=43.
The tabu list is the empty set.
Let us randomly generate the following solutions in the neighborhood of the initial state:

Solution	Move	f
1 1 1 0 0 1 1	Add_2	54
1 0 1 1 0 1 1	Add_4	66
1 0 1 0 1 1 1	Add_5	72
0 0 1 0 0 1 1	$Delete_1$	1
1 0 0 0 0 1 1	$Delete_3$	7
1 0 1 0 0 0 1	$Delete_6$	12

The solution with minimum f is 0 0 1 0 0 1 1 which is better than the initial state, so this solution will be the new current state.

The tabu list is now {1}

Iteration 2
We will now generate successors in the neighborhood of this new state.

Solution	Move	f
0 1 1 0 0 1 1	Add_2	46
0 0 1 1 0 1 1	Add_4	58
0 0 1 0 1 1 1	Add_5	64
0 0 1 0 0 1 0	$Delete_7$	9
0 0 0 0 0 1 1	$Delete_3$	23
0 0 1 0 0 0 1	$Delete_6$	24

There is no solution obtaining a better value for f at this step. Thus, a new set of neighboring solutions will be generated. The process will iterate this way (and we leave the remaining as an exercise). The solution with the optimal value is 0 0 1 1 0 1 0, with items 3, 4 and 6 considered and with a value f = -7. It might take quite a lot time to reach this solution by allowing only one move at a time. If we will allow at least two moves to be performed at a step, the solution will be reached much faster.

Let us start with the same initial state 1 0 1 0 0 1 1 (f=43) and consider a maximum of two moves at a time.

At the first iteration we might get the following set of neighboring solutions:

Solution	Move	f
0 1 1 0 0 1 1	Add_2 $Delete_1$	46
1 0 1 1 0 0 1	Add_4 $Delete_6$	49
1 0 1 1 0 1 0	Add_4 $Delete_7$	60
1 0 1 0 0 0 1	$Delete_6$	12
1 1 1 0 0 1 0	Add_2 $Delete_7$	48
1 0 1 0 1 1 1	Add_5	72

The best solution obtained now is 1 0 1 0 0 1 0 whose f is 12. This solution is better than the current solution thus this will be the new current solution in the next iteration.

Iteration 2
Current solution: 1 0 1 0 0 0 1
Tabu list: {6}

Solution	Move	f
0 0 1 0 1 0 1	Add_5 $Delete_1$	44
0 0 1 1 0 0 1	Add_4 $Delete_1$	-7
1 0 0 1 0 0 1	Add_4 $Delete_3$	-1
1 1 1 0 0 0 1	*Add2*	-1
0 1 1 0 0 0 1	Add_2 *Delete1*	11
1 0 0 0 0 0 1	$Delete_3$	30

The best solution among all the successors is 0 0 1 1 0 0 1 with the value -7 for f. This will be the next current solution.
The tabu list is updated and will be {6, 4, 1}.
This solution is actually the final solution and it is easy to observe that no improvements will occur if the iterations are continued.

Remarks

- a good starting (non-optimal solution) can be found quickly using a greedy approach and then the tabu search can be applied;
- number of iterations could be sometimes very large even for simple problems;
- in certain situation, the global optimum may not be found, depends on parameter settings.

4.5 Means Ends

The idea of this algorithm belongs to Newell and Simon [37]. It is very similar with hill-climbing and it works by reducing the difference between states, and so approaching the goal state. The algorithm is summarized in Algorithm 4.5:

```
Algorithm 4.5 Means ends search
set the current_state node to initial state;
loop
      if the goal state has been reached
      then return success and exit the algorithm
      else    find    the    difference    between    the
           current_state and the goal   state;
           choose   a   procedure   that   reduces   this
           difference
           apply it to the current_state to produce the
           new current state;
end
```

Means-Ends Analysis has some disadvantages such as:

- failure to find an operator to reduce a difference;
- sometimes must return to the initial state.

4.6 Summary

This chapter presented a number of heuristic search methods. The distinction between search techniques is related to the distinction between weak and strong methods of problem solving. A weak method uses blind search or a heuristic that is broadly applicable to many kinds of problems - e.g. means-ends-analysis as used with GPS.

A strong method uses a heuristic that incorporates significant knowledge about the specific problem - e.g. the examples we will look at in conjunction with using heuristic search with the 8 puzzle shortly. Three important methods are presented in this chapter: hill-climbing, simulated annealing and tabu search.

Hill-climbing only works if the heuristic is accurate (i.e., distinguishes "closer" states) and if the path to the goal is direct (i.e., state improves on every move). There are ways to generalize hill-climbing to continue even if the successor states look worse. For instance, always choose best successor and don't stop unless reaches the goal or there are no successors; dead-ends are still possible (and likely if the heuristic is not perfect).

Simulated annealing is a random-search technique, which exploits an analogy between the way in which a metal cools and freezes into a minimum energy crystalline structure (the annealing process) and the search for a minimum in a more general system. Simulated annealing has the advantage (over other search methods) to avoid becoming trapped in local minima. The algorithm employs a random search which not only accepts changes that decrease the objective function f (assuming a minimization problem), but also some changes that increase it (which are accepted with a probability).

It is an optimization technique for combinatorial and other problems[7][8][9].

A disadvantage of thc simulated annealing is that the methods are computation-intensive. There exists faster variants of basic simulated annealing, but these apparently are not as quite easily coded and so they are not widely used. Simulated Annealing guarantees a convergence upon running sufficiently large number of iterations.

Hill climbing suffers from problems in getting stuck at local minima (or maxima). Several ways can be tried to overcome these problems but none of them have proved satisfactory in practice when using a simple hill climbing algorithm.

Simulated annealing solves this problem by allowing worse moves (lesser quality) to be taken some of the time. That is, it allows some uphill steps so that it can escape from local minima. Unlike hill climbing, simulated annealing chooses a random move from the neighbourhood (recall that hill climbing chooses the best move from all those available – at least when using steepest descent (or ascent)).

Simulated annealing has been proved to converge to the best solution but it might take (in the worse case) more time than exhaustive search. Although it may not be practical to find the best solution using simulated annealing, simulated annealing does have this important property which is being used as the basis for future research. The efficiency of iterative solution methods depends mostly on the modeling. A fine tuning of parameters will never balance a bad choice of the neighborhood structure or of the objective function. On the opposite, an effective modeling should lead to robust techniques that are not too sensitive to different parameter settings [22].

Tabu search allows non-improving solution to be accepted in order to escape from a local optimum. Often it gives the global optimum, if correctly implemented, in a very short time. Tabu search obtains solutions that rival and often surpass the best solutions previously found by other approaches but there are too many parameters to be determined and also the number of iterations could be very large in several situations. Depending on the parameter settings, the global optimum may not always be reached. An important thing to notice in all these techniques is that only one solution is kept and tried to improve all the way during the search process. We will see in some of the forth coming chapters that there are several other heuristics which play not only with a single starting (and current solution) but with a whole set of solutions.

References

1. Lin, S., Kernighan, B.W.: An effective heuristic algorithm for the Traveling-Salesman Problem. Operations Research 21(2), 498–516 (1973)
2. Lin, S.: Computer Solutions of the Traveling Salesman Problem. Bell System Tech. J. 44, 2245–2269 (1965)
3. Laporte, G.: The traveling salesman problem: An overview of exact and approximate algorithms. European Journal of Operational Research 59(2), 231–247 (1992)
4. Metropolis, N., Rosenbluth, A.W., Rosenbluth, M.N., Teller, A.H., Teller, E.: Equation of State Calculations by Fast Computing Machines. Journal of Chemistry and Physics 21, 1087–1092 (1953)

5. Luke, B.T.: Simulated annealing, Technical Report, `http://members.aol.com/btluke/simann1.htm`
6. Kirkpatrick, S., Gelatt, C.D., Vecchi, M.P., Optimization, M.P.: by Simulated Annealing. Science 220(4598), 671–680 (1983)
7. Ingber, L.: Simulated annealing: practice versus theory. Mathl. Comput. Modelling 18(11), 29–57 (1993)
8. Ingber, L., Wehner, M.F., Jabbour, G.M., Barnhill, T.M.: Application of statistical mechanics methodology to term-structure bond-pricing models. Mathl. Comput. Modelling 15(11), 77–98 (1991)
9. Ingber, L.: Very fast simulated re-annealing. Mathl. Comput. Modelling 12(8), 967–973 (1989)
10. Moore, A., Climbing, H.: Simulated Annealing and Genetic Algorithms, Auton Lab Tutorials, Carnegie Mellon University, `http://www.autonlab.org/tutorials/hillclimb02.pdf`
11. Ingber, L., Rosen, B.: Genetic algorithms and very fast simulated reannealing: A comparison, Mathl. Comput. Modelling 16(11), 87–100 (1992)
12. Press, W.H., Flannery, B.P., Teukolsky, S.A., Vetterling, W.T.: Numerical recipes in C: the art of scientific computing. Cambridge University Press, Cambridge (1992)
13. Kendall, G.: Simulated Annealing: course notes, AI Methods course, University of Nottingham, `http://www.cs.nott.ac.uk/~gxk/`
14. Cĕrny, V.: A Thermodynamical Approach to the Travelling Salesman Problem; An Efficient Simulation Algorithm. J. of Optimization Theory and Applic. 45, 41–55 (1985)
15. Connolly, D.T.: An Improved Annealing Scheme for the QAP. European Journal of Operations Research 46, 93–100 (1990)
16. Dowsland, K.A.: Simulated Annealing. In: Reeves, C.R. (ed.) Modern Heuristic Techniques for Combinatorial Problems. McGraw-Hill, New York (1995)
17. Hajek, B.: Cooling Schedules for Optimal Annealing. Mathematics of Operations Research 13(2), 311–329 (1988)
18. Johnson, D.S., Aragon, C.R., McGeoch, L.A.M., Schevon, C.: Optimization by Simulated Annealing: An Experimental Evaluation; Part II. Graph Coloring and Number Partitioning. Operations Research 39, 378–406 (1991)
19. Lundy, M., Mees, A.: Convergence of an Annealing Algorithm. Math. Prog. 34, 111–124 (1986)
20. Mitra, D., Romeo, F., Sangiovanni-Vincentelli, A.: Convergence and Finite Time Behavior of Simulated Annealing. Advances in Applied Probability 8, 747–771 (1986)
21. Rayward-Smith, V.J., Osman, I.H., Reeves, C.R., Smith, G.D.: Modern Heuristic Search Methods. John Wiley & Sons, Chichester (1996)
22. Hertz, A., Taillard, É.D., de Werra, D.: Tabu Search. In: Aarts, E., Lenstra, J.K. (eds.) Local Search in Combinatorial Optimization, pp. 121–136. J. Wiley & Sons Ltd., Chichester (1997)
23. `http://en.wikipedia.org`
24. Laguna, M., Glover, F.: What is Tabu Search? Colorado Business Review LXI(5) (1996)
25. Glover, F.: Heuristics for integer programming using surrogate constraints. Decision Sciences 8(1), 156–166 (1977)
26. Glover, F.: Future paths for Integer Programming and Links to Artificial Intelligence. Computers and Operations Research 5, 533–549 (1986)
27. Glover, F.: Tabu Search Methods in Artificial Intelligence and Operations Research. ORSA Artificial Intelligence 1(2), 6 (1987)

28. Glover, F.: Tabu Search - Part I. ORSA Journal on Computing 1(3), 190–206 (1989)
29. Glover, F.: Tabu Search - Part II. ORSA Journal on Computing 2(1), 4–32 (1990)
30. Glover, F.: Tabu Search: A Tutorial. Interfaces 20(4), 74–94 (1990)
31. Glover, F.: Tabu Search: New Options for Optimization. ORSA Computer Science TS Letters 15(2), 13–20 (1994)
32. Glover, F., Laguna, M.: Tabu Search. Kluwer Academic Publishers, Boston (1997)
33. Glover, F.: A template for scatter search and path relinking. In: Hao, J.-K., Lutton, E., Ronald, E., Schoenauer, M., Snyers, D. (eds.) AE 1997. LNCS, vol. 1363, pp. 13–54. Springer, Heidelberg (1998)
34. Glover, F.: Scatter Search and Path Relinking. In: Corne, D., Dorigo, M., Glover, F. (eds.) New Methods in Optimisation, McGraw-Hill, New York (1999)
35. Glover, F., Laguna, M.: Tabu Search. In: Pardalos, P.M., Resende, M.G.C. (eds.) Handbook of Applied Optimization, pp. 194–208. Oxford University Press, Oxford (2002)
36. Glover, F., Laguna, M., Martí, R.: In: Ghosh, A., Tsutsui, S. (eds.) Advances in Evolutionary Computation: Theory and Applications, pp. 519–537. Springer, New York (2003)
37. Simon, H.A., Newell, A.: Heuristic problem solving: The next advance in operations research. Operations Research 6, 1–10 (1958)

Verification Questions

1. What are the issues with hill-climbing?
2. Which are the three dangerous situations of hill climbing?
3. What are the advantages and disadvantages of hill climbing?
4. Describe the analogy between simulated annealing and metal cooling (or annealing) in physics.
5. What are the main parameters of simulated annealing?
6. How you define a neighborhood for a current state in simulated annealing?
7. What are the main steps in the cooling process) in the case of simulated annealing)?
8. How can a high initial temperature influence the search process?
9. How can a low initial temperature influence the search process?
10. Is there an optimal way to decrease the temperature?
11. Define at least two stopping criteria for the simulated annealing algorithm.
12. Will simulated annealing always reach the optimum?
13. How you compare simulated annealing and hill-climbing?
14. What are the advantages and disadvantages of simulated annealing?
15. Which type of problems is simulated annealing fit for?
16. Define the main parameters of tabu search
17. Explain the differences between tabu search and hill climbing.
18. What is the meaning of a tabu list?
19. How can the search re-visit some of the states which are in the tabu list?
20. What are the advantages and disadvantages of tabu search?

Exercises

4.1 Describe the way hill-climbing can be used to solve n-queens problem.

4.2. Consider the simple TSP instance with 4 nodes given in Figure 1. Apply hill-climbing, tabu search and simulated annealing. Compare the results and the performances.

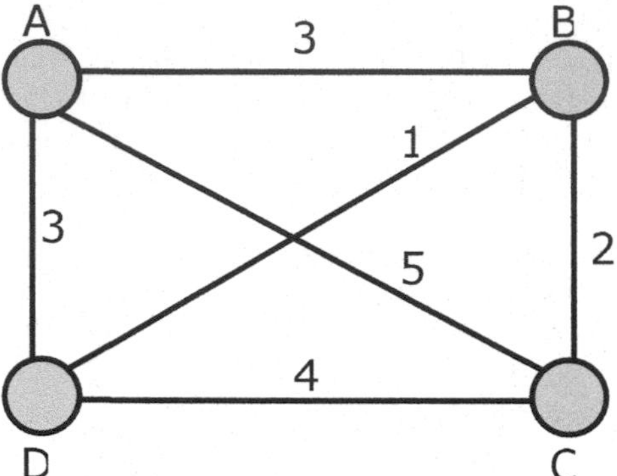

Fig. 1 Graph for the problem 4.2.

4.3 Find a simple problem for which hill-climbing works better (faster) than simulated annealing (and give explanations).

4.4 Find an example for which simulated annealing works better than hill climbing (or build such an example and explain the process).

4.4 Consider the following problem:

A set of n jobs and a set of m machines are given. Each job is composed on a number of units and for each machine the capacity is known (the number of job units processed in a time unit). A machine can only process a single job at a time. The task is to allocate each job to a single machine such as the total time required to process all the jobs is minimal.

a) design a simulated annealing algorithm to solve it;
b) change the annealing schedule defined above;
c) use a different stopping criterion from the one defined initially;
d) change the initial starting point;
e) design a tabu search algorithm to solve it;
f) compare the results of simulated annealing and tabu search.

4.5. Consider the optimization problem $f(x)=x^2$, $x\in[-2, 2]$. The graphic of this function is depicted in Figure 2 (left).

a) Apply hill climbing for finding the minimum of this function;
b) Apply simulated annealing for finding the minimum of f;

c) Apply tabu search for optimizing the function.
d) Compare the results;
e) Consider the function $f(x)= 5+0.5 \cdot x^2-5 \cdot \cos(3 \cdot x)$ (the graphic is depicted in Figure 2 (right).

    1. Apply again hill-climbing and simulated annealing for the optimization (minimization of this problem). In case hill-climbing fails, explain why it happens like that and try with several other starting points.
    2. Try simulated annealing with different starting point, different temperature schedules, different initial temperature and different stopping criteria.
    3. Report the values of the parameters required by simulated annealing to approximate the solution with an error of 0.0001.

(Minimum for both test functions is reached at x=0, and the value of f in this point is also 0 in both cases.)

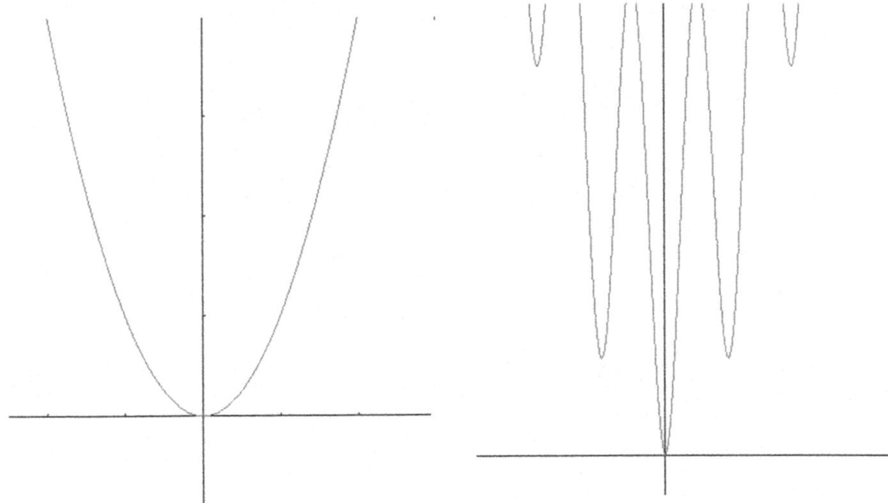

Fig. 2 Illustration of the functions for Exercise 4.5

Chapter 5
Adversarial Search

5.1 Introduction

In the Chapters 2-4, we presented single agent search methods, that is, we only have one player, which has to move, without depending on the moves of another player (or players) and without competing or collaborating with any other players. This type of search is single agent search, and, naturally, multi-agent search is there in its turn.

In this chapter we formulate a multi-player game as a search problem[9][13][14][17][19] and also illustrate how multi-agent search works. We then consider games in which the players alternatively making moves. The goal is to maximize and respectively minimize a scoring function (also called utility function). We only consider the following type of games:

- two player games;
- zero sum - one player's win is the other's loss; there are no cooperative victories.

We also focus on games of perfect information. A perfect game is a game with the following characteristics:

- deterministic and fully observable;
- turn taking: the actions of two players alternate;
- zero sum: the utilities values at the end of the game are equal and opposite.

Examples of perfect games are chess, checkers, go, Othello. As in the case of uninformed and informed search, we can define the problem by its four basic elements:

- *Initial state*: the initial board (or position);
- *Successor function* (or operators): defines the set of legal moves from any position;
- *Goal test*: determines when the game is over;
- *Utility function* (or *evaluation function*): gives a numeric outcome for the game.

C. Grosan and A. Abraham: Intelligent Systems, ISRL 17, pp. 111–129.
springerlink.com

Adversarial search is used in games where one player (or multiple players) tries to maximize its score but it is opposed by another player (or players).

5.2 MIN-MAX Algorithm

Proposed by John von Neumann in 1944, the search method called *minimax* maximizes your position whilst minimizing your opponent's position. The search tree in adversarial games consists of alternating levels where the moving player tries to maximize the score (or fitness) and this player it is called MAX and then the opposing player tries to minimize it (this player is called MIN). MAX always moves first and MIN is the opponent. An action by one player is called a *ply*, two ply (an action and a counter action) is called a *move*.

Remark

The utility function has a similar role as the heuristic function (as illustrated in the previous chapters), but it evaluates a node in terms of how good it is for each player. Figure 5.1 shows an example of utility function for tic-tac-toe. Positive values indicate states advantageous for MAX and negative values indicate states advantageous for MIN.

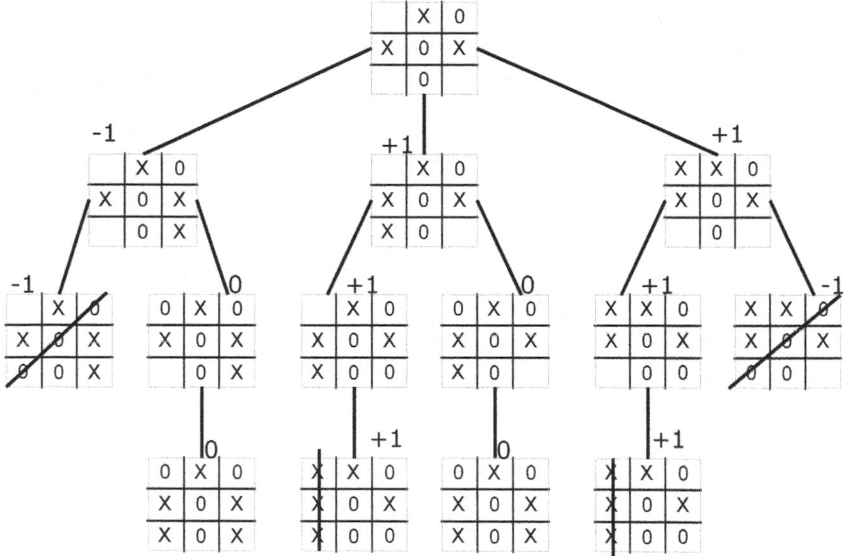

Fig. 5.1 Example of utility function for tic-tac-toe. Positive values indicate states advantageous for MAX and negative values indicate states advantageous for MIN.

To find the best move, the system first generates all possible legal moves, and applies them to the current board. In a simple game this process is repeated for each possible move until the game is won, lost, or drawn. The fitness of a top-level move is determined by whether it eventually leads to a win. The generation of all moves is possible in simple games such as tic-tac-toe, but for complex games (such as chess) it is impossible to generate all the moves in reasonable time. In this case only the state within a few steps ahead may be generated.

In its simplest form, the MIN-MAX algorithm is outlined as Algorithm 5.1:

Algorithm 5.1 MIN-MAX Algorithm

```
Step 1. Expand the entire tree below the root.
Step 2. Using the utility (evaluation) function,
        evaluate the terminal nodes as wins for the mi-
        nimizer or maximizer.
Step 3. Select a node all of whose children have been
assigned values.
      Step 3.1. if there is no such node
                 then the search process is finished.
                 Return the value assigned to the root.
      Step 3.2 if the node is a minimizer move
                then assign it a value that is the
                minimum of the values of its children.
      Step 3.3 if the node is a maximizer move assign
                it a value that  is the maximum of the
                values of its children.
Step 4. Return to Step 3.
end.
```

5.2.1 Designing the Utility Function

A suitable design for the utility function will influence the final result of the search process. Thus, it is not an easy task to design an adequate utility function. We provide few examples to illustrate the ways the utility functions may be designed for a couple of problems. The utility function is applied at the leaves of the tree. In what follows, we refer to a node n for which we calculate the utility function.

Utility function for tic-tac-toe

Let us suppose MAX is using X and MIN is using 0.

The utility function can be defined as:

- if n is win for MAX then $f(n)= +\infty$
- if n is win for MIN then $f(n) - \infty$
- else count how many rows, columns and diagonals are occupied by each player and subtract.

Let us consider the states given in Figure 5.2. For simplicity denote by the triplet (r, c, d) the number of rows, columns and diagonals respectively occupied by either X or 0.

For the state in Figure 5.2 (a), we have (1, 1, 2) for X which means X is occupying 1 row (the middle row), one column (the middle column) and 2 diagonals. For 0 we have (1, 1, 0). Thus the utility function for this state will have the value 1-1 + 1-1 +2-0 =2.

For the state in Figure 5.2 (b) we have (1, 2, 2) for X and (1, 1, 0) for 0. The value of f in this case is 3. The value of f for the state in Figure 5.2 (c) is 1+1+2 = 4; we have (2, 2, 2) for X and (1, 1, 0) for 0. The values of f for the cases in Figure 5.2 (d), (e), (f) and (g) are 1, 1, 2 and 1 respectively.

For the cases depicted in Figure 5.2 (i)-(l) we obtain +∞ for (i) (X is the winner with 3 on a row), and the values 3, 3, and 1 for (j), (k) and (l) respectively. For the case (m) the value of f is -∞ (0 is the winner with 3 on a row). For the case (n) the utility function value is 1 (we have the values (2, 3, 2) for X and (2, 3, 1) for 0).

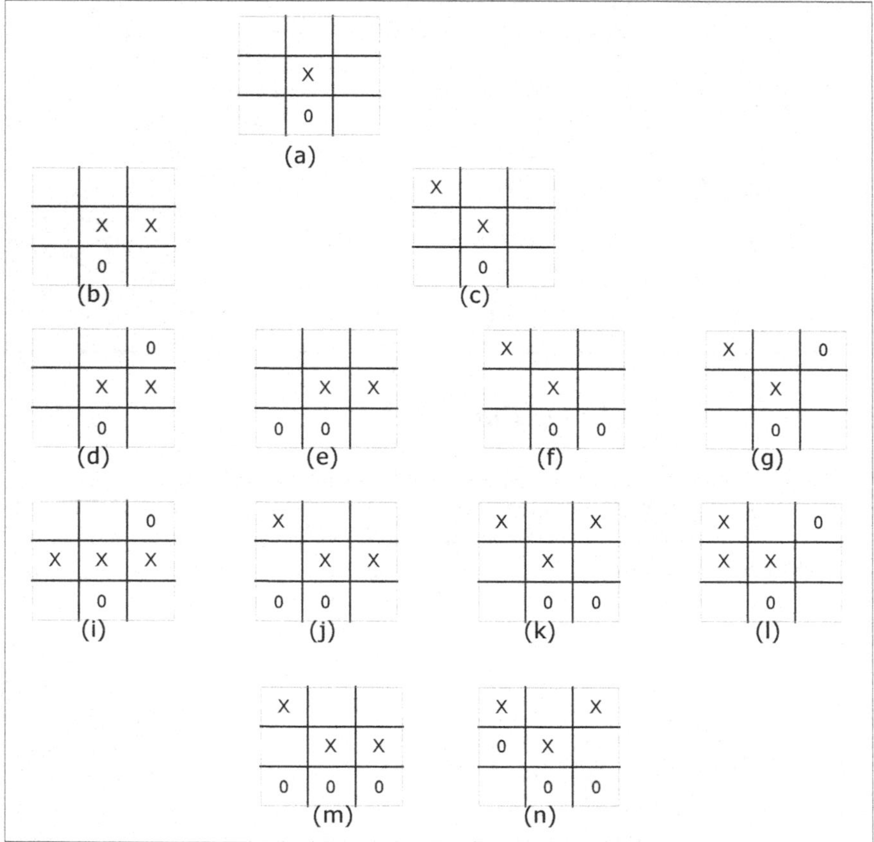

Fig. 5.2 Different utility function values corresponding to different states for the tic tac toe game.

Utility function for Othello

Othello game (also known as reverse) consists on an 8 × 8 board (like chess board) and 64 pawns (32 black and 32 white). There are two players which move alternately, by placing their pawns on the board. The pawns placed on the board are not allowed to be moved. The only thing players can do is to change their color. The board starts with the configuration given in Figure 5.3 and the black moves first. When the player's pawn lies near the enemy one, and the player puts a new pawn behind the enemy one, it will change its color into the player's color. It is called capturing. A player can capture any number of enemy pawns provided that the pawns are in one row between the two player's pawns. Furthermore capturing during making a move is absolutely obligatory. Actually, when a player cannot capture during his move, he must resign from the move and the other player will move.

At a time, one player can have more than one possibility of capturing the enemy pawns and he can freely choose any of them. The objective of the game is to cover all squares on the board and have more pawns in your color than the opponent.

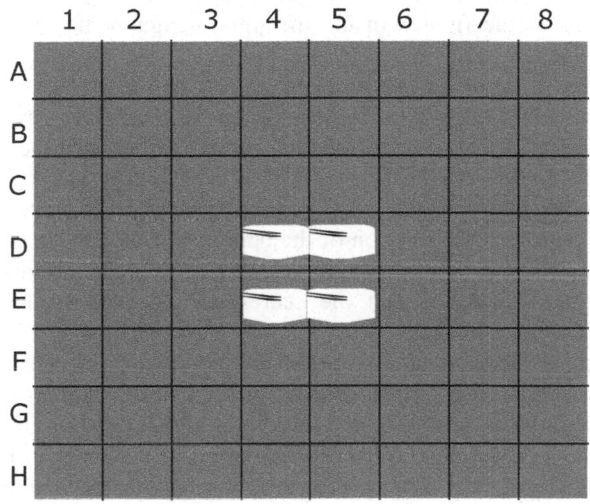

Fig. 5.3 Initial board configuration for Othello game.

The utility function for the Othello game can be defined by calculating the number of black pawns and the number of white pawns on the board and then subtracting them.

Utility function for chess game

For the chess game, and example of utility function may be build as follows:

each piece on the board is assigned a value; for instance:

pawn = 1;
bishop = 3;
knight = 5;
rook = 7;
queen = 9;

Then the total value of all black pieces and the total value of all white pieces on the board are calculated and then subtracted.

MIN-Max Example 1: NIM game

In the NIM game several piles of sticks are given. A player may remove, in one turn, any number of sticks from one pile of his/her choice. The player who takes the last stick loses.

For instance, if we have 4 piles with 1, 2, 3 and 5 sticks respectively, we can denote a state by (1 2 3 5). After a move (for instance one player is taking 2 sticks from the third pile), the configuration can be expressed as (1 2 1 5) or (1 1 2 5).

Let us consider the very simple NIM game (1 1 2). The tree is depicted in Figure 5.4 (look just at the figures inside the squares, ignore the digit above each square at this step).

Suppose MAX is the player who makes the first move. MAX takes one or two sticks. After this, it is MIN's turn to move. Then the opponent moves one or two sticks and the status is shown in the next nodes and so on until there is one stick left.

The MAX nodes represent the configuration before MAX makes a move and the MIN nodes represent the position of the opponent. Since the goal of this game is that the player who removes the last stick loses, the scores are assigned to 0 if the leaves are at MAX nodes and the scores are assigned to 1 if the leaves are MIN.

Then we back up the scores to assign the internal nodes from the bottom nodes. At MAX nodes we take the maximum score of the children and at MIN nodes the minimum score of the children respectively. In this manner, the scores (or utility) of non leaf nodes are computed from the bottom up. If we analyze the Figure 5.4, the root node is 1, and thus corresponds to a win for the MAX player. The first player should pick a child position that corresponds to a 1.

For real games, search trees are much bigger and deeper than NIM and one cannot possibly evaluate the entire tree; there is a need to put a bound on the depth of the search.

MIN-MAX Example 2

For the tree in Figure 5.5 the utility values of the leaves nodes are known. Use MIN-MAX search to assign utility values for each internal node and indicate which path is the optimal solution for the MAX node at the root of the tree.

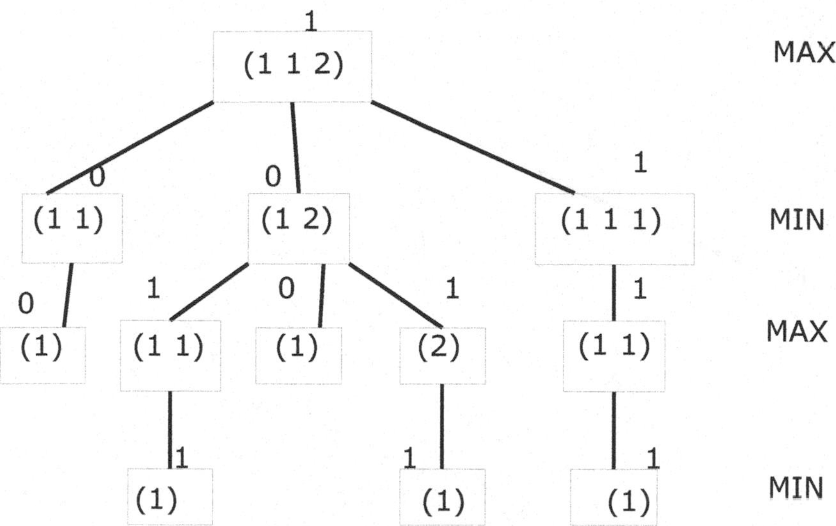

Fig. 5.4 Game tree for the (1 1 2) NIM.

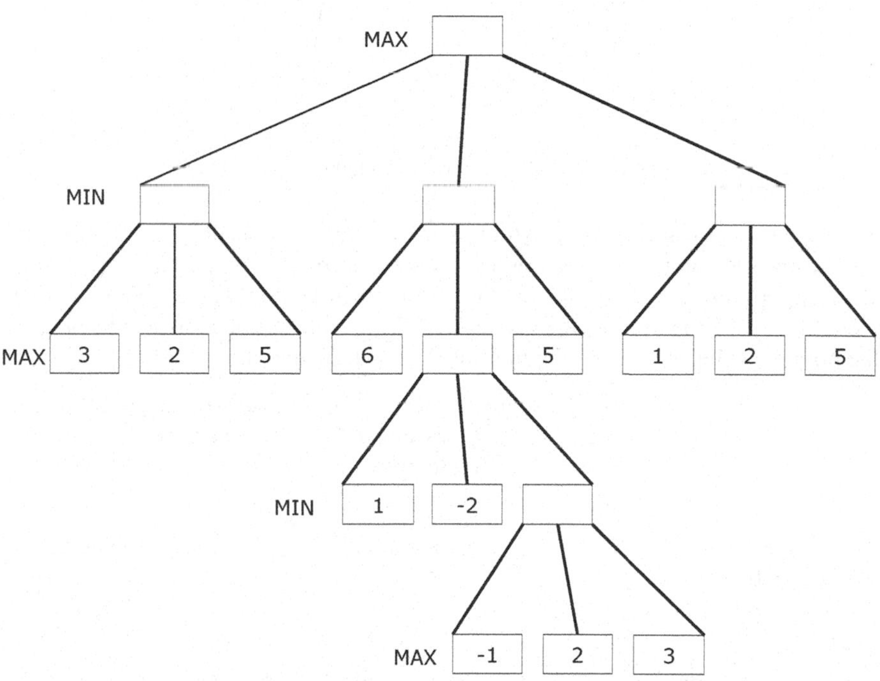

Fig. 5.5 The tree for the MIN-MAX search example.

The solution is depicted in Figure 5.6 with the heavy black line showing the path. The node's values are written for each internal node.

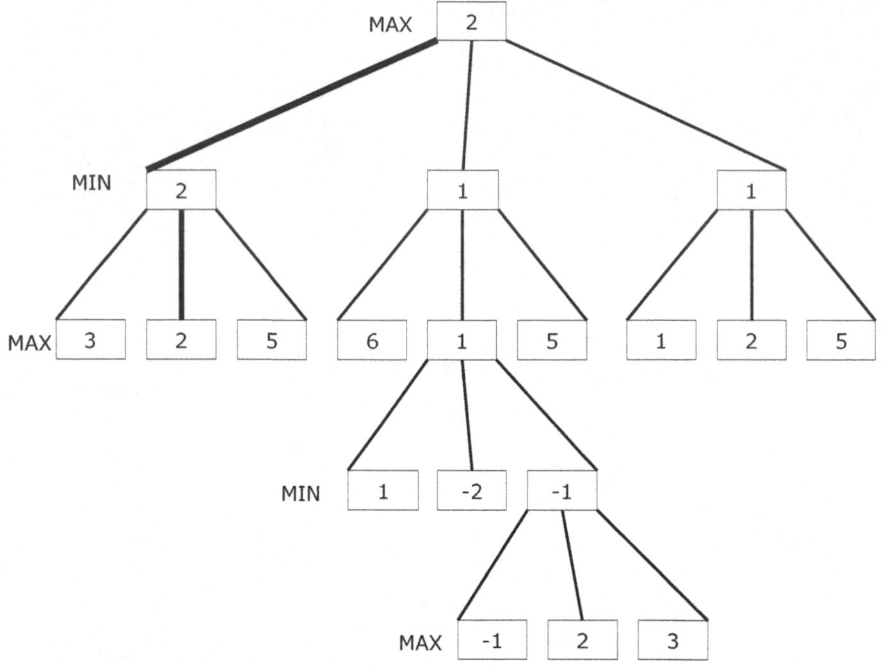

Fig. 5.6 Solution for the tree depicted in Figure 5.5.

If the terminal states are not definite win, loss or draw, or they actually are but with reasonable computer resources we cannot determine this, we have to heuristically/approximately evaluate the quality of the positions of the states.

Evaluation of the utility function is expensive if it is not a clear win or loss. One possible solution is to do depth limited Minimax search.

- search the game tree as deep as possible can in the given time;
- evaluate the fringe nodes with the utility function;
- back up the values to the root;
- choose best move, repeat.

This optimization is known as alpha-beta cutoffs and the algorithm in presented in the next Section.

Remarks

(i) alpha-beta principle: If you know it's bad, don't waste time finding out HOW bad;
(ii) may eliminate some static evaluations;
(iii) may eliminate some node expansions.

5.3 Alpha-beta Pruning

One of the most elegant of AI search algorithms is alpha-beta pruning. Apparently Jon McCarthy came up with the original idea in 1956 but didn't publish it. It first appeared in print in an MIT technical report and a thorough treatment of the algorithm can be found in [12].

The idea, similar to branch and bound, is that the minimax value of the root of the game tree can be determined without examining all the nodes at the search frontier.

Why the algorithm is called alpha-beta? Alpha is the value of the best (i.e., highest-value) choice found so far at any choice point along the path for MAX. If there is a value worse than alpha, MAX will avoid it and will prune that branch.

Beta is defined similarly but for MIN (or the opponent).

Shortly, we can express alpha and beta as:

- Alpha: value of the best (highest value) choice for MAX
- Beta: value of the best (lowest value) choice for min

If we are at MIN node and the value is less than or equal to alpha, then we can stop looking further at the children because MAX node will ignore. If we are at MAX node and value is greater or equal than beta we can stop looking further at the children because MIN node will ignore. The alpha-beta pruning algorithm is provided in Algorithm 5.2.

Algorithm 5.2 Alpha-beta pruning

```
Step 1. Have two values passed around the tree nodes:
      - the alpha value which holds the best MAX value
      found (set to -∞ at the beginning);
      - the beta value which holds the best MIN value
found (set to +∞ at the beginning);.
Step 2. If terminal state, compute the utility function
and return the result;
Step 3. Otherwise:
      At MAX level:
      Repeat
            Step 3.1 Use the alpha-beta procedure, with
            the current alpha and beta value, on a
            child and note the value obtained.
            Step 3.2 Compare the value reported with
            the alpha value;
              if the obtained value is larger, reset
            alpha to the new value.
        Until all children are examined with alpha-beta
      or alpha is equal to or greater than beta

      At MIN level:
```

```
Repeat
        Step 3.1 Use the alpha procedure, with the
        current alpha and beta value, on a child
        and note the value obtained.
        Step 3.2 Compare the value reported with
the beta value;
        if the obtained value is smaller, reset be-
        ta to the new value.
Until all children are examined with alpha-beta
or beta is equal to or lesser than alpha.

Step 4. Go to step 2.
end.
```

Remarks

(i) At MAX level, before evaluating each child path, compare the returned value of the previous path with the beta value. If the value is greater than it, abort the search for the current node;

(ii) At MIN level, before evaluating each child path, compare the returned value of the previous path with the alpha value. If the value is lesser than it, abort the search for the current node.

Alpha- beta pruning Example 1

Consider the tree given in Figure 5.5. For simplicity, we have assigned a label to each node as it can be seen in Figure 5.7 which represents the result of alpha-beta pruning for this tree.

First, the nodes E, F and G are evaluated and their minimum value (2) is backed up to their parent node B. Node H is then evaluated at 6 and since there are more nodes to evaluate the nodes N, O and P are the next ones to be evaluated. Node N is evaluated. Its value is 1. Node O is evaluated and its value is -2. We still need to have some information for the node P (it is of interest whether the value of node I is less than 6 and greater than what we already have, 2). It is enough to analyze Q since P is at a MIN level and we obtain a value ≤ -1. We can now label the node I with 1. Since the value of node A will be maximum between B, C and D and we already have the value 2 for the node B, it is meaningless to search further for the node C because the value we already have (<=1) is lower than 2. Then the backed up value for the node C is <=1. Thus, we can abort searching the children R and S of node P and node J and we have the first cutoffs.

Node K is further evaluated. Its value is 1 which is again less than the minmax value of node B. We ca then back up the value <=1 for the node D because it is meaningless to search further for values lower than 1 in the children of D. A lower value for this node will not change the situation. The portions of the tree, which are pruned are shown with heavy black lines in Figure 5.7.

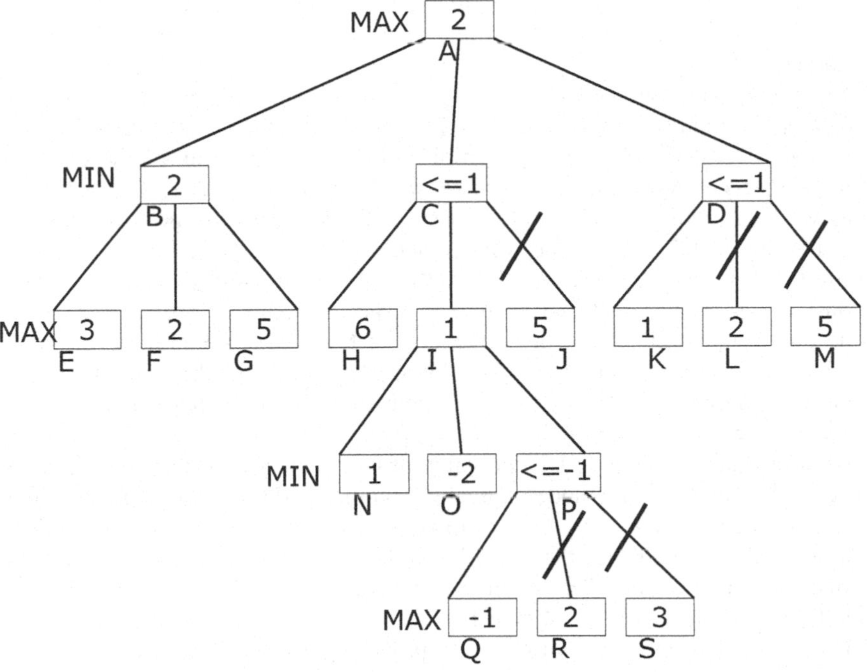

Fig. 5.7 Alpha-beta pruning for the tree depicted in Figure 5.5.

Alpha- beta pruning Example 2

Let us consider a second example for which we show how alpha-beta search works. The tree structure is given in Figure 5.8.

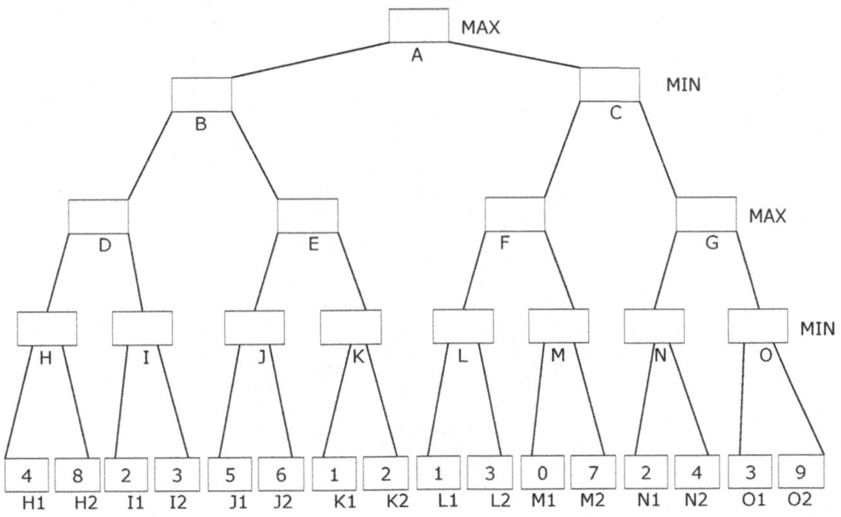

Fig. 5.8 Tree for the alpha-beta example 2.

We now follow the way in which alpha-beta pruning works in Figure 5.9. First, nodes H1 and H2 are evaluated and their minimum value – 4 – is backed up to the parent node H.

Node I1 is then evaluated at 2 and its parent node I must be less than or equal to 2 since it is the minimum of 2 and an unknown value (on its right child). Thus, we label node I by <=2. The value of node D is then 4 (as maximum between 4 and something less or equal than 2). Since we can determine the value of node D from what we have until now, there is no need to further evaluate the other child of node I (which is I2). We further evaluate nodes J1 and J2. The node J will get the minimum of J1 and J2 which is 5. This tells us that the minimax value of the node E must be greater or equal than 5 since it is the maximum of 5 and an unknown value for its right child. Thus, the value of node B is 4 as the minimum between 4 and a value greater of equal to 5. We got another cutoff for the right child of E. We have examined half of the tree at this stage and we know that the value of the root is greater than or equal to 4.

After evaluating the node L1, the value of its parent is less than or equal to 1. Since the value of the root node is greater than or equal to 4, the value of node L cannot propagate to the root. After evaluation of node M1, the value of M is less than or equal to 0 and hence the backed up value for node F is less than or equal to 1. Since the value of node C is minimum between the values of nodes F and G and node F has a value less of equal than 1, node C will also have less than or equal to 1. This means the right child of C can be pruned. Thus, the minimax value of the root is 4.

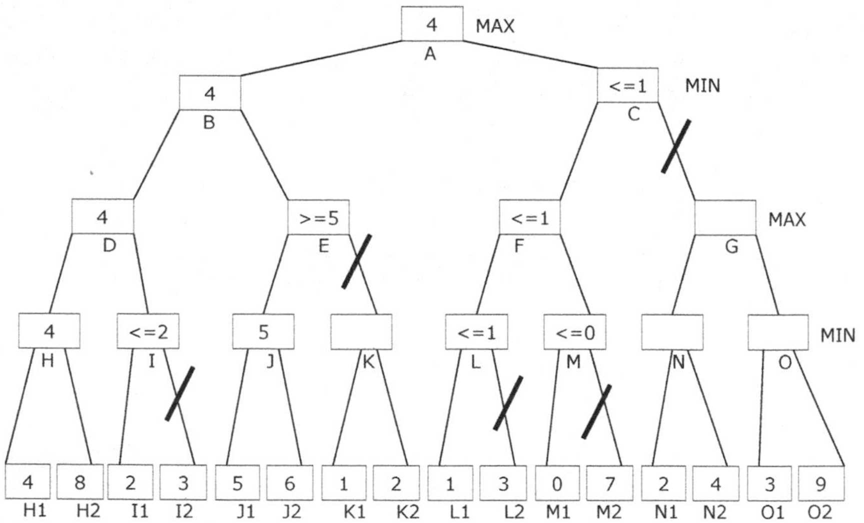

Fig. 5.9 Alpha-beta pruning results for Example 2.

5.4 Comparisons and Discussions

As we used to do for the other uninformed and informed search techniques, we will also compare MIN-MAX search and alpha-beta pruning. The results of comparison in terms of completeness, time complexity, space complexity and optimality are given in Table 1 where:

- b: maximum branching factor of the search tree;
- d: number of ply;
- m: maximum depth of the state space.

	MIN-MAX	**Alpha-beta**
Complete	Yes	Yes
Time complexity	$O(b^m)$	With perfect ordering $O(b^{m/2})$
Space complexity	$O(b^m)$	Best case $O(2b^{d/2})$ Worse case $O(b^d)$
Optimal	yes	Yes

Alpha-Beta is guaranteed to compute the same minimax value for the root node as computed by MIN-MAX. In the worst case alpha-beta does no pruning, examining b^d leaf nodes (where each node has *b* children and a *d*-ply search is performed). In the best case, alpha-beta will examine only $2b^{d/2}$ leaf nodes. Hence if the number of leaf nodes is fixed then one can search twice as deep as MIN-MAX. The best case occurs when each player's best move is the leftmost alternative (i.e., the first child generated). So, at MAX nodes the child with the largest value is generated first, and at MIN nodes the child with the smallest value is generated first[8][9][10][11][15].

MIN-MAX performs a depth first search exploration. For instance, for the chess game, if the branching factor b is approximately 35 and m is approximately 100, this gives a complexity of 35^{100} which is about 10^{154}. Thus, the exact solution is completely infeasible.

Summary

This chapter presented another kind of search – adversarial search –that is of great interest in game playing. Two well-known algorithms are presented for one player and two-player games: MIN-MAX search and alpha-beta pruning. Although the MIN-MAX algorithm is optimal, the time complexity is $O(b^m)$ where *b* is the effective branching factor and *m* is the depth of the terminal states. (Space complexity is only linear in *b* and *m*, because we can do depth first search).

Alpha-beta pruning brings an improvement for the MIN-MAX search.

The basic idea of alpha-beta pruning is that is possible to compute the correct minimax decision without looking at every node in the search tree pruning (allows us to ignore portions of the search tree that make no difference to the final choice)

The pruning does not affect final result. Also, it is important to note that a good move ordering improves effectiveness of pruning. With perfect ordering, time complexity is $O(b^{m/2})$. In games theory there is a huge need for effective and efficient searching techniques due to the complexity of these problems. Some of the well known games have the following complexity:

- Chess[6]
  - b~35(average branching factor)
  - d~100(depth of game tree for typical game)
  - b^d~35^{100}~10^{154} nodes
- Tic-Tac-Toe
  - ~5 legal moves, total of 9 moves
  - 5^9=1,953,125
  - 9!=362,880 (Computer goes first)
  - 8!=40,320 (Computer goes second)
- Go
  - b starts at 361 (19 x 19 board)

The line of perfect play leads to a terminal node with the same value as the root node. All intermediate nodes also have that same value. Essentially, this is the meaning of the value at the root node.

Adversary modeling is of general importance and some of the application domains including certain economical situations and military operations[2][3][4][5].

In practice, there are a few important situations where "machines" were able to compete (and defeat) world champions for certain well known games.

- For checkers game, there exist Chinook. After 40-year-reign of human world champion Marion Tinsley, Chinook defeated it in 1994. Chinook used a pre-computed end game database defining perfect play for all positions involving 8 or fewer pieces on the board, a total of 444 billion positions.
- For Chess game, there exists Deep Blue. Deep Blue defeated human world champion Garry Kasparov in a six-game match in 1997. Deep Blue searches 200 million positions per second, uses very sophisticated evaluation, and undisclosed methods for extending some lines of search up to 40 ply. In the chess program Deep Blue, they found empirically that alpha-beta pruning meant that the average branching factor at each node was about 6 instead of about 35-40[16].
- Othello game: human champions refuse to compete against computers, who are too good.
- Go game: human champions refuse to compete against computers, who are too bad. In Go, the branching factor b is greater than 300, so most programs use pattern knowledge bases to suggest plausible moves.
- Backgammon game: program has beaten the world champion, but was lucky.

References

1. http://www.cs.ucr.edu/~eamonn/205/
2. Owen, G.: Game theory, 3rd edn. Academic Press, San Diego (2001)
3. Nillson, N.J.: Principles of Artificial Intelligence. Tioga Publishing Co. (1980)
4. Rich, E., Knight, K.: Artificial Intelligence. McGraw-Hill, New York (1991)
5. Luger, G.F., Stubblefield, W.A.: Artificial Intelligence: Structures and strategies for complex problem solving. The Benjamin/Cummings Publishing Co. (1993)
6. Russell, S., Norvig, P.: Artificial Intelligence, a Modern Approach. Prentice-Hall, Englewood Cliffs (1995)
7. Norvig, P.: Paradigms of Artificial Intelligence Programming. Morgan Kaufmann, San Francisco (1992)
8. Stockman, G.: A minimax algorithm better than alpha-beta? Artificial Intelligence 12, 179–196 (1979)
9. Newborn, M.M.: The efficiency of the alpha-beta search in trees with branch dependent terminal node scores. Artificial Intelligence 8, 137–153 (1977)
10. Marsland, T.A., Campbell, M.: A survey of enhancements to the alpha-beta algorithm. In: Proceedings of the ACM National Conference, Los Angeles, CA, pp. 109–114 (1981)
11. Griffith, A.K.: Empirical exploration of the performance of the alpha-beta tree-searching heuristic. IEEE Transactions on Computers 25(1), 6–11 (1976)
12. Knuth, D., Moore, R.: An analysis of alpha-beta pruning. Artificial Intelligence 6, 293–326 (1975)
13. Marsland, T.A., Rushton, P.G.: A study of techniques for game playing programs. In: Rose, J. (ed.) Advances in Cybernetics and Systems, vol. 1, pp. 363–371. Gordon and Breach, London (1971)
14. Pearl, J., Korf, R.E.: Search techniques. Annual Review of Computer Science 2 (1987)
15. Pearl, J.: The solution for the branching factor of the Alpha-beta pruning algorithm and its optimality. Communications of the Association of Computing Machinery 25(8), 559–564 (1982)
16. Keene, R., Jacobs, B., Buzan, T.: Man v machine: The ACM Chess Challenge: Garry Kasparov v IBM's Deep Blue, B.B. Enterprises, Sussex (1996)
17. Kanal, L., Kumar, V. (eds.): Search in Artificial Intelligence. Springer, New York (1988)
18. Hart, T.P., Edwards, D.J.: The alpha-beta heuristic. MIT Artificial Intelligence Project Memo. MIT, Cambridge (1963)
19. Korf, R.E.: Artificial intelligence search algorithms. In: Algorithms and Theory of Computation Handbook, CRC Press, Boca Raton (1999)

Verification Questions

1. What is the importance of adversarial game and what are the practical applications of it?
2. Name some problems for which MIN-MAX search is optimal.
3. What are the advantages of alpha-beta pruning while compared to MIN-MAX search?

4. Find an example for which both alpha-beta pruning and MIN-MAX perform same. In which situations is alpha-beta better?
5. Find some examples (other than the ones given in this chapter) in which machines can beat humans for different games.

Exercises

5.1 For the tree in Figure 2 use MIN-MAX search to assign utility values for each internal node (i.e., non-leaf node) and indicate which path is the optimal solution for the MAX node at the root of the tree.

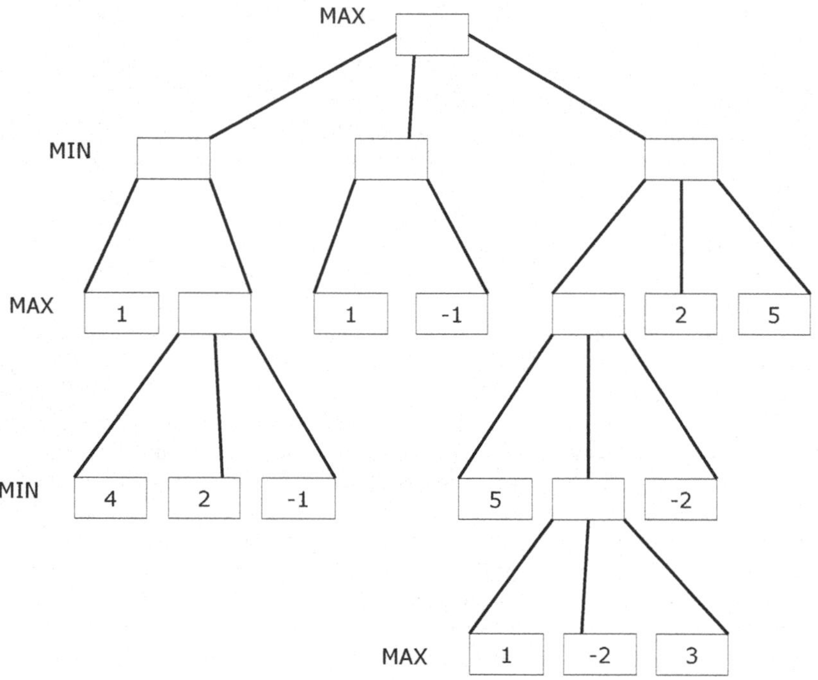

Fig. 1 Tree for the problem 5.1.

5.2 Use alpha-beta pruning for the (1 1 2) NIM game. How you compare with MIN-MAX search? Now consider the (1 2 2) NIM and apply both alpha-beta pruning and MIN-MAX search. Does alpha-beta reduces more the search in this case while compared with the previous one?

5.3 Use alpha-beta pruning and MIN-MAX search for each of the trees given in Figures 2-4.

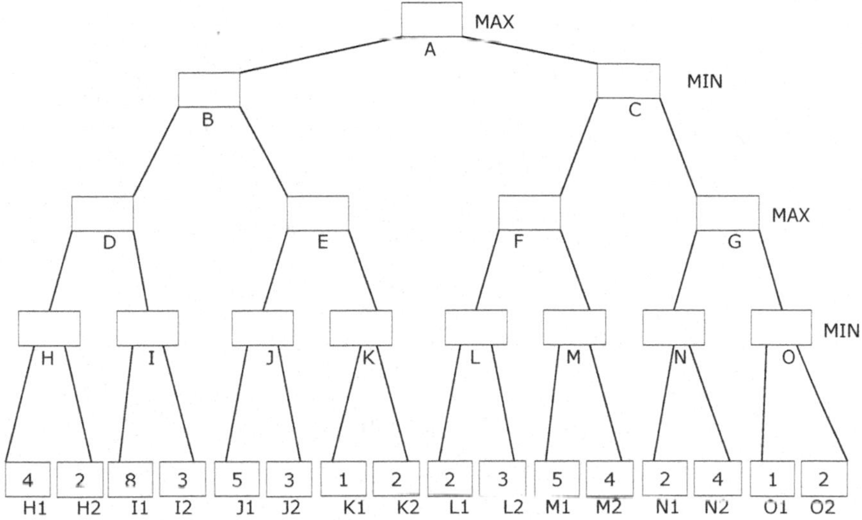

Fig. 2 First tree example for the problem 5.2.

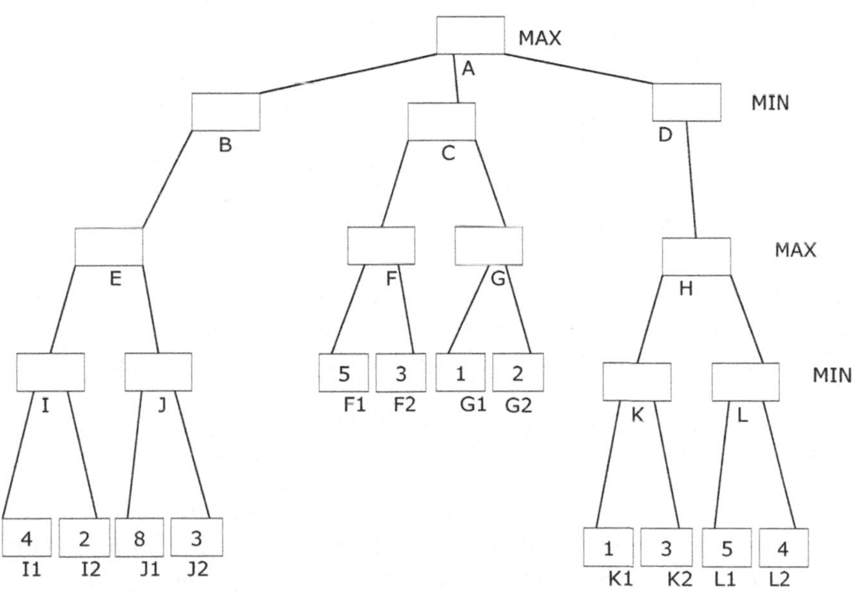

Fig. 3 Second tree example for the problem 5.2.

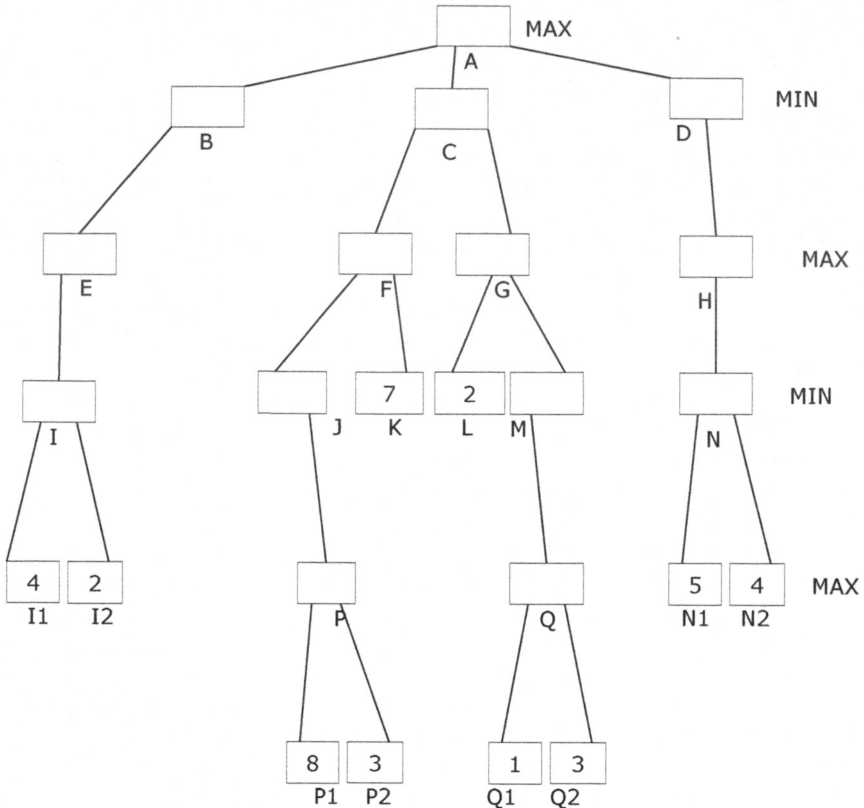

Fig. 4 Third tree example for the problem 5.2.

5.4. Use both alpha-beta pruning and MIX_MAX for the tic-tac-toe problem and compare the results. Consider starting with the empty board but also analyze the behavior of the two techniques on a given non-empty board configuration.

5.5 Consider the connect-4 game (also known as 4 in a line) which is a two player game stated as follows:

A 7x6 (7 rows and 6 columns) rectangular board placed vertically is given. 21 red and 21 yellow tokens are to be placed on this board by two players which alternate their moves by dropping a token into one of the seven columns. The token falls down to the lowest unoccupied square. A player wins if connects four token vertically, horizontally or diagonally. If the board is filled and no player has aligned four tokens the game ends in a draw (see Figure 1 for example).

a) Design the min-max-search algorithm for connect-4 game;
b) Design a proper utility function for connect-4;

c) Design and implement a game playing program for the deterministic two player game Connect-4

This game is centuries old, Captain James Cook used to play it with his fellow officers on his long voyages, and so it has also been called "Captain's Mistress".

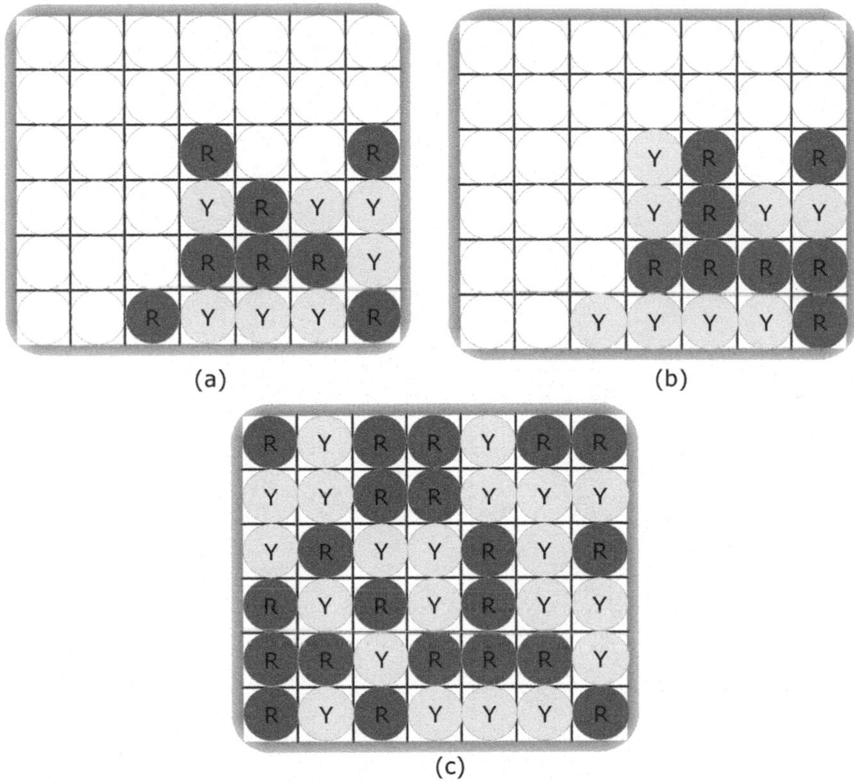

Fig. 5 Connect-4 example: (a) red won, (b) yellow won, (c) draw.

5.6. Design an implement and alpha-beta pruning for the Othello game.

Chapter 6
Knowledge Representation and Reasoning

6.1 Introduction

What is knowledge?
In a very generic way, knowledge can be defined as information (which can be expressed in the form of propositions) from the environment.

What is knowledge representation?
Again, in simple words, can be defined as symbols used to represent the propositions.

What is knowledge representation and reasoning?
One way to define it is as the manipulation of symbols encoding propositions to produce representations of new propositions.

The question of representing knowledge is a key issue in artificial intelligence: how can human knowledge of all kinds be represented by a computer language, and in such a way that computers can use this knowledge for purposes of reasoning? Modern computer applications have led to generalized use of knowledge representations in various contexts, including information search, simulation, web semantic ontology description.

Knowledge representation is of immense importance in the field of Artificial Intelligence. An intelligent agent should be able to acquire information (or knowledge) from environment, to represent and understand it, and to be able of reasoning, that is to infer the implications of what it knows and of the choices it has.

The main component of a knowledge-based agent is its knowledge-base. A knowledge-base is a set of sentences, each of them being expressed in a language called the knowledge representation language. Sentences represent some assertions about the world. The mechanism for deriving new sentences from old ones is known as inferencing or reasoning. Inference must obey the primary requirement that the new sentences should follow logically from the previous ones.

Logic is widely used in Artificial Intelligence as a representational method. The advantage of using formal logic as a language of AI is that it is precise and definite and allows reason about negatives and disjunctions.

C. Grosan and A. Abraham: Intelligent Systems, ISRL 17, pp. 131–147.
springerlink.com

This allows programs to be written which are declarative - they describe what is true and not how to solve problems. This is of importance for automated reasoning techniques for general purpose inferencing. A large amount of the reasoning carried out by humans depends on handling knowledge that is uncertain. Logic cannot represent this uncertainty well. Similarly, natural language reasoning requires inferring hidden state, namely, the intention of the speaker. Humans can cope with virtually infinite variety of utterances using a finite store of commonsense knowledge. Formal logic has difficulty with this kind of ambiguity[3][4][5].

Facts are claims about the world and can be true or false. A representation is an expression (sentence) in some language that can be encoded in a computer program and stands for the objects and relations in the world. The representation has to be consistent with reality.

A logic consists of two parts:

- *language*: has two aspects: syntax and semantics.
    - *Syntax* represents the atomic symbols of the logical language, and the rules for constructing well formed, non-atomic expressions (symbol structures) of the logic. Syntax specifies the symbols in the language and how they can be combined to form sentences. Hence facts about the world are represented as sentences in logic.
    - *Semantics* gives meanings to the atomic symbols of the logic. It specifies what facts in the world a sentence refers to. Hence, also specifies how you assign a truth value to a sentence based on its meaning in the world.
- *method of reasoning*: consists of the rules for determining a subset of logical expressions, called theorems of the logic. It refers to mechanical method for deriving new sentences from existing sentences[7][8][9][11].

There are a number of logical systems with different syntax and semantics. In what follows we will refer to two of them:

- propositional logic;
- first order predicate logic.

6.1 Propositional Logic

Propositional logic is the simplest logical system. In propositional logic the user defines a set of propositional symbols, like P and Q. User defines the semantics of each of these symbols. For example:

P means "Sunday is a holiday".
Q means "Today is Sunday".

In reasoning about truth values we have to use a set of operators which can be applied to truth values. We will present in what follows some of them.

6.2.1 Logical Operators

Any of us is familiar with these operators from our everyday language:

"I am going for shopping **and** for a walk".
"John likes apples **or** oranges".
"Mary is **not** slim". etc.

We will use the following operator's symbols even tough sometimes some other symbols may be used:

And $\wedge$
Or $\vee$
Not$\neg$
Implies$\Rightarrow$
Iff (if and only if) $\Leftrightarrow$

It is first necessary to convert facts and rules about the real world into logical expressions using the logical operators. For more details on how to translate logical propositions into English and vice versa, see [1].

Let us consider the following logical propositions:

R: It is raining
D: It is dark
C: It is cold.

Then a *sentence* (which is also called a *formula*) can be defined by:

a) A symbol (any of $\wedge$, $\vee$, $\neg$, $\Rightarrow$, $\Leftrightarrow$);
b) If P is a sentence, then $\neg$P is also a sentence;
c) If P and Q are sentences, then P$\wedge$Q, P$\vee$Q, P$\Rightarrow$Q and P$\Leftrightarrow$Q are also sentences;
d) A sentence contains a finite number of applications of (a)-(c).

Examples:

R$\wedge$C – It is raining *and* it is cold.
R$\Rightarrow$C – *If* it is raining *then* it is cold.
(R$\wedge$D) $\Rightarrow$C – *If* it is raining *and* it is dark *then* it is cold.

The behavior of the logical operators is usually represented using *truth tables*. A truth table shows the possible values that can be generated by applying an operator to truth values.

In what follows we will show the truth tables for the five logical operators we presented above.

Not Operator ($\neg$)
Not is a unary operator, which means it is applied only to one variable (all other operators above are binary operators). Its behavior is very simple:

$\neg$true is false
$\neg$false is true

For a given variable P we then have the truth table:

P	¬P
true	false
false	true

And Operator **(∧)**

And is a binary operator (it acts on two variables). It is also called the *conjunctive operator* and P ∧ Q is the conjunction of P and Q. The truth table of *and* operator is given below.

P	Q	P ∧ Q
true	true	true
true	false	false
false	true	false
false	false	false

We can observe that P ∧ Q is true only if both P and Q are true. If one or tem is false or if both are false then P ∧ Q is also false.

Or Operator **(∨)**

Or is another binary operator. It is also called the *disjunctive operator* and P ∧ Q is the disjunction of P and Q. The truth table of *or* operator is given below.

P	Q	P ∨ Q
true	true	true
true	false	true
false	true	true
false	false	false

We can observe that P ∨ Q is true if either P or Q are true and it is false only if both P and Q are false.

Implies operator **(⇒)**

This form of implication is also known as *material implication*. In a P⇒Q statement, P is the *antecedent*, and Q is the *consequent*.

P⇒Q it usually reads as "P implies Q" but can also be read as "If P then Q" or "If P is true then Q is true". Hence, if P is *false*, the statement is not really saying anything about the value of Q, so Q can take on any value (*true* or *false*).

The truth table is given below:

P	Q	P ⇒ Q
true	true	true
true	false	false
false	true	true
false	false	true

For example, if P is "I like chocolate" and Q is "I eat chocolate", then if P is true and Q is true, P⇒Q is also true (If I like chocolate then I eat chocolate). If P is true and Q is false then P⇒Q is also false (If I like chocolate then I don't eat chocolate). A natural implication is the one where P and Q are false and P⇒Q is true: (If I don't like chocolate then I don't eat chocolate).

Iff Operator (⇔)

Iff operator takes true values if both variables are true or if both are false and takes false values otherwise as it can be see in the truth table below:

P	**Q**	**P⇔Q**
true	true	true
true	false	false
false	true	false
false	false	true

In the truth tables we have seen until now only the values of applying a single operator are presented. But truth tables can display the values of expressions having more variables and more operators. The truth table for a complex expression given by (¬P∧Q)∨R is given in the table below. Instead of displaying only the value of the final expression we are also presenting the values of all sub-expressions which are ¬P and ¬P∧Q.

P	**Q**	**R**	**¬P**	**¬P∧Q**	**(¬P∧Q)∨R**
true	true	True	false	false	true
true	true	False	false	false	false
true	false	False	false	false	false
true	false	True	false	false	true
false	true	True	true	true	true
false	true	False	true	true	true
false	false	True	true	false	true
false	false	False	true	false	false

6.2.2 Terminology

A *tautology* is an expression whose value is *true* regardless of the value of the variables. For instance, P∨¬P is a tautology.

If P is a tautology then it is denoted by ⊨.

A *valid* expression is an expression that is true under any interpretation. No matter what meanings and values we assign to the variables in a valid expression, it will still be true.

If an expression is false in any interpretation, it is described as being *contradictory* (also called *inconsistent* sentence or unsatisfiable).

Two expressions that always have the same value for given values of their variables and *logically equivalent*.

For instance, the expressions:

$P \wedge Q$
$Q \wedge P$

are equivalent and we denote this by $P \wedge Q \equiv Q \wedge P$.

Going back to the syntax of logical systems, this can be defined using the alphabet:

$\{\neg, \wedge, \vee, \Rightarrow, \Leftrightarrow, (,), \text{true}, \text{false}, P, Q, \ldots R, ..\}$

We should note that any number of propositional symbols (like P, Q, R) can be allowed.

An expression is referred to as a *well-formed formula* or a sentence if it is constructed correctly, according to the rules of the syntax of propositional calculus. For example, if we have:

P, Q, R,...
$\neg P$
$P \wedge Q$
$P \vee Q$
$P \Rightarrow Q$
$P \Leftrightarrow Q$

Thus:

$(P \wedge Q) \vee (\neg R \Rightarrow P) \Leftrightarrow (P \vee Q) \wedge (R \vee Q)$

is a well-formed formula. A sentence is defined recursively, in terms of other sentences.

If we have a set of assumptions {P1, P2, ..., Pk}, and from those assumptions a conclusion C can be derived, then we say that we have *deduced C* from the assumptions and this can be written as:

$\{P1, P2, \ldots Pk\} \vdash C$

If *C* can be concluded without any assumptions, then we write:

$\vdash C$

To derive a conclusion from a set of assumptions, a set of *inference rules* are applied.

Sentence P *entails* sentence Q, written $P \models Q$, means that whenever P is True, Q is also true. In other words, all models of P are also models of Q.

6.2.3 Inference

To derive a conclusion from a set of assumptions, a set of *inference rules* are applied. Instead of using the formula $P \vdash C$ we can simply use $\frac{P}{C}$.

Given two sentences P and Q we say Q is inferred from P ($P \vdash Q$) if there is a sequence of rules of inference that apply to P and allow Q to be added.

Inference is not directly related to truth; i.e. we can infer a sentence provided we have rules of inference that produce the sentence from the original sentences.

However, if rules of inference are to be useful we wish them to be related to entailment. Thus, would be ideal that:

$$P \vdash Q \Leftrightarrow P \models Q.$$

Remarks

(i) If Q was inferred by applying rules of inference to P but there is some model in which P holds but Q does not hold then the rules of inference have inferred too much and we have:
$P \vdash Q$ but $P \not\models Q$

(ii) If Q is a sentence which holds in all models in which P holds but we cannot find rules of inference that will infer Q from P then the rules of inference are insufficient to infer the things we wish and we have:
$P \models Q$ but $P \vdash Q$.

In inference procedure $\vdash$ is *sound* if whenever $P \vdash Q$ then also $P \models Q$. A sound inference procedure infers things that are valid consequences.

An inference procedure $\vdash$ is complete if whenever $P \models Q$ then it is also the case that $P \vdash Q$. A complete inference procedure is able to infer anything that is a valid consequence.

The best inference procedures are both sound and complete but this is computationally expensive (especially the completeness part). Even if an inference is not complete it is desirable that it is sound.

A logical system is *decidable* if there is a procedure that is guaranteed to terminate having determinate whether the logical expressions in that system are valid or not (will determine whether any well-formed formula is a theorem).

Propositional logic is decidable; propositional logic is complete and we can prove that any well-formed formula is a theorem by showing that it is a tautology (and this can be deduced from the truth tables).

A logical system is *monotonic* if a valid proof in the system cannot be made invalid by adding additional premises or assumptions.

For example, if we have proved a conclusion C by applying rules of deduction to a premise Q with assumptions P, by adding additional assumptions and premises will not stop us from being able to deduce C.

Propositional logic is monotonic.

Some of the most useful inference rules for propositional logic are presented below. In these rules, P, Q, R and C stand for any logical expressions.

6.2.3.1 Introduction

Example: And (∧) Introduction

$$\frac{P \quad Q}{P \wedge Q}$$

This rule says that given P and Q we can deduce P∧Q.

Example Or (∨) Introduction

$$\frac{P}{P \vee Q}, \frac{Q}{P \vee Q}$$

This rule says that from P we can deduce the disjunction of P with any expression (P ∨ Q is true for any value of Q).

Example: Implies (⇒) Introduction

$$\frac{\begin{array}{c} P \\ \vdots \\ C \end{array}}{P \Rightarrow C}$$

This rule says that if we start from an assumption P and derive a conclusion C, then we can conclude that P⇒C.

6.2.3.1 Elimination

Example: And (∧) Elimination

$$\frac{P \wedge Q}{P}, \frac{P \wedge Q}{Q}$$

These rules say that given P∧Q , we can deduce P and we can also deduce Q separately.

Example: Implies (⇒) Elimination

$$\frac{P \quad P \Rightarrow Q}{Q}$$

This rule says that if P is true and P implies Q is true, then we know that Q is true.

This rule is usually known as *modus ponens* and is one of the most commonly used rules in logical deduction. This kind of reasoning is clearly valid.

For instance, if we have:

P: I like chocolate.
Q. I eat chocolate.

If we replace in the expression above we get: I like chocolate. If I like chocolate I eat chocolate. Thus, I eat chocolate.

Example: Not Not (¬¬) Elimination

$$\frac{\neg\neg P}{P}$$

This rule says that if we have a sentence that is negated twice, we can conclude the sentence itself, without the negation.

An important inference rule is known as the *deduction theorem* and is stated as:

if $P \cup \{Q\} \vdash C$ then $P \vdash (Q \Rightarrow C)$

The reverse also holds:

if $P \vdash (Q \Rightarrow C)$ then $P \cup \{Q\} \vdash C$

6.3 First Order Predicate Logic (FOPL)

6.3.1 Predicate Calculus

In predicate calculus, we use *predicates* to express properties of objects. Predicate calculus allows us to reason about the object's proprieties and the relationships between them. In propositional calculus, we can express sentence "I like chocolate" by P. We can also construct ¬ P from here which means "I do not like chocolate".

But the thing is this does not allow us to extract any information about me, or I like or chocolate.

In predicate logic, we can express the sentence P by

L(me, chocolate)

where L is a predicate that represents "liking." This statement also expresses a relationship between me and chocolate.

Predicate calculus can be generalized and used for more general statements. For instance, we can extend the statement "I like chocolate" to "Everyone likes chocolate" and we might express this as:

$\forall x\ P(x) \Rightarrow L(x, c)$

where ∀ means "for all" and it is called *universal quantifier*.

This can be read as "for every x it is true that if P holds for x then the relationship L holds between x and c.

We can also make another statement using another quantifier, ∃, to express that only some values have a certain propriety, not all of them.
For instance:

$\exists x\ L(x, c)$

Can be read as "there exists an x such as x likes chocolate". The quantifier ∃ is called *existential quantifier* and in the example above can be interpreted as there is at least one value of x for which L(x, c) holds.

Remarks

(i) $\forall x\ L(x, c) \Rightarrow \exists x\ L(x, c)$ is true;
(ii) $\exists x\ L(x, c) \Rightarrow \forall x\ L(x, c)$ is false.

We can express an object that relates to another object in a specific way using functions with the same meaning they have in mathematics. For example, to represent the statement "My brother likes chocolate," we might use:

L(B(me), chocolate)

where the function *b*(*x*) means the brother of *x*. Functions can take more than one argument. A general function with *n* arguments is represented as in mathematics by $f(x_1, x_2, \ldots, x_n)$.

A first-order logic is one in which the quantifiers ∀ and ∃ can be applied to objects or terms, but not to predicates or functions.

6.3.2 FOPL Alphabet

FOPL alphabet is a bit more complex than the alphabet used by propositional logic. It consists of:

- Logical Symbols: ∧, ∨, ¬, ⇒, ⇔, ∀, ∃, true, false;
- Non-Logical Symbols which can be variables and constants:
  - Any identifier might be considered as a variable;
  - Constants can be predicates and functions. 0-ary functions are also called individual constants.

The identifiers used for predicates, functions, and variables must be easily distinguishable by using some appropriate convention.

A *term* is any of the following:

- a constant;
- a variable;
- a function $f(x_1, x_2, \ldots, x_k)$ where $x_1, x_2, \ldots, x_k$ are all terms.

An *atomic formula* is either false or an n-ary predicate applied to n terms: $P(x_1, x_2, \ldots, x_n)$.

A *literal* can be defined as:

- *positive literal*: an atomic formula;
- *negative literal*: the negation of an atomic formula;
- *ground literal*: a variable-free literal.

A *clause* is a disjunction of literals. There are a few types of closes:

- *ground clause:* is a variable-free clause;
- *Horn clause*: is a clause with at most one positive literal;
- *definite clause*: is a Horn Clause with exactly one positive literal.

Depending on the logical operators used, a formula can be:

- an atomic formula;
- *negation* (the NOT of a formula);
- *conjunctive* formula (the AND of formulae);
- *disjunctive* formula (the OR of formulae);
- an *implication* (a formula of the form formula1 $\Rightarrow$ formula2);
- an *equivalence* (a formula of the form formula1 $\Leftrightarrow$ formula2);
- *universally quantified* formula ($\forall$ variable formula; occurrences of variable are *bound* in formula);
- *existentially quantified* formula ($\exists$ variable formula; occurrences of variable are *bound* in formula).

A formula that is the disjunction of clauses is said to be in *clausal form*. For convenience, the terms and formulae are referred as *form* or *expression*.

Substitution can be seen as a map from terms to terms and from formulae to formulae.

Given a term s, the *substitution of a term* t *in* s *for a variable* x, denoted by s[t/x], is:

- t, if s is the variable x;
- y, if s is the variable y different from x;
- $F(s_1[t/x]\ s_2[t/x] \,..\, s_n[t/x])$, if s is $F(s_1\ s_2 \,..\, s_n)$.

Given a formula P, *substitution of a term* t *in* P *for a variable* x, denoted P[t/x], is:

false, if P is false;

- $F(t_1[t/x]\ t_2[t/x] \,..\, t_n[t/x])$, if P is $F(t_1\ t_2 \,..\, t_n)$;
- $(Q[t/x] \wedge R[t/x])$ if P is $(Q \wedge R)$, and similarly for the other relationships;
- $(\forall\ x\ Q)$ if P is $(\forall\ x\ Q)$, (similarly for $\exists$),
- $(\forall\ y\ Q[t/x])$, if P is $(\forall\ y\ Q)$ and y is different from x (similarly for $\exists$).

Given two substitutions $S = [t_1/x_1 \,..\, t_n/x_n]$ and $V = [u_1/y_1 \,..\, u_m/y_m]$, the *composition* of S and V – denoted S.V – is the substitution obtained by:
Applying V to $t_1, \ldots, t_n$ (this is called *concatenation*), and
adding any pair u_j/y_j such that y_j is not in $\{x_1 \,..\, x_n\}$.

For example [2]:

[G(x y)/z].[A/x B/y C/w D/z]

is

[G(A B)/z A/x B/y C/w].

Remark

Composition is an operation that is associative and non commutative.

A set of forms f1,f2, ..., fn is *unifiable* iff there is a substitution S such that:

f1.S = f2.S = ... = fn.S.

S is said to be a *unifier* of the set.

For example:

{P(x F(y) B) P(x F(B) B)}

is unified by

[A/x B/y]

and also unified by [B/y].

A *Most General Unifier* (MGU) of a set of forms f1, f2, ... fn is a substitution S that unifies this set and such that for any other substitution T that unifies the set there is a substitution V such that S.V = T. The result of applying the MGU to the forms is called a *Most General Instance* (MGI).

Facts about FOPL:

(*i*) FOPL is not decidable (while compared to propositional logic which is). It is not possible to develop an algorithm that will determine whether an arbitrary well-formed formula in FOPL is logically valid.

(*ii*) FOPL is monotonic.

6.4 Resolution in Propositional Logic and FOPL

We have introduced above the inference rule modus ponens. *Resolution* is another important inference rule.

6.4.1 Resolution in Propositional Logic

Resolution inference rule can be stated as follows:

given the clauses C1 and C2 as premises, where C1 contains the literal L and C2 contains the literal (¬L), infer the clause C, where C is the union of (C1 - {L}) ∪(C2 -{¬L}).

This is also called the *resolvent* of C1 and C2.

In other words, two clauses can be combined together, and *L* and ¬ *L* can be removed from those clauses.

This can be written as:

$$\frac{\{C1, C2\}}{(C1-\{L\})\cup(C2-\{\neg L\})}$$

A simple example is:

$$\frac{\{(P,Q)\ \ (\neg Q,R)\}}{(P,R)}$$

Another example:

$$\frac{\{(P,Q,R),S,(\neg Q,S,T),(\neg S,V)\}}{\{(P,R,S,T),S,(\neg S,V)\}}\text{can be resolved to :}$$

$$\frac{\{(P,R,S,T),S,(\neg S,V)\}}{\{(P,R,T,V),S\}\ \text{or}\ \{(P,R,S,T),V\}}\text{can be further resolved to :}$$

We also have one more choice at the first step (involving S and ¬S). But we leave this resolution as an exercise.

If the resolution of a set of clauses leads to falsum means that the closes are inconsistent. The original closes are *refuted* using *resolution refutation*.

For instance, if we have the original clauses:

{(P, Q), (¬P, ¬S), S, ¬Q}

By elimination P and ¬P we obtain:

{(Q, ¬S), S, ¬Q}

And by eliminating Q and ¬Q we obtain:

{S, ¬S}

⊥ (falsum).

6.4.2 Resolution in FOPL

Given clauses C1 and C2, a clause C is a resolvent of C1 and C2 if the following conditions are fulfilled:

1) There is a subset C1' = {P_1, .., P_m} of C1 of literals of the same sign, say positive, and a subset C2' = {Q_1, .., Q_n} of C2 of literals of the opposite sign, say negative,
2) There are substitutions s1 and s2 that replace variables in C1' and C2' so as to have new variables,
3) C2" is obtained from C2 removing the negative signs from P_1, P_2, …, P_n
4) There is an Most General Unifier s for the union of C1'.s1 and C2".s2

and C is:

((C1 - C1').s1 ∪ (C2 - C2').s2).s

Example [2]:

C1 = {(P z (F z)) (P z A)}
C2 = {(NOT (P z A)) (NOT (P z x)) (NOT (P x z))
C1' = {(P z A)}
C2' = {(NOT (P z A)) (NOT (P z x))}
C2" = {(P z A) (P z x)}
s1 = [z1/z] s2 = [z2/z]
C1'.s1 UNION
C2'.s2 = {(P z1 A) (P z2 A) (P z2 x)}
s = [z1/z2 A/x]
C = {(NOT (P A z1)) (P z1 (F z1))}

This application of Resolution has eliminated more than one literal from C2, i.e. it is not a binary resolution.

To apply resolution to FOPL expressions, we first need to deal with the quantifiers ∀ and ∃. The method that is used is to move these quantifiers to the beginning of the expression, resulting in an expression that is in *prenex normal form.*

The following rules to move the quantifiers to the front [1]:

1. $\neg(\forall x)P(x) \equiv (\exists x)\neg P(x)$
2. $\neg(\exists x)P(x) \equiv (\forall x)\neg P(x)$
3. $(\forall x)P(x) \wedge Q \equiv (\forall x)(P(x) \wedge Q)$
4. $(\forall x)P(x) \vee Q \equiv (\forall x)(P(x) \vee Q)$
5. $(\exists x)P(x) \wedge Q \equiv (\exists x)(P(x) \wedge Q)$
6. $(\exists x)P(x) \vee Q \equiv (\exists x)(P(x) \vee Q)$
7. $(\forall x)P(x) \wedge (\forall y)Q(y) \equiv (\forall x)\ (\forall y)\ (P(x) \wedge Q(y))$
8. $(\forall x)P(x) \wedge (\exists y)Q(y) \equiv (\forall x)\ (\exists y)\ (P(x) \wedge Q(y))$
9. $(\exists x)P(x) \wedge (\forall y)Q(y) \equiv (\exists x)\ (\forall y)\ (P(x) \wedge Q(y))$
10. $(\exists x)P(x) \wedge (\exists y)Q(y) \equiv (\exists x)\ (\exists y)\ (P(x) \wedge Q(y))$

In converting a well-formed formula to prenex normal form, we use the rules [1]:

1. $P \Leftrightarrow Q \equiv (P \Rightarrow Q) \wedge (Q \Rightarrow P)$
2. $(P \Rightarrow Q) \equiv \neg P \vee Q$
3. $\neg(P \wedge Q) \equiv \neg P \vee \neg Q$
4. $\neg(P \vee Q) \equiv \neg P \wedge \neg Q$
5. $\neg\neg P \equiv P$
6. $P \vee (Q \wedge R) \equiv (P \vee Q) \wedge (P \vee R)$

Before resolution can be carried out on a well-formed formula, we need to eliminate all the existential quantifiers (∃).

This is done by replacing a variable that is existentially quantified by a constant, for instance:

$\exists (x) P(x)$

would be converted to:

$P(c)$

where c is a constant *that has not been used elsewhere in the well-formed formula*. Although P(c) is not logically equivalent to ∃ (x) P(x), we are able to make this substitution in the process of resolution because we are interested in seeing whether a solution exists. If there exists some x for which P(x) holds, then we may as well select such an *x* and name it c.

This process is called *skolemization*, and the variable *c* is called a *skolem constant*.

In order to produce an automated system for generating proofs using resolution on FOPL expressions, we can prove (given a set of assumptions and a conclusion) whether the assumption logically follows from the assumptions as in the steps below:

1. negate the conclusion and add it to the list of assumptions.
2. convert the assumptions into prenex normal form.
3. skolemize the resulting expression.
4. convert the expression into a set of clauses.

Summaries

This chapter presented the fundamentals of two logical systems: propositional logic and first-order predicative logic. Operators, syntax and semantics were introduced for both systems as well as the inference and resolution.

The behavior of the logical operators can be expressed in truth tables. Truth tables can also be used to solve complex problems. Propositional logic deals with simple propositions while first-order predicate logic allows us to reason about more complex statements using the quantifiers ∀ and ∃ (for all, and there exists).

Propositional logic is sound, complete, and decidable while first-order predicate logic is sound and complete, but not decidable. Resolution can be applied to a set of clauses that have been skolemized. This process can be automated because each step can be expressed algorithmically. Resolution can be used to automate the process of proving whether a conclusion can be derived in a valid way from a set of premises or not. Resolution may help in telling us whether a solution exists or not for combinatorial optimization problems.

References

1. Coppin, B.: Artificial intelligence illuminated. Jones and Bartlett, USA (2004)
2. `http://www.onlinefreeebooks.net/free-ebooks-computer-programming-technology/artificial-intelligence/artificial-intelligence-course-material-pdf.html` (accessed on December 22, 2010)
3. Robinson, J.A.: Logic, Form and Function: The Mechanization of Deductive Reasoning. Elsevier Science, Amsterdam (1980)
4. Burckert, H.J.: A Resolution Principle for a Logic With Restricted Quantifiers. Springer, Heidelberg (1992)
5. Jardine, L., Silverthorne, M. (eds.): Francis Bacon: The New Organon. Cambridge University Press, Cambridge (2002)
6. Büning, H.K., Lettmann, T.: Propositional Logic: Deduction and Algorithms. Cambridge University Press, Cambridge (1999)
7. Epstein, R.L.: Predicate Logic: The Semantic Foundations of Logic. Wadsworth Publishing, Belmont (2000)
8. Epstein, R.L.: Propositional Logics: The Semantic Foundations of Logic. Wadsworth Publishing, Belmont (2000)
9. Kelly, J.: The Essence of Logic. Prentice-Hall, Englewood Cliffs (1997)
10. Pospesel, H.: Introduction to Logic: Propositional Logic. Prentice-Hall, Englewood Cliffs (1999)
11. Reeves, S., Clarke, M.: Logic for Computer Science. Addison-Wesley, Reading (1993)

Verification Questions

1. Explain the meaning of the following terms: validity, truth, equivalent, tautology, satisfiable, sound, complete, decidable.
2. What are the components of the propositional logic alphabet?
3. What type of inferences can be performed? Explain each of them.
4. What are the elements of the first-order predicative logic alphabet?
5. Explain the role of the quantifiers $\exists$ and $\forall$ for the predicative calculus.
6. What does it means that logic is monotonic? Which of the propositional logic and first-order predicative logic is monotone?
7. What does it means that logic is decidable? Why first-order predicative logic is not decidable?
8. Which is the most famous inference rule? Explain it and give an example.
9. What entailment means?

10. What is the meaning of equivalence? Give an example.
11. Explain the algorithm for resolution in first-order predicate logic.

Exercises

6.1 Prove the following:

a) $\{P \wedge Q\} \vdash P \vee Q$
b) $\vdash (\neg P \Rightarrow Q) \Rightarrow (\neg Q \Rightarrow P)$

6.2 Construct the truth table for the expression:

$(P \wedge Q) \vee (\neg Q \vee R) \Rightarrow (P \wedge R)$

6.3 Prove the following:

$\vdash (P \Rightarrow Q) \Rightarrow ((Q \Rightarrow R) \Rightarrow ((R \Rightarrow S) \Rightarrow (P \Rightarrow S)))$

6.4 Write expressions in first-order predicate logic to represent the following statements:

1. All computer science students love artificial intelligence.
2. Everyone who knows programming loves artificial intelligence.
3. Therefore, all computer science students know programming.

Prove whether the conclusion follows from the premises or not.

6.5 Convert the following English statements in first-order logic:

1. Every apple or pear is a fruit.
2. Every fruit has a yellow or a green or a red color.
3. No pear has red color.
4. No fruit which is sweet is green.
5. Pear is a fruit.

Construct a proof of the statement:

If pear is not yellow then it is not sweet.

6.6 Determine whether each of the following are valid, satisfiable (but not valid), or unsatisfiable.

1. (rich $\Rightarrow$ happy) $\wedge$ rich $\wedge$ $\neg$happy
2. (rich $\Rightarrow$ happy) $\wedge$ rich $\wedge$ unhappy
3. $\neg P \wedge (\neg(Q \Rightarrow P) \vee (\neg P \wedge Q))) \vee (\neg Q \wedge \neg P) | \neg(\neg P \wedge \neg Q)$
4. $\neg(\neg R \vee T \vee \neg S) \vee \neg(R \Rightarrow S) \vee (R \Rightarrow T)$
5. (e) $(\neg R \wedge (S \Leftrightarrow \neg(Q \vee R))) \Leftrightarrow (\neg S \vee \neg((R \wedge Q) \Rightarrow T))$

6.7 Using propositional linear resolution, show the following propositional sentence is unsatisfiable. Convert this sentence to clausal form and derive the empty clause using resolution:

$(P \vee Q \vee \neg R) \vee ((\neg R \vee Q \vee P) \Rightarrow ((R \vee Q) \wedge \neg Q \wedge \neg P))$

Chapter 7
Rule-Based Expert Systems

7.1 Introduction

Rule-based systems (also known as *production systems* or *expert systems*) are the simplest form of artificial intelligence. A rule based system uses rules as the knowledge representation for knowledge coded into the system [1][3][4][13][14][16][17][18][20]. The definitions of rule-based system depend almost entirely on expert systems, which are system that mimic the reasoning of human expert in solving a knowledge intensive problem. Instead of representing knowledge in a declarative, static way as a set of things which are true, rule-based system represent knowledge in terms of a set of rules that tells what to do or what to conclude in different situations.

A rule-based system is a way of encoding a human expert's knowledge in a fairly narrow area into an automated system. A rule-based system can be simply created by using a set of assertions and a set of rules that specify how to act on the assertion set. Rules are expressed as a set of if-then statements (called IF-THEN *rules* or *production rules*):

IF P THEN Q

which is also equivalent to:

P⇒Q.

A rule-based system consists of a set of IF-THEN rules, a set of *facts* and some *interpreter* controlling the application of the rules, given the facts. The idea of an expert system is to use the knowledge from an expert system and to encode it into a set of rules. When exposed to the same data, the expert system will perform (or is expected to perform) in a similar manner to the expert. Rule-based systems are very simple models and can be adapted and applied for a large kind of problems. The requirement is that the knowledge on the problem area can be expressed in the form of if-then rules. The area should also not be that large because a high number of rules can make the problem solver (the expert system) inefficient.

C. Grosan and A. Abraham: Intelligent Systems, ISRL 17, pp. 149–185.
springerlink.com

7.2 Elements of a Rule-Based System

Any rule-based system consists of a few basic and simple elements as follows:

1. A set of *facts*. These facts are actually the assertions and should be anything relevant to the beginning state of the system.
2. A set of *rules*. This contains all actions that should be taken within the scope of a problem specify how to act on the assertion set. A rule relates the facts in the IF part to some action in the THEN part. The system should contain only relevant rules and avoid the irrelevant ones because the number of rules in the system will affect its performance.
3. A termination criterion. This is a condition that determines that a solution has been found or that none exists. This is necessary to terminate some rule-based systems that find themselves in infinite loops otherwise.

Facts can be seen as a collection of data and conditions. Data associates the value of characteristics with a thing and conditions perform tests of the values of characteristics to determine if something is of interest, perhaps the correct classification of something or whether an event has taken place.

For instance, if we have the fact:

temperature <0

then temperature is the data and the condition is <0.

Rules do not interact directly with data, but only with conditions either singly or multiple (joined by logical operators as shown below). Figure 7.1 contains an example showing the parts of a rule based systems and the interactions between them.

DATA	CONDITIONS	RULES
		Premises
Season winter		IF temperature<0
temperature	<0, >0	AND
wind blushing	strongly, gently	IF wind blushing is strongly
road	slippery, not slippery	OR
weather	cold, warm, hot	IF the road is slippery
		Conclusion
		THEN the weather is cold

Fig. 7.1 An example showing the parts of a rule based systems and the interactions between them.

There are two ways for a rule to set new values for the data [2]:

- by assignment, where the value is directly set, and
- by assertion. The assertion does not in itself assign a value to the data, but the condition acts like a constraint upon the data value, saying it must be the value specified by the condition.

7.2.1 Rules

A rule consists of two parts: the IF part and the THEN part. The IF part is called *antecedent* or *premise* (or *condition*) and the THEN part is called *consequent* or *conclusion* (or *action*).

Thus, a simple rule can be expressed as:

IF antecedent
THEN consequent.

Example:
IF the season is winter
THEN it is cold.

The rule tests the logical expression in the premise, and, if the expression evaluates to true, it then asserts that a fact about a thing or a class of things is true.

A general rule can have multiple antecedents joined by any of the logical operators AND, OR (or by a mixture of both of them).

Example 1 (multiple antecedents combined by AND*)*

IF antecedent1
AND antecedent2
⋮
AND antecedentN
THEN consequent

IF the season is winter
AND the temperature is <0 degrees
AND it is windy
THEN the weather is cold

Example 2 (multiple antecedents combined by OR*)*

IF antecedent1
OR antecedent2
⋮
OR antecedentN
THEN consequent

IF the season is winter
OR the temperature is <0 degrees
OR it is windy
THEN it is cold

Example 3 (multiple antecedents combined by AND and OR*)*

IF antecedent1
AND antecedent 2
⋮
OR antecedentN
THEN consequent

IF the season is winter
AND the temperature is <0 degrees
OR the weather is windy
THEN it is cold

The consequent can also have multiple clauses, for instance:

IF antecedent
THEN consequent1
consequent2
⋮
consequentN

Example:

IF the season is winter
THEN the temperature is low
the road is slippery
the forecast is snow

The antecedent of a rule has two parts:

- an object (also called a linguistic object);
- the value of the linguistic object.

The object and its value are linked by an operator (like, for example, *is*, *are* or mathematical operators) which identifies the object and assigns the value.

Examples:

IF x>0
THEN x is positive

IF the temperature is high
THEN the weather is hot

A consequent also combines an object and a value connected by an operator. The operator assigns a value to the linguistic object. Numerical objects as well as arithmetical expressions can be used as rule consequent.

7.2.1.1 Rules Classification

There are several ways to classy the rules. Some of them are based on their:

- function,
- structure and
- behavior.

Functional classifications are the easiest to use and understand. Often, in practice, rules can be associated with a single subject. Functional categorization often remains the default classification scheme for rules.

Behavioral classification of rules can be complex and is often encountered more in the implementation and debugging of rules rather than their specification. Behavioral classification and rule behavior in general can be a very complex topic. Rule behavior can arise either from interactions between rules or between the values of the data being used to test conditions within a rule.

Structural classification of rules has the advantage of classifying rules precisely and unambiguously. However, the structural classification of a rule will not describe important aspects of that rule, unlike functional classification. Structural rules can be in their turn classified into three groups:

- logic rules,
- definitions and
- constraints.

Logic rules have a standard IF-THEN structure that is easy to follow and understand. If the conditions of the premise are met, then the conclusion about relationships between data entities must be true. The value of some data element is set according to the conclusion of the rule.

From a structural standpoint, a logic rule in the restricted sense has a distinct test condition in the IF clause and a conclusion that changes something in the THEN clause [2].

Example:

IF x=1 Test
AND y=2
THEN z=3 Assign

Only a logic rule has a clearly recognizable IF condition and a THEN conclusion. The conclusion changes the value of something in the system.

Definition is a THEN conclusion with no IF condition. It is unconditional, except for trivial validations of variable bindings and divides by zero conditions.
A definition constructs values from other values. Values may be the result of the application of previous definitions, or they may be the result of executing the conclusions of logic rules [2].

Example:

IF P (something which is always true)
THEN Q=x Assign or compute a value

Definitions often comprise the majority of rule entities in most applications. Definitions are unconditional so they are often implemented as procedural code or logical view of the database.

Constraints are IF conditions with no THEN conclusion. A constraint describes a violation of a relationship between data entities. There is no change of value within the data. A constraint will often trigger an exception, such as sending a message (for instance “x should be a positive number”). There is no change in the value or state of the any entity.

Example:

IF x= -1
AND y=2
THEN Raise an exception – such as sending the message “x should be a positive value” (Do not allow a change to a value).

The exception referred to in the example is usually an error process. The process will either prompt the user that there is an error condition or log the error condition to an error file. The exception may alter or interrupt the flow of process steps or of the rule engine itself [2].

Based on the conclusion or the consequent of a rule, rules can express:

- *Relation*

  IF x >0
  THEN x is positive

- *Recommendation*

  IF it is rainy
  THEN take an umbrella

- *Directive*

  IF Phone battery signal
  AND phone battery empty
  THEN charge the phone

- *Heuristic*

  IF phone light is off
  THEN battery is flat

7.3 Structure of a Rule-Based Expert System

A rule-based expert systems has the structure given in diagram in Figure 7.2 and consists of the following main elements (the five most important ones are marked in bold in the figure):

- ***Knowledge base***

  Contains the domain knowledge which is represented as rules (IF-THEN rules) about subject at hand [5][19].
- ***Database***

  Consists of predicate calculus facts that match against the IF parts of the rules in the knowledge base.
- ***Inference Engine***

  Consists of all the processes that manipulate the knowledge base to deduce information requested by the user and carries the reasoning required by the expert system to reach a solution.
- ***Explanation subsystem***

  Analyzes the structure of the reasoning performed by the system and explains it to the user, giving the user the possibility to enquire the systems about the way in which a conclusion has been reached or about the facts used.
- ***User interface***

  Refers to the communication between a user looking for a solution and the expert system and consists of some kind of natural language processing system or graphical user interfaces with menus.

- ***Knowledge engineer***
Is usually a computer scientist with AI training which works with an expert in the field of application in order to represent the relevant knowledge of the expert in a forms that can be entered into the knowledge base.
- ***Knowledge acquisition subsystem***
Checks and updates the growing knowledge base for possible inconsistencies and incomplete information.

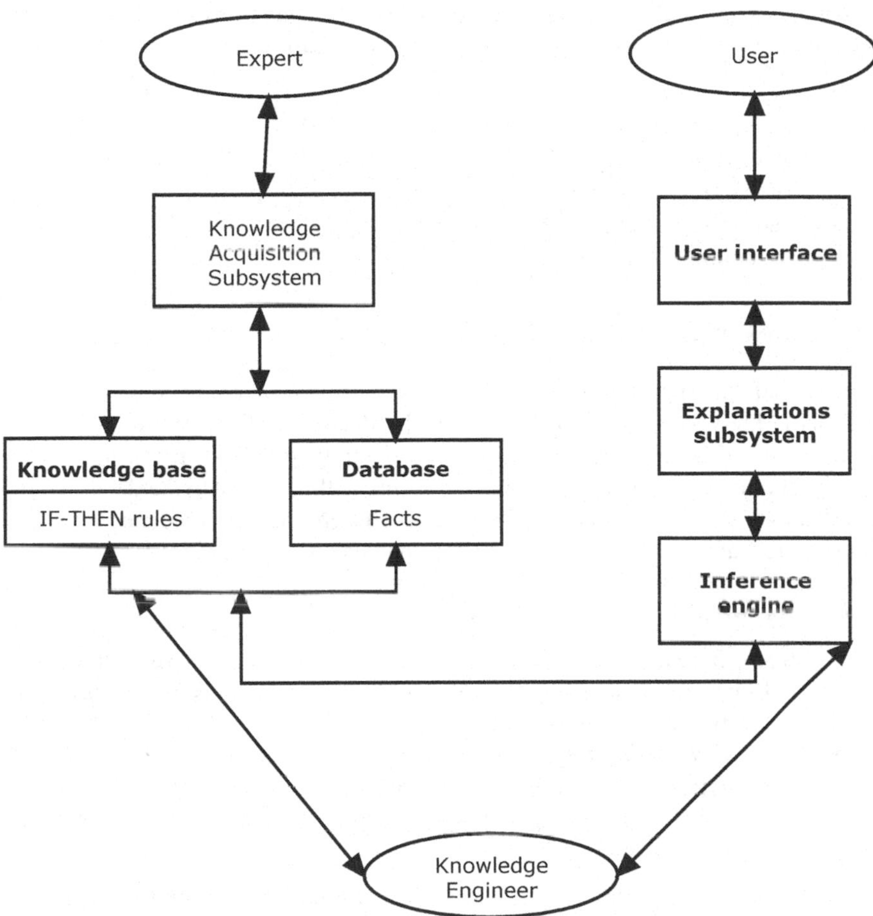

Fig. 7.2 The structure of a rule-based expert system

The rule-based system works in a very simple way: it starts with a rule-base, which contains all of the appropriate knowledge encoded into IF-THEN rules, and a working memory, which may or may not initially contain any data, assertions or initially known information. The system examines all the rule conditions (IF) and determines a subset, the conflict set, of the rules whose conditions are satisfied based on the working memory. Of this conflict set, one of those rules is triggered (fired). Which one is chosen is based on a conflict resolution strategy. When the rule is fired, any actions specified in its THEN clause are carried out. These actions can modify the working memory, the rule-base itself, or do just about anything else the system programmer decides to include. This loop of firing rules and performing actions continues until a termination criterion is met. This termination criterion can be given by the fact that there are no more rules whose conditions are satisfied or a rule is fired whose action specifies the program should terminate.

Reasoning is the way in which rules are combined to derive new knowledge. Reasoning is how humans work with knowledge, facts and problem solving strategies to draw conclusions. There are a few types of reasoning:

- inductive reasoning;
- deductive reasoning;
- abductive reasoning;
- analogical reasoning;
- common-sense reasoning;
- non-monotonic reasoning.

7.4 Types of Rule-Based Expert Systems

A rule-based expert system works as follows: the inference engine compares each rule in the knowledge base with facts in the database. If the IF part of a rule matches a fact then the THEN part is executed and the rule fires. By firing a rule a new result (a new fact) may be obtained and this will be added to the database. By firing rules inference chains are obtained. An inference chain indicates how an expert system applies the rules to reach the conclusion or the goal.

There are two main ways in which rules are executed and this conducts to the existence of two main rule systems:

- *forward chaining* systems. A forward chaining system starts with the initial facts and keep using the rules to draw new conclusions (or take certain actions) given those facts.
- *backward chaining* systems. A backward chaining system starts with some hypothesis (or goal) to prove, and keep looking for rules that would allow concluding that hypothesis, by setting new subgoals to prove as the process advances.

Forward chaining systems are primarily data-driven, while backward chaining systems are goal-driven.

Example

Consider the following expert systems whose database consists of the facts A, B, C, D, E and whose knowledge base is given by the rules below:

Rule 1: IF A is true
AND C is true
THEN B is true

Rule 2: IF C is true
AND D is true
THEN F is true
Rule 3: IF C is true
AND D is true
AND E is true
THEN X is true
Rule 4: IF A is true
AND B is true
AND X is true
THEN Y is true
Rule 5: IF D is true
AND Y is true
THEN Z is true

The inference chain for this example is given in Figure 7.3.

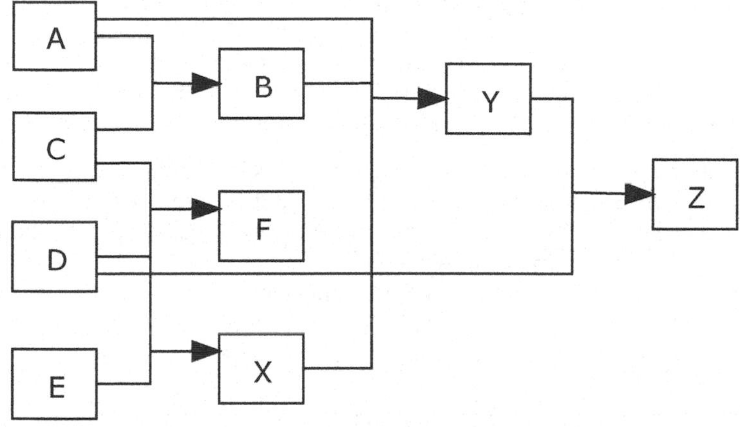

Fig. 7.3 An inference chain example.

Rule-based systems differ from standard procedural or object-oriented programs in that there is no clear order in which code executes. The knowledge of the expert is captured in a set of *rules*, each of which encodes a small piece of the expert's knowledge. Each rule has a left hand side and a right hand side (IF part and THEN part respectively).

The IF part contains information about certain facts and *objects* which must be true in order for the rule to potentially fire (or execute). Any rules whose IF part match in this manner at a given time are placed on an *agenda*. One of the rules on the agenda is picked (there is no way of predicting which one), and its THEN part is executed, and then it is removed from the agenda.

The agenda is then updated (generally using a special algorithm called the *Rete algorithm* which helps in reducing the number of comparisons that need to be made between rules and facts in the database), and a new rule is picked to execute. This continues until there are no more rules on the agenda. The Rete is a directed, acyclic, rooted graph (or a search tree). Each path from the root node to a leaf in the tree represents the left-hand side of a rule (IF part). Each node stores details of which facts have been matched by the rules at that point in the path.

As facts are changed, the new facts are propagated through the Rete from the root node to the leaves, changing the information stored at nodes appropriately. This means either adding a new fact or deleting an old fact, or changing information about an old fact. In this way, the system only needs to test each new fact against the rules, and only against those rules to which the new fact is relevant, instead of checking each fact against each rule [12]. A general form of expert systems is an expert system shell. An expert system shell is actually and expert system whose knowledge is removed. Thus, the user can just add its own knowledge in the form of rules and provide information to solve the problem. Expert system shells are commercial versions of the expert systems.

7.4.1 Forward Chaining Systems

The forward chaining works as follow: given a certain set of facts in the working memory, use the rules to generate new facts until the desired goal is reached. The steps below are followed (see Figure 7.4 for the illustration of forward chaining):

1. *Match* the IF part of each rule against facts in working memory.
2. If there is more than one rule that could be used (more than one rule which fires), *select* which one to apply by using conflict resolution (described in the following section).
3. *Apply* the rule. If new facts are obtained add them to working memory.
4. *Stop* (or *exit*) when the conclusion is added to the working memory or if there is a rule which specifies to end the process.

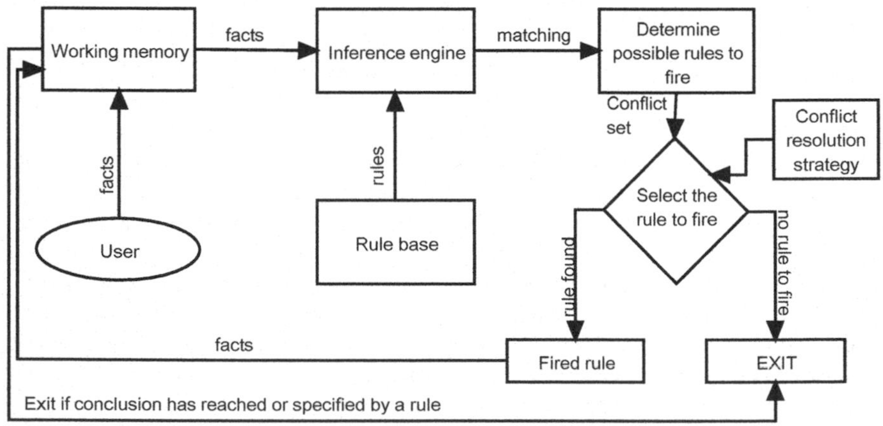

Fig. 7.4 Forward chaining diagram.

Example 1

Let us consider the example given in Section 7.4 with the database consisting of facts A, B, C, D and E and with the knowledge base consisting of the 5 given rules. We will now show how forward chaining can be applied to this system to reach the conclusion Z. If multiple rules can fire at a time then we will choose to fire the first of the rules, which was not fired before.

Let us now follow step by step the forward chaining as presented above.

1. Match the IF part of each rule against facts in working memory.

The following rules can be selected:

Rule 1: IF A AND C THEN B (since both A and C are in the database);
Rule 2: IF C AND D THEN F (since both C and D are in the database);
Rule 3: IF C AND D AND E THEN X (since all C, D and E are in the database)
Rules 4 and 5 cannot be selected because their IF part cannot be matched (X in the case of Rule 4 and Y in the case of Rule 5 respectively are not in the database at this moment).

2. If there is more than one rule that could be used (more than one rule which fires), select which one to apply.

As we can see, there are 3 rules that can be used: Rule 1, Rule 2 and Rule 3. And we will always select the first one, which can be applied and was not applied earlier. Thus, Rule 1 will be the rule fired first.

3. Apply the rule. If new facts are obtained add them to working memory.

At this step we are applying the Rule 1:

Rule 1: IF A AND C THEN B

The consequent of this rule is B is true. But B is already in the database, so no new facts are obtained by applying this rule.

4. Stop (or exit) when the conclusion is added to the working memory or if there is a rule which specifies to end the process.

Our conclusion is Z and was not reached yet, so we will go again to the first step.

1. At this time, the rules which can be selected are the same: Rule 1, Rule 2 and Rule 3.
2. Since Rule 1 already fired, the next selected rule is Rule 2.
3. Rule 2 is fired: Rule 2: IF C AND D THEN F.

A new fact if obtained, F, which is not already in the database, so F will be added to the database which is now: A, B, C, D, E, F.

4. Conclusion Z was not reached yet, so we will start the process again.

1. Rules which can fire now are still the same: Rule 1, Rule 2, Rule 3.
2. Since rules Rule 1 and Rule 2 have been already used, the only remaining one to be selected is Rule 3:

Rule 3:

3. Rule 3 is fired and a new fact is obtained, X, which is not in the database, so will be added.
4. X is not the conclusion, so we should still restart the process.

1. The database contains now the facts A, B, C, D, E, F, and X. Thus, the first 4 rules can be matched at this point.
2. Since the first 3 rules have been already used, the only one which can be selected to fire is Rule 4:

Rule 4: IF A AND B AND X THEN Y.

3. Rule 4 is fired and a new term, Y, is added to the database.
4. Still, Y is not the conclusion, so we have to continue.

1. All the 5 rules match the IF condition.
2. The only remaining rule to use is Rule 5:

Rule 5: IF D AND Y THEN Z.

3. Rule 5 is fired and a new fact, Z, is obtained.
4. Z represents our conclusion so the process may stop here.

The diagram showing the whole forward chaining process to obtain the conclusion Z is depicted in Figure 5.

Example 2

Let us now consider a practical example. Given a set of facts containing various information about flowers and given a set of rules, the task is to produce the solution, which indicates which flower is a white lily. The linguistic variables (objects) and their possible values allowed by the experts systems and contained into the database are given in Table 1. The knowledge base consists of the following rules:

Rule 1: IF size > 10
AND size <50
THEN height is small

Rule 2: IF size > 50
AND size <150
THEN height is medium

Rule 4: IF size > 150
THEN height is tall

Rule 5: IF life cycle is one year
THEN life type is annual

Rule 6: IF life cycle is more than one year
THEN life type is perennial

Rule 7: IF season is summer
AND color is blue
OR color is purple
OR color is yellow
AND life type is perennial
AND root type is bulb
THEN flower name is iris

Rule 8: IF season is autumn
AND color is white
OR color is pink
OR color is pinkish-red
THEN flower name is anemone

Rule 9: IF season is autumn
AND height is medium
AND color is yellow
OR color is while
OR color is purple
OR color is red
THEN flower name is Chrysanthemum

Rule 10: IF season is spring
AND root type is bulbs
AND color is white
OR color is yellow
OR color is orange
OR color is purple
OR color is red
OR color is blue
AND perfumed is true
THEN flower is Freesia

Rule 11: IF life type is perennial
AND height is tall
AND root type is bulbs
AND season is summer
THEN flower name is Dahlia

Rule 12: IF season is spring
AND root type is bulbs
AND color is yellow
OR color is white
THEN flower name is Narcissus

Rule 13: IF soil is acidic
AND color is white
OR color is pink
OR color is red
AND life type is perennial
AND root type is roots
THEN flower name is Camellia

Rule 14: IF season is spring
AND root type is bulbs
AND perfumed is true
AND height is small
AND life type is perennial
THEN flower name is Lily

Rule 15: IF height is small
AND life type is annual
AND soil is rich
OR soil is loose
OR soil is fertile
THEN flower name is Begonia

Rule 16: IF season is winter
AND color is white
OR color is pink
OR color is red
THEN flower name is Azalea

Rule 17: IF life type is perennial
AND root type is root
AND color is white
OR color is red
OR color is blue
OR color is yellow
THEN flower is Anemone

Rule 18: IF life type is perennial
AND root type is roots
AND color is white
OR color is pink

OR color is red
OR color is yellow
AND perfumed is true
AND soil is well-drained
THEN flower is rose

Rule 19: IF flower name is Lily
AND perfumed is true
THEN flower name is White lily

Table 1 The objects (linguistic variables) and their values are used in Example 2.

Object	**Value**	**Object**	**Value**
Flower name	Iris Anemone Chrysanthemums Freesia Dahlia Narcissus Camellias Lily Begonia Azaleas Anemone Roses White lily	color	blue purple yellow red white pink orange violet pinkish-red
Season	Autumn Summer Spring winter	Size	10-50 cm 50-150 cm >150 cm
Root type	Bulb root	Perfume	True False
Life type	Perennial Annual	Soil	Acidic Loose Fertile Rich Well-drained
Life cycle	One year More than one year	Height	Small Medium Tall

Suppose we have the following facts in the database: season: spring, root type: bulbs, perfumed: true, size: 16-18 cm, life cycle more than one year, color: orange, red, white, pink. We wish to infer the white lily flower. The forward reasoning process is carried out as follows:

Cycle 1

Matching: Rule 1 and Rule 6 are applicable.
Rule selection: Select Rule 1
Rule application: height is small will be added to the working memory.

Matching: Rule 1 and Rule 6 are applicable.
Rule selection: Can only select Rule 6 since Rule 1 already fired.
Rule application: Life type perennial will be added to the working memory.

Working memory consists now of: season: spring, root type: bulbs, perfumed: true, size: 16-18 cm, life cycle more than one year, color: orange, red, white, pink, height: small, life type: perennial.

Cycle 2

Matching: Rule 1, Rule 6, Rule 10, Rule 12 and Rule 14 are applicable.
Rule selection: Can only select Rule 10 Rule 12 or Rule 14 since Rule 1 and Rule 6 already fired. Select Rule 10.
Rule application: Flower name is Freesia will be added to the working memory.

Matching: Rule 1, Rule 6, Rule 10, Rule 12 and Rule 14 are applicable.
Rule selection: Can only select Rule 12 or Rule 14 since Rule 1, Rule 6 and Rule 10 already fired. Select Rule 12.
Rule application: Flower name is Narcissus will be added to the working memory.

Matching: Rule 1, Rule 6, Rule 10, Rule 12 and Rule 14 are applicable.
Rule selection: Can only select Rule 14 since the other rules already fired. Select Rule 14.
Rule application: Flower name is Lily will be added to the working memory.

Working memory consists now of: season: spring, root type: bulbs, perfumed: true, size: 16-18 cm, life cycle more than one year, color: orange, red, white, pink, height: small, life type: perennial, flower name: freesia, narcissus, lily.

Cycle 3

Matching: Rule 1, Rule 6, Rule 10, Rule 12, Rule 14 and Rule 19 are applicable.
Rule selection: Can only select Rule 19 since the others already fired. Select Rule 19.

Rule application: Flower name is White lily be added to the working memory and this is also the goal and the inference process will stop.

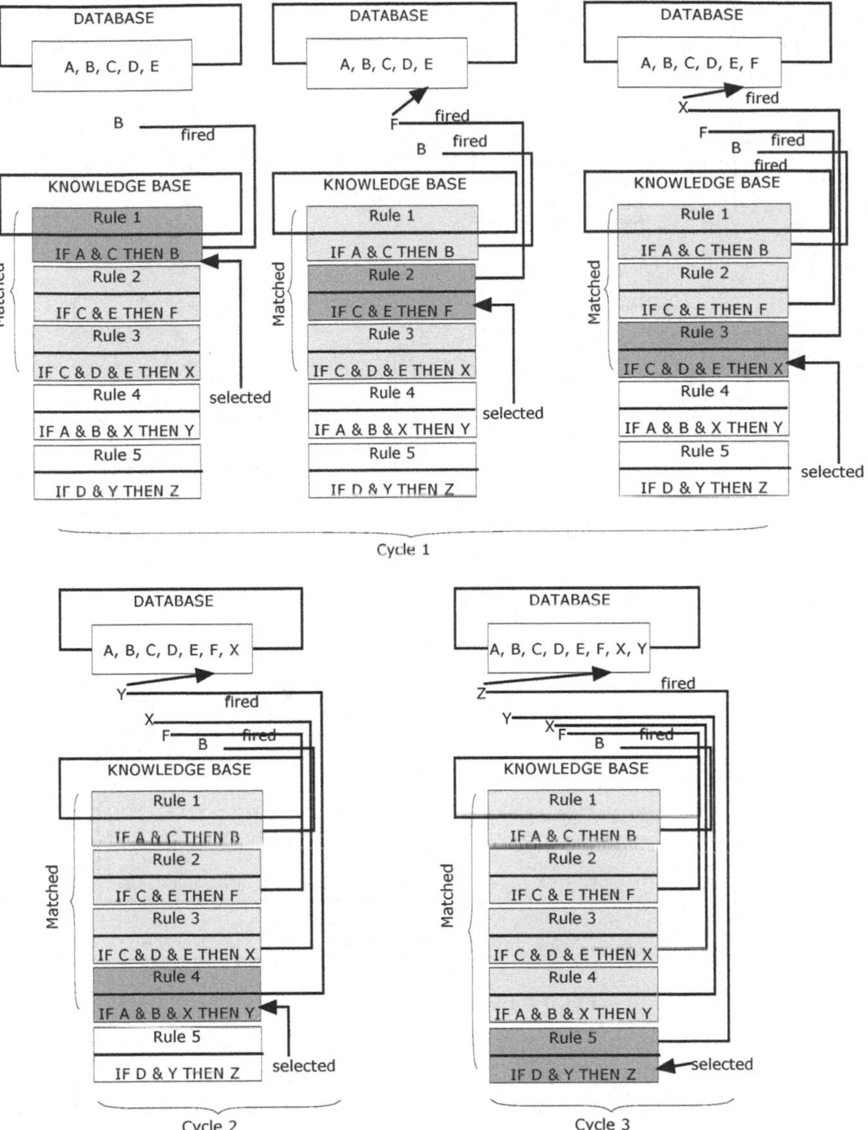

Fig. 7.5 Forward chaining for the example considered. The goal is to reach Z.

7.4.2 Backward Chaining Systems

In the backward chaining we first state a hypothesis. Then, the inference engine tries to find evidence to prove it. If the evidence doesn't match then we have to start over with a new hypothesis. If the evidence matches then the correct hypothesis has been made. Figure 7.6 presents the diagram and the working model of a backward chaining system.

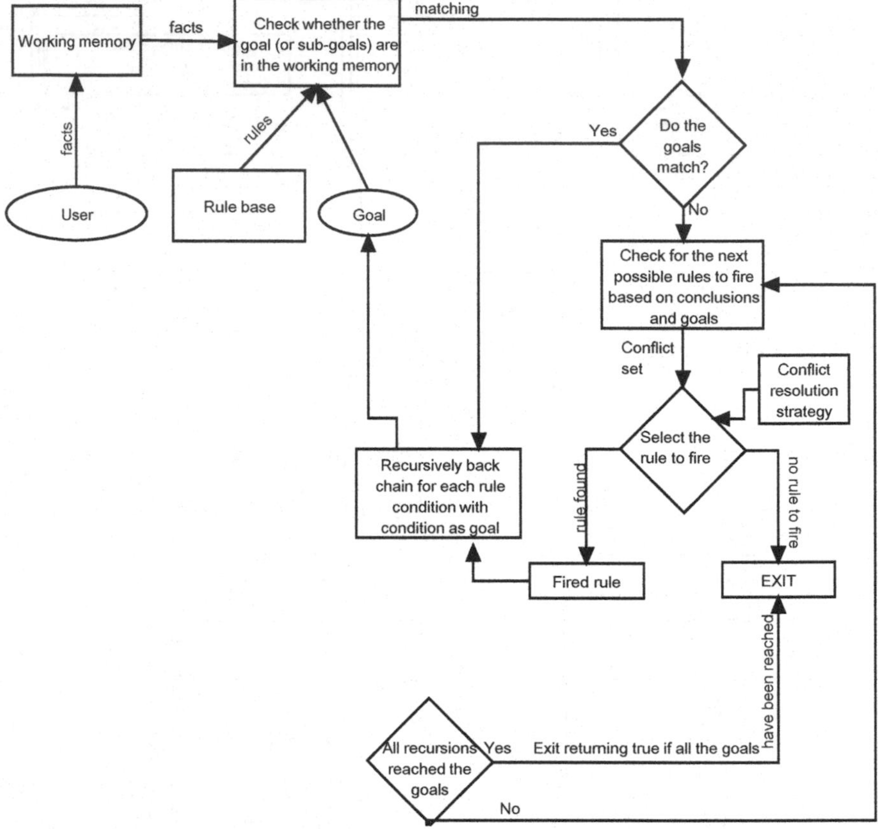

Fig. 7.6 Backward chaining diagram.

The backward chaining systems work backwards from a hypothesized goal, attempting to prove it by linking the goal to the initial facts. To backward chain from a goal in the working memory the inference engine must follow the steps:

1. Select rules with conclusions matching the goal.
2. Replace the goal by the rule's premises. These become sub-goals.
3. Work backwards until all sub-goals are known to be true. This can be achieved either:

- they are facts (in the working memory) or
- the user provides the information.

Example 1

Let us now consider the same example as for forward chaining: the database consists of the facts A, B, C, D, E and whose knowledge base is given by the rules below:

Rule 1: IF A is true
AND C is true
THEN B is true

Rule 2: IF C is true
AND D is true
THEN F is true

Rule 3: IF C is true
AND D is true
AND E is true
THEN X is true

Rule 4: IF A is true
AND B is true
AND X is true
THEN Y is true

Rule 5: IF D is true
AND Y is true
THEN Z is true

The goal of the system is Z.

For this example, backward chaining works as follows:

Step 1

1. *Select rules with conclusions matching the goal.*

   The only rule with conclusion matching the goal is Rule 5.

2. *Replace the goal by the rule's premises. These become sub-goals.*

D is in the database but we don't have Y. So, the first sub-goal is Y.

3. *Work backwards until all sub-goals are known to be true.*

We don't have all sub-goals as true, so we back-chain again.

Step 2

1. *Select rules with conclusions matching the goal.*

Our goal is Z but our sub-goal is Y. Rule 4 has Y as conclusion.

2. *Replace the goal by the rule's premises. These become sub-goals.*

Among the Rule's 4 premises we have A and B in the database but we don't have X. Thus, our current sub-goal is X

3. *Work backwards until all sub-goals are known to be true.*

We don't have all sub-goals as true, so we back-chain again.

Step 3

1. *Select rules with conclusions matching the goal.*

Our goal is Z but our sub-goals are Y and X. Most recent one is X. Rule 3 has X as conclusion.

2. *Replace the goal by the rule's premises. These become sub-goals.*

Rule's 3 premises are C, D and E and we can find all of them in the database. Thus, our sub-goal X can be obtained by first firing Rule 3.

3. *Work backwards until all sub-goals are known to be true.*

We obtained one of the sub-goals. We don't have all sub-goals as true, so we recursively back-chain to obtain the other sub-goals. First on the agenda is Y.

Going back to Step 2, we now have all the premises for Rule 4. Thus, rule 4 can fire and Y is obtained and added to the database.

Going back to Step 1, we now have all the premises for Rule 5 whose conclusion is our desired goal. Thus, we can fire Rule 5 and obtain the goal Z.

The diagram of the backward chaining for this example is depicted in Figure 7.7.

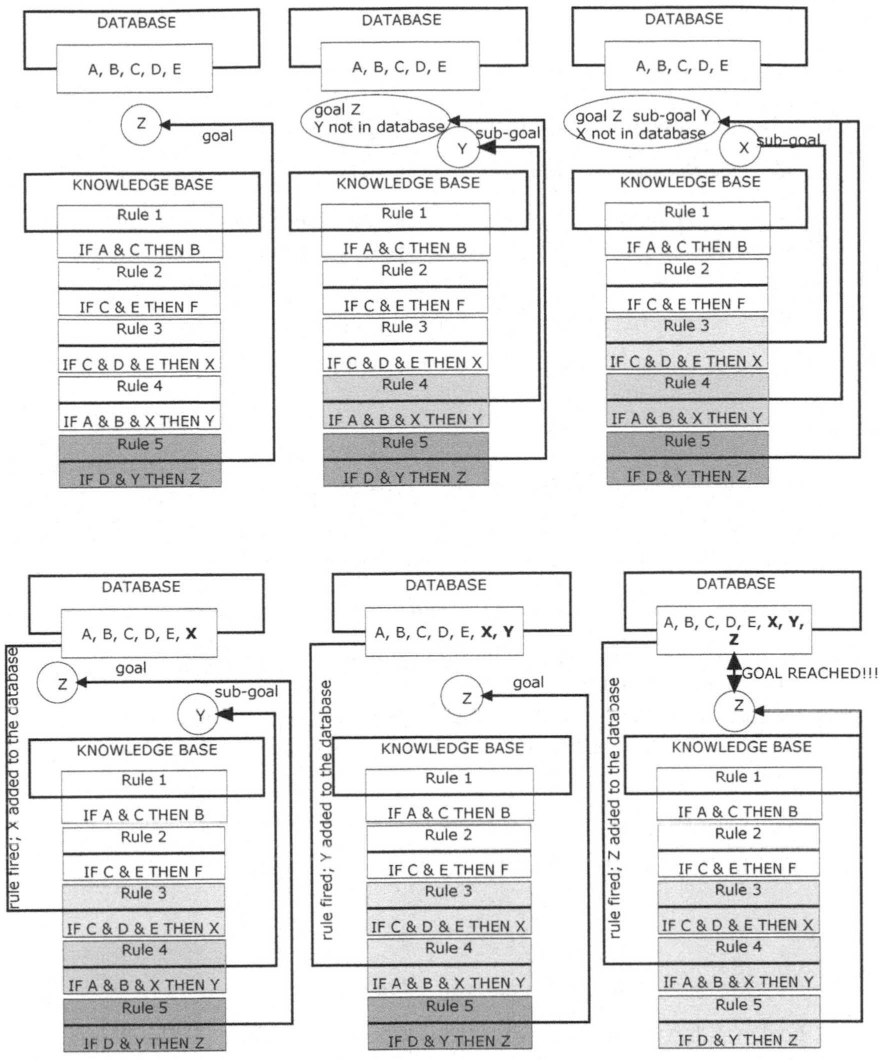

Fig. 7.7 Backward chaining diagram.

Example 2

Consider now the same example as the one used for forward chaining in Example 2. The initial facts in the database are: season: spring, root type: bulbs, perfumed: true, size: 16-18 cm, life cycle more than one year, color: orange, red, white, pink. We wish to infer the white lily flower.

Step 1

1. *Select rules with conclusions matching the goal.*

The only rule with conclusion matching the goal is Rule 19.

2. *Replace the goal by the rule's premises. These become sub-goals.*

The premise "perfumed is true" is in the database but the premise "flower name is Lily" is not. The conclusion "flower name is Lily" is our sub-goal and this is the consequent of Rule 14.

3. *Work backwards until all sub-goals are known to be true.*

We don't have all sub-goals as true, so we back-chain again.

Step 2

1. *Select rules with conclusions matching the goal.*

Our goal is "flower name is white lily" (the conclusion of Rule 19) and our sub-goal is "flower name is lily" the conclusion of Rule 14.

2. *Replace the goal by the rule's premises. These become sub-goals.*

Among the premises of Rule 14, the following are already in the database: "season is spring", "root type is bulbs", "perfumed is true".
We need to get two more premises: "height is small" and "life type is perennial".
We consider now "height is small" "life type is perennial" as the following sub-goals.

3. *Work backwards until all sub-goals are known to be true.*

We don't have all sub-goals as true, so we back-chain again.

Step 3

1. *Select rules with conclusions matching the goal.*

Our goal is "flower name is white lily" (the conclusion of Rule 19) and our sub-goals are "flower name is lily" the conclusion of Rule 14, "height is small", the conclusion of Rule 1 and "life type is perennial" the conclusion of Rule 6.

2. *Replace the goal by the rule's premises. These become sub-goals.*

All the premises of Rule 1 are already in the database. Thus, Rule 1 can fire and the fact "height is small" will be added to the database.

3. *Work backwards until all sub-goals are known to be true.*

We don't have all sub-goals as true, so we back-chain again.

Step 4

1. *Select rules with conclusions matching the goal.*

Our goal is "flower name is white lily" the conclusion of Rule 19 and our sub-goals are "flower name is lily" the conclusion of Rule 14 and "life type is perennial" the conclusion of Rule 6.

2. *Replace the goal by the rule's premises. These become sub-goals.*

The premise of Rule 1 is in the database. Thus, Rule 6 can fire and the fact "life type is perennial" will be added to the database.

3. *Work backwards until all sub-goals are known to be true.*

We don't have all sub-goals as true, so we back-chain again.

Step 4

1. *Select rules with conclusions matching the goal.*

Our goal is "flower name is white lily" the conclusion of Rule 19 and our sub-goal is "flower name is lily" the conclusion of Rule 14.

2. *Replace the goal by the rule's premises. These become sub-goals.*

We now have all the premises to fire Rule 14. The fact "flower name is lily" is obtained and introduced in the database.

3. *Work backwards until all sub-goals are known to be true.*

We don't have yet the goals so we continue to back-chain.

Step 4

1. *Select rules with conclusions matching the goal.*

Our goal is "flower name is white lily" the conclusion of Rule 19. There are no subgoals at this stage and all the premises of Rule 19 are true. Thus, Rule 19 can now fire and the final goal is obtained.

Remark

As we can observe from our particular examples, the backward chaining reaches the goal in fewer steps than forward chaining does. The semantics of creating a backward-chaining expert system can be very similar to the forward-chaining systems. It's the engine that handles the systems differently. Some engines do both forward and backward chaining with the same expert system. The benefit of a backward-chaining inference engine is that it doesn't have to evaluate all of the evidence to arrive at a conclusion. If your expert system contains a huge number of possible hypothesis, it may perform substantially better using backward chaining, than having to evaluate all of the evidence sequentially, essentially having to process all of the cases.

7.4.3 *Forward Chaining or Backward Chaining? Which One Should Apply?*

In order to choose one of the forward chaining or backward chaining, we should observe how a domain expert solves the problem. If the expert need to first cumulate all the possible information and infer from there whatever can be inferred, then the forward chaining is suggested. If the expert starts with a hypothetical solution and then tries to find facts to prove it then the backward chaining system is suitable.

Forward chaining is generally suggested if:

- All or most of the data is given in the problem statement;
- There exist a large number of potential goals but only a few of them are achievable in a particular problem instance;
- It is difficult to formulate a goal or hypothesis.

Generally, backward chaining is suggested in the flowing situations:

- A goal or hypothesis is given in the problem statement or can be easily formulated;
- There are a large number of rules that match the facts, producing a large number of conclusions - choosing a goal prunes the search space;
- Problem data are not given (or easily available) but must be acquired as necessary (in certain systems).

Forward chaining (data-driven) search can appear aimless but produces all solutions to a problem (if desired).

Mixed reasoning is also possible - facts get added to the working memory and sub-goals get created until all the sub-goals are present as facts.

7.5 Conflict Resolution

It is important to define a way or an order in which rules are firing during the inference process. There are several different strategies such as [2][11]:

- ***First applicable:*** If the rules are in a specified order, firing the first applicable one is the easiest way to control the order in which rules fire. From a practical perspective the order can be established by ordering the rules in the knowledge base by placing them in the preferred order (but this only works for small systems of up to 100 rules).

  This is the simplest strategy and has a potential for a large problem: that of an infinite loop on the same rule. If the database remains unchanged (nothing new is introduces or nothing is removed), as does the rule-base, then the conditions of the first rule have not changed and it will fire again and again. To solve this, it is a common practice to suspend a fired rule and prevent it from re-firing until the data that satisfied the rule's conditions, has changed.

Example

Suppose we have the following rules in the knowledge base:

Rule 1: IF color is yellow
THEN fruit is apple;

Rule 2: IF color is yellow
AND shape is long
THEN fruit is banana

Rule 3: IF shape is round
THEN fruit is apple.

And the data base consisting of yellow (color) and round (shape). Then two rules can fire: Rule 1 and Rule 3 and the order will be Rule 1 first and then Rule 3.

Random: A random strategy simply chooses a single random rule to fire from the conflict set. It is also advantageous even though it doesn't provide the predictability or control of the first-applicable strategy.

Example

If we consider the same example as above, then one of the rules Rule 1 and Rule 3 will be chosen at random to fire first.

Most Specific: This strategy is based on the number of conditions of the rules. From the conflict set, the rule with the most conditions is chosen. This is based on the assumption that if it has the most conditions then it has the most relevance to the existing data. It can also be called *longest matching strategy* and it is based on the assumption that a specific rule process more information than a general one.

Example

Rule 1: IF the weather is cold
THEN the season is winter

Rule 2:
IF the weather is cold
AND the temperature is low
AND the wind is blushing
AND the forecast is snow
THEN the season is winter.

Among the two rules, most significant one is Rule 2. Thus, Rule 2 is selected if the most specific strategy is used.

- **Least Recently Used:** Each of the rules has a time or step stamp (or time and date) associated, which marks the last time it was used. This maximizes the number of individual rules that are fired at least once. If all rules are needed for the solution of a given problem, this is a perfect strategy.

There might also be situations that the new rules have been added by an expert whose opinion is less trusted than that of the expert who added the earlier rules. In this case, it clearly makes more sense to allow the earlier rules priority.

Example:

Rule 1: IF color is yellow [28.02.2009, 13:45]
THEN fruit is apple;

Rule 2: IF color is yellow [01.03.2009, 12:00]
AND shape is long
THEN fruit is banana

Rule 3: IF shape is round [05.03.2009, 20:00]
THEN fruit is apple.

In this example, the Rule 1 is the least recently introduces, thus this will be the one selected to fire.

- **"Best" rule:** In the case of this strategy, each rule is given a 'weight,' which specifies how much it should be considered over the alternatives. The rule with the most preferable outcomes is chosen based on this weight.

Example:

Rule 1: IF color is yellow 30%
THEN fruit is apple;

Rule 2: IF color is yellow 30%
AND shape is long
THEN fruit is banana

Rule 3: IF shape is round 40%
THEN fruit is apple.

In this example Rule 3 will be selected because it is having the highest weight among all the three rules.

In many expert systems, the order in which rules are used affects the conclusion. But there's no explicit knowledge saying why the rules are in that order. Maybe the author ordered them in order of specificity, or execution time. As of Goodall [8], the conflict-resolution strategy is not explicit.

To improve the performance of a knowledge system, the system may be supplied some extra knowledge about the knowledge is posses. This kind of knowledge is called meta knowledge—knowledge about knowledge. The rules that define how conflict resolution will be used, and how other aspects of the system itself will run, are called meta rules.

7.6 Benefits and Capabilities of Rule Based Expert Systems

One of the advantages of the rule based expert systems is their ability to automatically generate explanations suitable for novices, without any extra work by the knowledge engineer.

We will outline below some of the main advantages of rule based expert systems as well as their disadvantages.

Advantages

- Allows the organizations to replicate their very best people. Expert systems carry the intelligence and information found in the intellect of experts and provides this knowledge to other members of the organization.
- Reduce the error due to automation of tedious, repetitive or critical tasks
- Reduce the manpower and time required for system testing and data analysis
- Reduce the costs through acceleration of fault observations
- Eliminate the work that people should not do (such as difficult, time-consuming or error prone tasks, jobs where training needs are large or costly).
- Eliminates work that people would rather not do (such as jobs involving decision making, which does not satisfy everyone; expert systems ensure fair decisions without favoritism in such cases).
- Expert systems perform better than humans in certain situations.
- Perform knowledge acquisition, process analysis, data analysis, system verification
- Increased visibility into the state of the managed system
- Develop functional system requirements
- Coordinate software development
- For simple domains, the rule-base might be simple and easy to verify and validate.
- Expert system shells provide a means to build simple systems without programming.
- Provide consistent answers for repetitive decisions, processes and tasks

- Hold and maintain significant levels of information
- Reduces creating entry barriers to competitors.

Disadvantages

- Expert knowledge is not usually easily codified into rules.
- Experts often lack access to their own analysis mechanisms.
- Validation/Verification of large systems is very difficult.
- When the number of rules is large, the effect of adding new rules can be difficult to assess.
- There is a lack of human common sense needed in some decision makings
- The creative responses human experts can respond to in unusual circumstances cannot be incorporated in an expert system.
- Domain experts are not always being able to explain their logic and reasoning
- There is a lack of flexibility and ability to adapt to changing environments as questions are standard and cannot be changed
- The expert system is not able to recognize when no answer is available

7.7 Types of Expert Systems

We can distinguish several types of expert systems or we can classify the expert systems based on several aspects [21]:

- *Nature of task to be done by and expert system.* From this aspect we can distinguish four classes:
    - Diagnosis (or classification): medical diagnosis expert systems are included into this category.
    - Design: The system has to build a solution to a given problem while satisfying certain constraints. The solution space is unknown and must be generated based on the given constraints.
    - Monitoring: The system starts with a given (known) space of solutions and iteratively analysis the behavior in order to detect possible failures. One iteration influences the following iteration(s).
    - Prediction (or simulation): the solutions space is unknown in the beginning and the system has to predict the changes due to some initial perturbations.
- *Role of the system in interaction with the user.* From this aspect we can distinguish three classes:
    - Advisory systems: which let the user themselves to make the final decision.
    - Dictatorial systems: which make decisions without consulting the user.

  - o Critical systems: which evaluate alternatives given by the user and give an opinion.

- *Time limitations.* There are two straightforward classes:
  - o Systems with limited time (such us real-time systems).
  - o Systems with unlimited time.
- *Nature of knowledge.* There are many types of knowledge. Most important ones are:
  - o Knowledge based on experience: human experts know the experiments and their outcomes (the knowledge is contained in them) but they do not know the exact causes of the outcomes.
  - o Causal knowledge: A classical logic analysis is possible in this case.
- *Temporary nature of knowledge.* Based on this aspect we can distinguish two classes:
  - o Static: the knowledge base is not altered during the expert system session.
  - o Dynamic: the knowledge base is altered during the expert system session. The changes include:
    - ▪ Predictable
    - ▪ Unpredictable
    - ▪ Increasing information
    - ▪ Change of information.
- *Certainty of information.* The degree of completeness or certainty of information involves a classification. Information can be:
  - o Incomplete: it is not sufficient for the expert system to make a decision.
  - o Imprecise: different terms are used with the same meaning or the same term has multiple meanings.

7.8 Examples of Expert Systems

One of the earliest expert systems was DENDRAL [7]. Developed at Stanford University, DENDRAL was designed to analyze mass spectra. A mass spectrum is a particular trace or analytical record formed when a molecule is bombarded with electrons. Each molecule has its own spectrum, determined by the way it breaks into fragments when hit. Chemists have charts of mass spectra for some common molecules. And they know some general rules that determine how a given type of molecule will break up, and what kind of spectrum it will give. But identifying a new molecule from its spectrum is not easy [6]. DENDRAL did contain rules, but it worked differently from most expert systems. It consists of two subprograms, Heuristic Dendral and Meta-Dendral and its developers believed that it can compete and experienced chemist (was marketed commercially in the United States) [9].

One of the first expert systems about which the scientists spent a lot of discussions is MYCIN [7][8][15]. The system was developed at Stanford University too to diagnose blood infections and recommend treatments, given lab data about tests on cultures taken from the patient. Although never put to practical use, MYCIN showed to the world the possibility of replacement of a medical professional by an expert system.

PROSPECTOR has been developed by NASA. It takes geological information about rock formations, chemical content, etc, and advises on whether there were likely to be exploitable mineral deposits nearby. Popular accounts of AI say that Prospector (in 1978-ish) discovered a hundred-million-dollar deposit of molybdenum [6].

XCON it is often mentioned because it was one of the most successful expert systems: it performed a task that couldn't be done manually or with a conventional computer program. It helped ``configure'' computer systems for DEC (who make VAXes amongst other things) [6][10]. XCON took about 3 years to get the job done, but Crevier joke that XCON may have replaced 75 people, but 150 were needed to keep it running.

PROSPECTOR and XCON/R1 are two literature examples of commercially systems. MYCIN has never been used for real diagnosis, perhaps partly because of fears over who'd be legally responsible for mistakes.

DENDRAL and XCON/R are data driven reasoning models (they use forward chaining) while MYCIN is a goal driven model (using backward chaining). Forward chaining is less efficient since we need to assume the disease and match their cause. Hence backward chaining is used in diagnosis expert systems.

These expert systems are the first ones and ones of the most important and most discussed. But there are several others which we will just enumerate:

- DELTA, a system for diagnosing electric loco repair problems [6][10].
- Tax Advisor: asks questions about its users' financial state and advises on how to minimize tax while maximizing investment [6][8].
- GASOIL: designing gas-oil separation systems for offshore oil platforms
- A program that planned where to site transistors on silicon chips so as to achieve the most compact and easy to fabricate layout.
- PATHFINDER IV: lymph node pathology
- Loan Probe is a microcomputer based expert system that evaluates a commercial loan, recommending the amount of reserves the bank should keep in case the borrower does not pay the loan.
- OncoLogic - A Computer System to Evaluate the Carcinogenic Potential of Chemicals
- Motorola's Helpdesk: Learns the general nature of the caller's problem. If can be addressed by some expert system component, it routes the call to that system. If cannot be routed, it logs the caller's problem, its characteristics, and its probable causes and solutions.

- Federal Express Corporation: helps customers weigh, label and document their own shipments. Decides which parts to stock by considering cost, importance and other such issues.
- Puff: diagnoses the results of pulmonary function tests. It is one of the very earliest medical expert systems in use.
- SETH: gives specific advice concerning the treatment and monitoring of drug poisoning.
- Building code checking
- Detecting credit card fraud
- AGREX: helps the Agricultural field personnel to give timely and correct advice to the farmers. Finds extensive use in the areas of fertilizer application, crop protection, irrigation scheduling, and diagnosis of diseases etc.
- Rice-Crop Doctor: expert system to diagnose pests and diseases for rice crop and suggest preventive/curative measures.
- Air craft design
- …

Summaries

Expert systems are a class of computer programs that can advise, analyze, categorize, communicate, consult, design, diagnose, explain, explore, forecast, form concepts, identify, interpret, justify, learn, manage, monitor, plan, present, retrieve, schedule, test, and tutor. They address problems normally thought to require human specialist for their solution.

Rule-based systems are a relatively simple model that can be adapted to any number of problems. As with any AI model, a rule-based system has its strengths as well as limitations that must be considered before deciding if it's the right technique to use for a given problem. Overall, rule-based systems are really only feasible for problems for which any and all knowledge in the problem area can be written in the form of if-then rules and for which this problem area is not large. If there are too many rules, the system can become difficult to maintain and can suffer a performance hit.

There are a couple of advantages in using expert systems. One is that the human expert's knowledge then becomes available to a very large range of people. Another advantage is that if you can capture the expertise of an expert in a field, then any knowledge, which they might have is not lost when they retire or leave the firm.

There are two main methods for rule-based expert systems: forward chaining or data driven and backward chaining or goal driven systems. The question is which one of the forward chaining or backward chaining we should use for a given problem. A simple answer is:

- If we are trying to prove a particular hypothesis then we should use backward chaining.
- If we are trying to find all possible conclusions then we should use forward chaining.

There are a couple of advantages and disadvantages in using rule based expert systems. Even thought the rule-based expert systems do not represent a main current research direction, rule-based expert systems are the most widely used and accepted AI in the world outside of games and there are still some more ideas to explore and exploit such as: systems that consult each other and maintain themselves or systems that do many things at once, or expert systems with eyes, ears and all the rest (Robot cooking or bringing things) or systems that already know the facts

References

1. Ligęza, A.: Logical Foundations for Rule-based Systems, 2nd edn. Springer, Heidelberg (2006)
2. http://www.billbreitmayer.com/rule_based_systems/rule_based_design.html (accessed on February 10, 2011)
3. Durkin, J.: Expert Systems: Design and Development. Prentice Hall, New York (1994)
4. Durkin, J.: Expert Systems: Catalog of Applications. Intelligent Computer Systems, Inc., Akron (1993)
5. Firebaugh, M.W.: Artificial Intelligence, A Knowledge-Based Approach. PWS-Kent Publishing Company (1993)
6. http://www.j-paine.org/students/lectures/ (accessed on February 10, 2011)
7. Winston, P.H.: Artificial Intelligence, 3rd edn. Addison-Wesley, Reading (1992)
8. Goodall, A.: Guide to Expert Systems. Information Today Inc. (1985)
9. Lederberg, J.: How Dendral was conceived and born. In: ACM Symposium on the History of Medical Informatics. National Library of Medicine, Rockefeller University (1987)
10. Crevier, D.: AI: The Tumultuous History of the Search for Artificial Intelligence. Basic Books, London (1993)
11. Negnevitsky, M.: Artificial intelligence: a guide to intelligent systems. Addison-Wesley, Reading (2002)
12. Coppin, B.: Artificial Intelligence Illuminated. Jones and Bartlett Publishers Inc., USA (2004)
13. Nikolopoulos, C.: Expert Systems – Introduction to First and Second Generation and Hybrid Knowledge Based Systems. CRC, Boca Raton (1997)
14. Lindsay, S.: Practical Applications of Expert Systems. John Wiley & Sons Inc., Chichester (1988)
15. Buchanan, B.G., Shortliffe, E.H.: Rule-Based Expert Systems: The MYCIN Experiments of the Stanford Heuristic Programming Project. Addison-Wesley, Reading (1984)
16. Giarratano, J.C.: Expert Systems: Principles and Programming. Brooks Cole, Pacific Grove (1998)
17. Jackson, P.: Introduction to Expert Systems. Addison-Wesley, Reading (1999)
18. Kidd, A.L.: Knowledge Acquisition for Expert Systems: A Practical Handbook. Plenum Publishing Corporation, New York (1987)
19. Levesque, H.J., Lakemeyer, G.: The logic of knowledge bases. MIT Press, Cambridge (2001)

20. Durkin, J.: Expert systems design and development. Prentice Hall, Englewood Cliffs (1994)
21. Castillo, E., Alvarez, E.: Expert systems: uncertainty and learning. Elsevier Science publishing Company, New York (1991)

Verification Questions

1. How is knowledge represented in a rule-based expert system?
2. What is a production rule? What is rule matching? When will a rule fire?
3. What are the main components of an expert system?
4. What is knowledge base and what is database?
5. Which type of relations rules can represents?
6. What do meta-rules represent?
7. What is Rete algorithm used for?
8. What does the working memory contains?
9. Explain how forward chaining works.
10. Explain how backward chaining works.
11. Give an example for which forward chaining is better than backward chaining.
12. Give an example for which backward chaining is better than forward chaining.
13. What is conflict resolution and which are the main strategies?
14. When we should use forward chaining and when backward chaining?
15. List some of the first expert systems.
16. What is an expert system shell?
17. What are the advantages and disadvantages of the rule-based expert systems?
18. Give some future rule-based expert systems ideas.

Exercises

7.1 The following knowledge may be used for recommending you to buy a car:

You should get a car of suitable size for your garage, and one that is suitable for your family. If your garage is small then small cars are of suitable size. If you have children then a 4 doors car is appropriate. VW polo, Renault Clio, Peougeot 206 are small while Ford Mondeo, BMW 3 series and Audi A4 are medium.

Represent the above knowledge as a set of production rules (and possibly initial facts). Briefly describe how the rules might be used to check on the suitability of a car - say Audi A4 using backward chaining.

7.2 Consider the following production system which identifies pets:

Rule 1	IF	the pet has hair
	THEN	it is a mammal
Rule 2	IF	the pet gives birth
	THEN	it is a mammal

Rule 3 IF the pet has feathers
THEN the pet is it is a bird

Rule 4 IF the pet flies
AND the animal lays eggs
THEN the pet is it is a bird

Rule 5 IF the pet is a mammal
AND the pet eats meat
THEN it is a carnivore

Rule 6 IF the pet is a mammal
AND the pet has pointed teeth
AND the pet has claws
AND the pet's eyes point forward
THEN it is a carnivore

Rule 7 IF the height is <10 cm
THEN the size is small

Rule 8 IF the height is >10 cm
AND the height is < 30 cm
THEN the size is medium

Rule 9 IF the height is >50 cm
THEN the size is big

Rule 10 IF the pet is a carnivore
AND the pet barks
AND the pet has long legs
AND the animal is big size
THEN the pet is a dog.

R11 IF the pet is a carnivore
AND the pet has soft hair
AND the pet size is medium
THEN the pet is a cat

R12 IF the pet is small
AND the pet has short legs
AND the pet has soft hair
THEN the pet is a mouse

R13 IF the pet has light hair
AND the pet size is medium
AND the pet is not carnivore
THEN then the pet is Guinea pig

1) Given these facts in working memory initially:
   the pet gives birth
   the pet has long legs
   the pet size is 70 cm

the pet eats meat
the pet barks

Establish by *forward chaining* that the pet is a dog.

2) Given the facts that:

the pet has soft hair
the pet has claws
the pet has pointed teeth
the pet's eyes point forward
the pet's height is 25 cm
the animal has dark spots

Establish by *backward chaining* that the animal is a cat.

7.3 Consider the following production system which identifies a treatment:

Rule 1	IF	temperature <37
	THEN	no fever
Rule 2	IF	temperature >37
	AND	temperature <38
	THEN	low fever
Rule 3	IF	if temperature > 38
	THEN	high fever
Rule 4	IF	light nasal breathing
	THEN	nasal discharge
Rule 5	IF	heavy nasal breathing
	THEN	sinus membranes swelling
Rule 6	IF	low fever
	AND	headache
	AND	nasal discharge
	AND	cough
	THEN	cold
Rule 7	IF	cold
	AND	not soar throat
	THEN	don't treat
Rule 8	IF	cold
	AND	soar throat
	THEN	treat
Rule 9	IF	don't treat
	THEN	don't give medication
Rule 10	IF	treat
	THEN	give medication

Rule 11 IF give medication
AND antibiotics allergy
THEN give Tylenol

Rule 12 IF give medication
AND not antibiotics allergy
THEN give antibiotics

Given these facts:

The patient has headache and nasal breathing. His fever is 37.5 and he is coughing. The patient denies that he is having allergy to antibiotics. Laboratory experiments show that he is also having soar throat.

1) Establish by forward chaining that the patient has to take antibiotics.
2) Use the backward chaining to show the things back (if the treatment is antibiotics then the patient has the given symptoms).

7.4. Consider the following production system:

Rule 1 IF shape is long
AND color is yellow
THEN fruit is banana

Rule 2 IF shape is round
AND color is red
AND size is medium
THEN then fruit is apple

Rule 3 IF shape is round
AND color is red
AND size is small
THEN then fruit is cherry

Rule 4 IF skin smell
THEN perfumed

Rule 5 IF fruit is lemon
OR fruit is orange
OR fruit is pomelo
OR fruit is grapefruit
THEN citrus fruit

Rule 6 IF size is medium
AND color is yellow
AND perfumed
THEN then fruit is lemon

Rule 7 IF size is medium
AND color is green
THEN fruit is kiwi

Rule 8	IF	size is big
	AND	perfumed
	AND	color is orange
	AND	citrus fruit
	THEN	fruit is grapefruit
Rule 9	IF	perfumed
	AND	color is orange
	AND	size is medium
	THEN	fruit is orange
Rule 10	IF	perfumed
	AND	color is red
	AND	size is small
	AND	no seeds
	THEN	fruit is strawberry
Rule 11	IF	diameter <2 cm
	THEN	size is small
Rule 12	IF	diameter >10 cm
	THEN	size is big
Rule 13	IF	diameter >2 cm
	AND	diameter <10 cm
	THEN	size is medium

The fruit has no seed, a 7 cm diameter, smelling skin, orange color

1) Establish by forward chaining that the fruit is a citrus fruit.
2) Use the backward chaining to show the things back.

7.5. Build a knowledge database with around 10 rules and 15 facts.

1) Perform a forward chaining
2) Perform a backward chaining
3) Create an inference chain.
4) Draw the chaining in a diagram.
5) Use your own creativity and expert knowledge in your chosen domain.

7.6. Implement an expert system shell at your choice and test it on any two of the examples given in this chapter or any of the exercises.

Chapter 8
Managing Uncertainty in Rule Based Expert Systems

8.1 What Is Uncertainty and How to Deal With It?

Uncertainty is essentially lack of information to formulate a decision. The presence of uncertainty may result in making poor or bad decisions. In our daily life, as human beings, we are accustomed to dealing with uncertainty – that's how we survive.

Dealing with uncertainty requires reasoning under uncertainty along with possessing a lot of common sense.

There are several sources of uncertainty [1][2][3][4][5][6][15]:

- Imprecise language: our (or expert's) natural language has to be transposed into IF-THEN rules. But sometimes our language is ambiguous and imprecise.
- Data (or information or knowledge) can be:
    - Incomplete
    - Incorrect
    - Missing
    - Unreliable
    - Imprecise
- Uncertain terminology
- Uncertain knowledge
- Incomplete information: Information is not sufficient for the expert system to make a decision.
- Imprecise data: different terms are used with the same meaning or a term has multiple (different) meanings.
- Many types of errors contribute to uncertainty:
    - Errors related to hypothesis
        - Type I Error – accepting a hypothesis when it is not true – False Positive.
        - Type II Error – Rejecting a hypothesis when it is true – False Negative

C. Grosan and A. Abraham: Intelligent Systems, ISRL 17, pp. 187–217.
springerlink.com

  - Errors related to measurement
    - Errors of precision – how well the truth is known
    - Errors of accuracy – whether something is true or not
    - Unreliability stems from faulty measurement of data – results in erratic data.
    - Random fluctuations – termed random error
    - Systematic errors result from bias
  - Errors in induction. Induction proceeds from specific to general (while compared to deduction which proceeds from general to specific). Expert systems may consist of both deductive and inductive rules based on heuristic information. When rules are based on heuristics, there will be uncertainty. Inductive arguments can never be proven correct (except in mathematical induction).
- Combination of different expert views: When huge expert systems require the presence of multiple experts, there is a low probability that all the experts will reach the same conclusion. They might have contradictory opinions and this will involve the production of conflicting rules.

Thus, uncertainty may be induces by the degree of validity of facts, rule conditions and rules themselves.

When dealing with uncertainty, we should be satisfied just with getting a good solution. There are a number of methods to pick the best solution in light of uncertainty.

General methods for dealing with uncertainty are:

- Probability-based methods which include:
  - objective probability
  - experimental probability
  - subjective probability
- Heuristic methods which include:
  - certainty factors
  - fuzzy logic

Since the Boolean approach to reasoning does not solve the problems in domains involving uncertainty, a number of theories have been developed. Some known theories to deal with uncertainty are:

- Bayesian Probability
- Hartley Theory
- Shannon Theory
- Dempster-Shafer Theory
- Markov Models
- Fuzzy Theory

In these theories, a scheme on how to introduce measure which numerically quantifies uncertainties and how to propagate and combine these measures of uncertainty during reasoning is usually proposed.

8.2 Bayesian Theory

Before explaining Bayesian reasoning, we will first review the classical probability theory, which will help to better understand the Bayesian reasoning concepts.

8.2.1 Classical Probability Theory

Classical probability has been proposed by Pascal and Fermat in 1654. It is also called *a priori* probability because it deals with ideal games or systems:

- assumes all possible events are known
- each event is equally likely to happen.

In propositional logic, primitives are propositions. In probabilistic reasoning, primitives are *random variables*. A random variable is not in fact a variable, but a function from a sample space to another space (often the real numbers). A probability of an event is the proportion of the cases this event occurs [15][19]. Mathematically, the probability is expressed as a real number between [0, 1], 0 representing absolute impossibility and 1 representing absolute certainty. Probabilistic information systems represent information with variables and their probability distributions. The value of a particular attribute A for a specific tuple t is a variable A(t) and this variable has an associated probability distribution P(A(t)). P(A(t)) assigns values in the range [0, 1] to the elements of the domain of attribute A, with the provision that the sum of all values assigned is 1.

For example, if *t* is rain and A is forecast, then we can write:

$$\text{P(forecast(rain))} = \begin{cases} \textit{true}, 0.3 \\ \textit{false}, 0.7 \end{cases}$$

The interpretation is that it will rain with a probability of 0.3 and it will not rain with a probability of 0.7.

Each event has at least two possible outcomes: success and failure. The probability of success is given by:

$$\text{P(success)} = \frac{\text{the number of successes}}{\text{the total number of possible outcomes}}$$

while the probability of failure is given by:

$$\text{P(failure)} = \frac{\text{the number of failures}}{\text{the total number of possible outcomes}}.$$

We have:

$$\text{P(success)} + \text{P(failure)} = 1.$$

Example: dice
If we consider the classical dice example, then for each of the sides 1 to 6 we will have:

$$P(\text{success}(1)) = P(\text{success}(2)) = \ldots = P(\text{success}(6)) = \frac{1}{6}$$

and

$$P(\text{failure}(1)) = P(\text{failure}(2)) = \ldots = P(\text{failure}(6)) = \frac{5}{6}.$$

There are two simple situations which can occur in our discussions and must be taken into account:

- *Events are independent and mutually exclusive:* this means that events cannot happen simultaneous (for instance, in the dice example, we cannot get a 6 and a 5 simultaneously).
- *Events that are not independent*: this means that one event (or multiple events) may affect the occurrence of the other event (events).

We have the following three axioms of formal theory of probability:

1. $0 \le P(E) \le 1$
2. $\sum_i P(E_i) = 1$
3. $P(E_1 \cup E_2) = P(E_1) + P(E_2)$

where E_1 and E_2 are mutually exclusive events.
For pairwise independent events:

$$P(A \cap B) = P(A)\, P(B)$$

The *additive low* states as:

$$P(A \cup B) = P(A) + P(B) - P(A \cap B)$$

$$P(A \cup B \cup C) = P(A) + P(B) + P(C) - P(A \cap B) - P(A \cap C) - P(B \cap C) + P(A \cap B \cap C).$$

Suppose A and B are two events which are not mutually independent. We can then define a conditional probability that event A occurs if event B occurs P(A|B). P(A|B) can be interpreted as the *conditional probability* of event A occurring given that event B has occurred and it is given by:

$$P(A|B) = \frac{\textit{the number of times both A and B can occur}}{\textit{the number of times B can occur}}.$$

The number of times both A and B can occur is called *joint probability* (or compound probability) of A and B and it is given by $P(A \cap B)$.

8.2.2 Bayes' Rules

We can write:

$$P(A|B) = \frac{P(A \cap B)}{P(B)}$$

From the above we can deduce:

$$P(A \cap B) = P(A|B)\, P(B).$$

Joint probability is commutative, thus we have:

$$P(A \cap B) = P(B \cap A)$$

and from

$$P(B \cap A) = P(B|A)\, P(A)$$

we obtain:

$$P(A|B) = \frac{P(B \mid A)\, P(A)}{P(B)} \quad (1)$$

This equation is known as *Bayesian rule* where:

- $P(A|B)$ is the probability that event A occurs given that event B has occurred;
- $P(B|A)$ is the probability that event B occurs given that event A has occurred;
- $P(A)$ is the probability that event A occurs;
- $P(B)$ is the probability that event B occurs.

The *sum rule* is given by:

$$P(A|B) + P(\neg A|B) = 1.$$

The Bayesian rule above has been only presented for two dependent events A and B but it can be further extended and generalized. We have three generalizations as given below.

1) *An event A is depended on a set of events B_1, B_2, ..., B_n which are mutually exclusive.*

We can thus derive generalized formula as follows:

$$P(A \mid B_1) = \frac{P(A \cap B_1)}{P(B_1)}$$

$$P(A \mid B_2) = \frac{P(A \cap B_2)}{P(B_2)}$$

$$\vdots$$

$$P(A \mid B_n) = \frac{P(A \cap B_n)}{P(B_n)}$$

Thus, by summing the above, we obtain:

$$\sum_{i=1}^{n} P(A \mid B_i) = \frac{\sum_{i=1}^{n} P(A \cap B_i)}{\sum_{i=1}^{n} P(B_i)}.$$

From

$$\sum_{i=1}^{n} P(A \cap B_i) = P(A)$$

we obtain:

$$P(A) = \sum_{i=1}^{n} P(A \mid B_i) P(B_i).$$

In the particular situation when we have the event A and two mutually exclusive events B and ¬B we obtain:

$$P(A) = P(A|B)\ P(B) + P(A|\neg B)\ P(\neg B)$$

and similarly:

$$P(B) = P(B|A)\,P(A) + P(B|\neg A)\,P(\neg A)$$

And from the above we get:

$$P(A|B) = \frac{P(B \mid A)\,P(A)}{P(B \mid A)\,P(A) + P(B \mid \neg A)\,P(\neg A)} \tag{2}$$

2) *A set of mutually exclusive events $A_1, A_2, \ldots, A_n$ is depended on an event B.*

In this situation, for each event A_i, the equation (2) becomes:

$$P(A_i \mid B) = \frac{P(B \mid A_i)\,P(A_i)}{\sum_{k=1}^{n} P(B \mid A_k)\,P(A_k)}.$$

3) *A set of mutually exclusive events $A_1, A_2, \ldots, A_m$ is depended on a set $B_1, B_2, \ldots, B_n$ which are mutually exclusive.*

In this case, for each event A_i we have:

$$P(A_i \mid B_1 B_2 \ldots B_n) = \frac{P(B_1 B_2 \ldots B_n \mid A_i) P(A_i)}{\sum_{k=1}^{m} P(B_1 B_2 \ldots B_n \mid A_k) P(A_k)}$$

which can be also written as:

$$P(A_i \mid B_1 B_2 \ldots B_n) = \frac{P(B_1 \mid A_i)\,P(B_2 \mid A_i)\ \ldots\ P(B_n \mid A_i)\ P(A_i)}{\sum_{k=1}^{m} P(B_1 \mid A_k)\,P(B_2 \mid A_k) \ldots\ P(B_n \mid A_k)\,P(A_k)}$$

8.2.3 Bayesian Reasoning

Consider the expert system whose rules in the knowledge base are represented in the following IF-THEN form:

IF A is true
THEN B is true [with probability P]

where A represents hypothesis and B represents evidences to support the hypotheses. In an expert system the probabilities required to solve the problem are provided by the experts which will determine (specify) right form the beginning the probabilities of all hypotheses as well all of their negations. Also, the expert will provide all the conditional probabilities for observing an evidence (or multiple evidences) if a hypotheses (or multiple hypothesis) is true and false respectively.

Users provide information about the evidence (evidences) observed and the expert system computes the conditional probabilities for the hypothesis using the evidences provided by the user. This probability is called *posterior probability*.

Example

Let us consider a simple expert system whose task is to see which of the three hypotheses will be finally considered say for a diagnosis problem.

The experts create three hypotheses A_1, A_2, A_3 (mutually exclusive) based on three independent evidences B_1, B_2, B_3. The experts also provide the conditional probabilities of observing each evidence for all the considered hypotheses. Suppose the ranking given to the hypotheses is in the order A_1, A_2, A_3, with A_1 having the highest probability (being the most trustful). We will now apply Bayesian reasoning to see if the order will be kept the same at the end of the process. Consider the data provided by the experts as given in Table 8.1.

Table 8.1 Prior probabilities and conditional probabilities for the example.

PROBABILITY					
$P(A_1)$	0.5	$P(A_2)$	0.3	$P(A_3)$	0.2
$P(B_1 \mid A_1)$	0.6	$P(B_1 \mid A_2)$	0.7	$P(B_1 \mid A_3)$	0.1
$P(B_2 \mid A_1)$	0.3	$P(B_2 \mid A_2)$	0.3	$P(B_2 \mid A_3)$	0.9
$P(B_3 \mid A_1)$	0.0	$P(B_3 \mid A_2)$	0.5	$P(B_3 \mid A_3)$	0.4

We will now calculate the posterior probabilities observing the evidences in the order B_1, B_2 and B_3.

We will first calculate $P(A_i|B_1)$, i= 1, 2, 3, using the formula:

$$P(A_i \mid B_1) = \frac{P(B_1 \mid A_i) P(A_i)}{\sum_{k=1}^{3} P(B_1 \mid A_k) P(A_k)}$$

$$P(A_1 \mid B_1) = \frac{0.6 \cdot 0.5}{0.6 \cdot 0.5 + 0.7 \cdot 0.3 + 0.1 \cdot 0.2} = \frac{0.3}{0.3 + 0.21 + 0.02} = \frac{0.3}{0.53} = 0.56$$

$$P(A_2 \mid B_1) = \frac{0.7 \cdot 0.3}{0.6 \cdot 0.5 + 0.7 \cdot 0.3 + 0.1 \cdot 0.2} = \frac{0.21}{0.3 + 0.21 + 0.02} = \frac{0.21}{0.53} = 0.39$$

$$P(A_3 \mid B_1) = \frac{0.1 \cdot 0.2}{0.6 \cdot 0.5 + 0.7 \cdot 0.3 + 0.1 \cdot 0.2} = \frac{0.02}{0.3 + 0.21 + 0.02} = \frac{0.02}{0.53} = 0.037$$

We now calculate $P(A_i|B_2B_1)$, i=1, 2, 3, using the formula:

$$P(A_i \mid B_2 B_1) = \frac{P(B_2 \mid A_i) P(B_1 \mid A_i) P(A_i)}{\sum_{k=1}^{3} P(B_2 \mid A_k) P(B_1 \mid A_k) P(A_k)}$$

$$P(A_1 \mid B_2 B_1) = \frac{0.3 \cdot 0.6 \cdot 0.5}{0.3 \cdot 0.6 \cdot 0.5 + 0.3 \cdot 0.7 \cdot 0.3 + 0.9 \cdot 0.1 \cdot 0.2} = \frac{0.09}{0.09 + 0.063 + 0.018} = \frac{0.09}{0.171} = 0.52$$

$$P(A_2 \mid B_2 B_1) = \frac{0.3 \cdot 0.7 \cdot 0.3}{0.3 \cdot 0.6 \cdot 0.5 + 0.3 \cdot 0.7 \cdot 0.3 + 0.9 \cdot 0.1 \cdot 0.2} = \frac{0.063}{0.09 + 0.063 + 0.018} = \frac{0.063}{0.171} = 0.36$$

$$P(A_3 \mid B_2 B_1) = \frac{0.9 \cdot 0.1 \cdot 0.2}{0.3 \cdot 0.6 \cdot 0.5 + 0.3 \cdot 0.7 \cdot 0.3 + 0.9 \cdot 0.1 \cdot 0.2} = \frac{0.018}{0.09 + 0.063 + 0.018} = \frac{0.018}{0.171} = 0.10$$

We now observe the last evidence, B_3, and we calculate $P(A_i|B_3B_2B_1)$, i=1, 2, 3, using the formula:

$$P(A_i \mid B_3 B_2 B_1) = \frac{P(B_1 \mid A_i) P(B_2 \mid A_i) P(B_3 \mid A_i) P(A_i)}{\sum_{k=1}^{3} P(B_1 \mid A_k) P(B_2 \mid A_k) P(B_3 \mid A_k) P(A_k)}$$

$$P(A_1 \mid B_3 B_2 B_1) = \frac{0.0 \cdot 0.3 \cdot 0.6 \cdot 0.5}{0.0 \cdot 0.3 \cdot 0.6 \cdot 0.5 + 0.5 \cdot 0.3 \cdot 0.7 \cdot 0.3 + 0.4 \cdot 0.9 \cdot 0.1 \cdot 0.2} =$$

$$= \frac{0}{0 + 0.0315 + 0.0072} = 0$$

$$P(A_2 \mid B_3 B_2 B_1) = \frac{0.5 \cdot 0.3 \cdot 0.7 \cdot 0.3}{0.0 \cdot 0.3 \cdot 0.6 \cdot 0.5 + 0.5 \cdot 0.3 \cdot 0.7 \cdot 0.3 + 0.4 \cdot 0.9 \cdot 0.1 \cdot 0.2} =$$

$$= \frac{0.0315}{0 + 0.0315 + 0.0072} = \frac{0.0315}{0.0387} = 0.814$$

$$P(A_3 \mid B_3 B_2 B_1) = \frac{0.4 \cdot 0.9 \cdot 0.1 \cdot 0.2}{0.0 \cdot 0.3 \cdot 0.6 \cdot 0.5 + 0.5 \cdot 0.3 \cdot 0.7 \cdot 0.3 + 0.4 \cdot 0.9 \cdot 0.1 \cdot 0.2} =$$

$$= \frac{0.0072}{0 + 0.0315 + 0.0072} = \frac{0.0072}{0.0387} = 0.186$$

Although the initial ranking of hypotheses was A_1, A_2, A_3, only the hypotheses A1 and A2 remain under consideration after all the evidences B_1, B_2 B_3 have been observed. The hypotheses A_1 which was the first one is not even under consideration after calculating the posteriori probabilities and the second hypothesis is now having a high credibility (and the highest one among the hypotheses).

8.2.4 Bayesian Networks

In real learning problems, we are typically interested in looking for relationships among a large number of variables. Bayesian networks are an ideal tool for doing it. Bayesian nets (BN) (also referred to as Probabilistic Graphical Models or Bayesian Belief Networks)[16][17][18][21] are directed acyclic graphs where each node represents a random variable. The meaning of an arrow from a parent to a child is that the parent directly influences the child. These influences are quantified by conditional probabilities. BNs are graphical representations of joint distributions.

A Bayesian network for a set of variables consists of the following elements (see Figure 8.1):

1) a network structure that encodes a set of conditional independence assertions about variables;
2) a set of local probability distributions associated with each variable.

Together, these components define the joint probability distribution for the set of given variables.

The probabilities encoded by a Bayesian network may be Bayesian or physical:

- when building Bayesian networks from prior knowledge alone, the probabilities will be Bayesian;
- when learning these networks from data, the probabilities will be physical (and their values may be uncertain).

Each node in a Bayesian network has an associated conditional probability table or CPT. This gives the probability values for the random variable at the node conditional on values for its parents. All the random variables are supposed to have only a finite number of possible values. If a node has no parents, then the CPT reduces to a table giving the probability of that random variable [7].

Example 1

The following example taken from [7] [8] is a very significant one. In Figure 8.1, the nodes are binary and have two possible values – true and false – denoted by T and F.

Given that it is cloudy with a probability of 0.5 and the conditional probabilities for:

- sprinkler being on (respectively off) given that it is cloudy (or being true or false)
- raining (or not) given that it is cloudy (also true or false)

we are getting the conditional probabilities for the grass being wet given that:

- it is raining (Rain is true) and the sprinkler is on (sprinkler on true);
- it is raining (Raining true) and the sprinkler is off (sprinkler on false);
- sprinkler is on (sprinkler on true) and it is not raining (raining false);
- sprinkler is off (sprinkler on false) and it is not raining (raining false).

The highest probability of the grass to be wet is in the given conditions that both sprinkler is on and it is raining. The probability is lower (and has the same value) when only one of the evidences sprinkler on true and raining true is observed.

The strength of these relationships is shown in the corresponding tables in Figure 8.1.

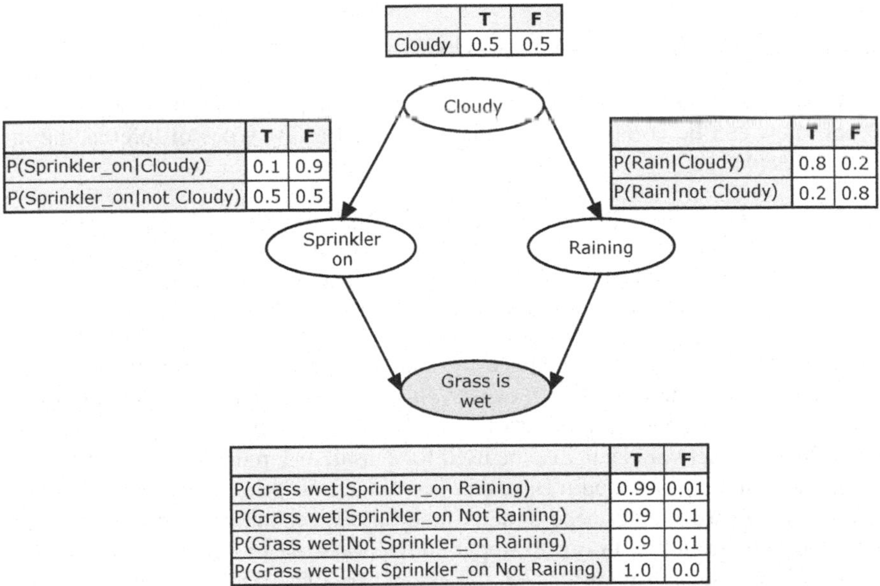

	T	F
Cloudy	0.5	0.5

	T	F
P(Sprinkler_on|Cloudy)	0.1	0.9
P(Sprinkler_on|not Cloudy)	0.5	0.5

	T	F
P(Rain|Cloudy)	0.8	0.2
P(Rain|not Cloudy)	0.2	0.8

	T	F
P(Grass wet|Sprinkler_on Raining)	0.99	0.01
P(Grass wet|Sprinkler_on Not Raining)	0.9	0.1
P(Grass wet|Not Sprinkler_on Raining)	0.9	0.1
P(Grass wet|Not Sprinkler_on Not Raining)	1.0	0.0

Fig. 8.1 The Bayesian network and the conditional probability tables for the example 1.

The conditional independence relationship encoded in a Bayesian network can be stated as follows: a node is independent of its ancestors given its parents, where the ancestor/parent relationship is with respect to some fixed topological ordering of the nodes [8].

By the chain rule of probability, the joint probability of all the nodes in the graph above is:

P(Cloudy, Sprinkler_on, Rain, Grass_wet) = P(Cloudy) * P(Sprinkler_on|Cloudy) * P(Rain|Cloudy,Sprinkler_on) * P(Grass_wet|Cloudy,Sprinkler_on,Rain)

By using conditional independence relationships, we can rewrite this as

P(Cloudy, Sprinkler_on, Rain, Grass_wet) = P(Cloudy) * P(Sprinkler_on|Cloudy) * P(Rain|Cloudy) * P(Grass_wet| Sprinkler_on, Rain)

where we were allowed to simplify the third term because R is independent of S given its parent C, and the forth term because W is independent of C given its parents S and R.

We can see that the conditional independence relationships allow us to *represent* the joint more compactly. Here the savings are minimal, but in general, if we had n binary nodes, the full joint would require $O(2^n)$ space to represent, but the factored form would require $O(n\ 2^k)$ space to represent, where k is the maximum fan-in of a node. And fewer parameters make learning easier.

8.2.4.1 Inference in Bayesian Networks

Given what we do know in the form of evidences, the distribution over what we do not know can be computed. For this, there exist a few types of inferencing in Bayesian networks:

- *Diagnostic Inferences*: infer from effects to causes.
- *Causal Inferences:* infer from causes to effects.
- *Intercausal Inferences*: between causes of a common event.
- *Mixed Inferences*: some causes and some effects known.

In the example given above, suppose we had evidence of an effect (that grass is wet), and inferred the most likely cause. This is called diagnostic, or bottom up, reasoning, since it goes from effects to causes; it is a common task in expert systems. Bayesian networks can also be used for causal, or top down, reasoning.

Once we have constructed a Bayesian network (from prior knowledge, data, or a combination), we usually need to determine various probabilities of interest from the model. For example, in our example above, we want to know the probability of wet grass given observations of the other variables. This probability is not stored directly in the model, and hence needs to be computed. In general, the computation of a probability of interest given a model is known as *probabilistic inference.*

Example 2 (taken from [10])

Consider the problem of detecting credit-card fraud and the following variables (together with the shortcut used in what follows): Fraud (F), Gas (G), Jewelry (J), Age (A), and Sex (S), representing whether or not the current purchase is fraudulent, whether or not there was a gas purchase in the last 24 hours, whether or not there was a jewelry purchase in the last 24 hours, and the age and sex of the card holder, respectively (represented in Figure 8.2; Y and N refers to true and false respectively (yes and no). Arcs are drawn from cause to effect. The local probability distribution(s) associated with a node are shown adjacent to the node.

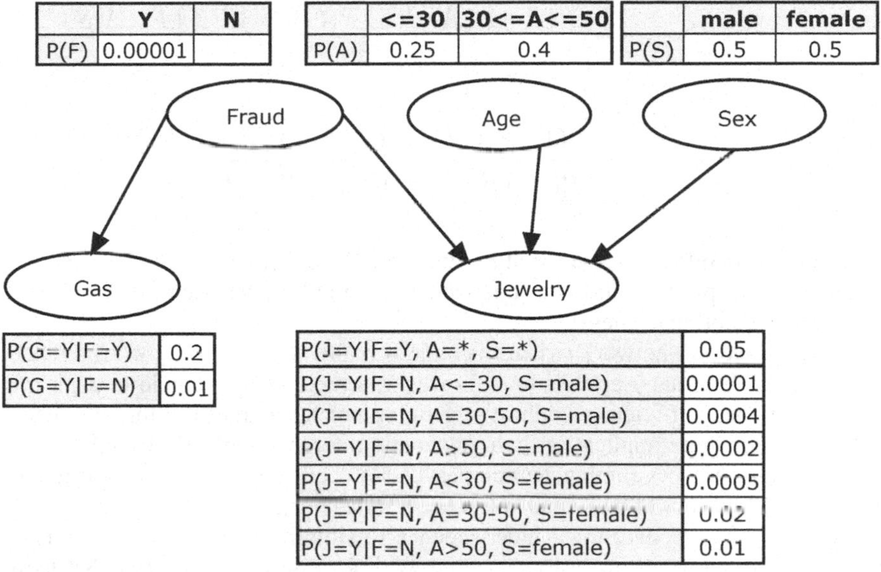

Fig. 8.2 Bayesian network for Example 2.

Using the ordering (F; A; S; G; J), we have the conditional independencies:

$$
\begin{aligned}
&P(A| F) = P(A)\\
&P(S| F; A) = P(S) \qquad (2)\\
&P(G| F; A; S) = P(GJF)\\
&P(J| F; A; S; G) = P(JJF; A; S)
\end{aligned}
$$

Because a Bayesian network for the set of all variables determines a joint probability distribution for the set of variables, we can use the Bayesian network to compute any probability of interest. For example, the probability of fraud given observations of the other variables can be computed as follows [10]:

$$P(F \mid A;S;G;J) = \frac{P(F;A;S;G;J)}{P(A;S;G;J)} = \frac{P(F;A;S;G;J)}{\sum_{F'} P(F';A;S;G;J)} \quad (3)$$

This direct approach is not practical for high number of variables. In the case that all variables are discrete, we can exploit the conditional independencies encoded in a Bayesian network to make this computation more efficient. In our example, given the conditional independencies in equations (2), we can re-write equation (3) as:

$$P(F \mid A;S;G;J) = \frac{P(F)P(A)P(S)P(G \mid F)P(J \mid F;A;S)}{\sum_{F'} P(F')P(A)P(S)P(G \mid F')P(J \mid F';A;S)}$$

$$= \frac{P(F)P(G \mid F)P(J \mid F;A;S)}{\sum_{F'} P(F')P(G \mid F')P(J \mid F';A;S)}$$

Several probabilistic inference algorithms for Bayesian networks with discrete variables that exploit conditional independence roughly have been proposed (see [10] for a description of them).

When a Bayesian-network structure contains many undirected cycles, inference is intractable. For many applications, however, structures are simple enough (or can be simplified sufficiently without sacrificing much accuracy) so that inference is efficient. For those applications where generic inference methods are impractical, researchers are developing techniques that are custom tailored to particular network topologies or to particular inference queries [10].

Although we use conditional independence to simplify probabilistic inference, exact inference in an arbitrary Bayesian network for discrete variables is NP-hard [11].

8.2.4.2 Variable Ordering in Bayesian Networks

The conditional independence assumptions expressed by a Bayesian network allow a compact representation of the joint distribution. Bayes network imposes a partial order of nodes. We can always break down the joint so that the conditional probability factor for a node only has non-descendants in the condition.

If we choose the variable order carelessly, the resulting network structure may fail to reveal many conditional independencies among the variables.

In the worst case, we have to explore n! variables orderings to find the best one.

There is another technique for constructing Bayesian networks that does not require an ordering. The approach is based on two observations:

(1) people can often readily assert causal relationships among variables;
(2) causal relationships typically correspond to assertions of conditional dependence.

In particular, to construct a Bayesian network for a given set of variables, we simply draw arcs from cause variables to their immediate effects. In almost all cases, doing so results in a network structure that satisfies the definition of the joint probability distribution.

Still, in practice there might be some problems. For example, judgments of conditional independence and/or cause and effect can influence problem formulation. Also, assessments of probability can lead to changes in the network structure.

8.2.4.3 Facts about Bayesian Networks

Bayesian networks can readily handle incomplete data sets. When one of the inputs is not observed, most models will produce an inaccurate prediction, because they do not encode the correlation between the input variables. Bayesian networks offer a natural way to encode such dependencies. Bayesian networks allow and facilitate learning about causal relationships. The process is useful when we are trying to gain understanding about a problem domain, and also, knowledge of causal relationships allows us to make predictions in the presence of interventions.

Bayesian networks have several issues, among others [12]:

- Require the knowledge of a large number of probabilities (for the hypotheses and then the conditional probabilities) which may not always be easy to estimate.
- The probabilistic approach assumes that the presence of evidence also affects the negation of a conclusion (for example, if P(A|B) = 0.6, then this implies that P(¬A|B)=0.4). But this is not necessary true in all domains.
- If the prior and posterior probabilities are based on frequency counts and statistics, then the samples must be of large enough size to derive accurate probabilities. If the probabilities are not based on frequencies, but are estimated by human domain experts, they may be inconsistent. They might not sum up to 1 for instance even if cases are exhaustive.

Bayesian networks in conjunction with Baycsian statistical techniques facilitate the combination of domain knowledge and data. Bayesian methods together with Bayesian networks and other types of models offers an efficient and principled approach for avoiding the over fitting of data. There is no need to hold out some of the available data for testing. Using the Bayesian approach, models can be smoothed in such a way that all available data can be used for training [10].

Remarks

Allan L. Yuille [9] has made a few interesting remarks about the Bayes theorem which we are reproducing below:

" Bayes Theorem is commonly ascribed to the Reverent Thomas Bayes (1701-1761) who left one hundred pounds in his will to Richard Price ``now I suppose Preacher at Newington Green." Price discovered two unpublished essays among Bayes's papers which he forwarded to the Royal Society. This work made little impact, however, until it was independently discovered a few years later by the

great French mathematician Laplace. English mathematicians then quickly rediscovered Bayes' work.

Little is known about Bayes and he is considered an enigmatic figure. One leading historian of statistics, Stephen Stigler, has even suggested that Bayes Theorem was really discovered by Nicolas Saunderson, a blind mathematician who was the fourth Lucasian Professor of Mathematics at Cambridge University. (Saunderson was recommended to this chair by Isaac Netwon, the second Lucasian Professor. Recent holders of the chair include the great physicist Paul Dirac and Stephen Hawking).

Bayes theorem and, in particular, its emphasis on prior probabilities has caused considerable controversy. The great statistician Ronald Fisher was very critical of the ``subjectivist" aspects of priors. By contrast, a leading proponent I.J. Good argued persuasively that ``the subjectivist (i.e. Bayesian) states his judgements, whereas the objectivist sweeps them under the carpet by calling assumptions knowledge, and he basks in the glorious objectivity of science"."

8.3 Certainty Factors

The MYCIN developers realized that a Bayesian approach was intractable, as too much data and/or estimates from the experts are required. Also, medical diagnosis systems based on Bayesian methods were not accepted due to lack of explanation facilities (the systems did not provide simple explanations of how it has reached its conclusion).

Doctors reason more in terms of gathering evidences that supports or contradicts a particular hypothesis. The MYCIN developers thus developed a logic which worked this way and this conducted to the raise of *certainty factors theory.* Certainty factors theory is an alternative to Bayesian reasoning. Certainty theory is an attempt to formalise the heuristic approach to reasoning with uncertainty. Human experts weight the confidence in their conclusions and reasoning steps in term of "unlikely", "almost certain", "highly probable", "possible". These are not probabilities but heuristics derived from experience.

A certainty factor is used to express how accurate, truthful, or reliable one judges a predicate to be. This judgment reflects how good the evidence is. A certainty factor is neither a probability nor a truth value. Certainty factors have been quantified using various different systems, including linguistics ones (certain, fairly certain, likely, unlikely, highly unlikely, definitely not) and various numeric scales, such as 0-1, 0-10, and -1 to 1.

Certainty factors may apply to:

- facts;
- rules (conclusion(s) of rules);
- both to facts and to rules.

When certainty factors apply to facts (evidences, premises) this represents the degree of belief (disbelief) associated to a given piece of evidence. When certainty factors apply to rules this represents the degree of confirmation (or disconfirmation) of a hypothesis given concrete evidence. A certainty factor value reflects confidence in given data, inferred data or hypothesis. The meaning of a certainty factor (CF) between -1 and 1 is:

- As the CF approaches 1 the evidence is stronger *for* a hypothesis.
- As the CF approaches -1 the confidence *against* the hypothesis gets stronger.
- A CF around 0 indicates that there is little evidence either for or against the hypothesis.

There is a similarity between certainty factors and conditional probabilities:

Certainty factors	***Conditional probabilities***
represent a measure of belief in the outcome	represent the degree of probability of the outcome
range from -1 (believed not to be the case) to 1 (believed to be the case)	range from 0 (false) to 1 (true)

Two examples of uncertain terms and their interpretation is given in Figure 8.3 [15].

In an expert system with certainty factors, the knowledge base consists of rule of the following form:

IF evidence
THEN hypothesis {CF}

where CF represents believe in the hypothesis given that the evidence occurs. We denote in what follows hypothesis by H (or H_1, H_2, … in case of multiple hypotheses) and evidence by E (respectively E_1, E_2, … in the case of multiple evidences).

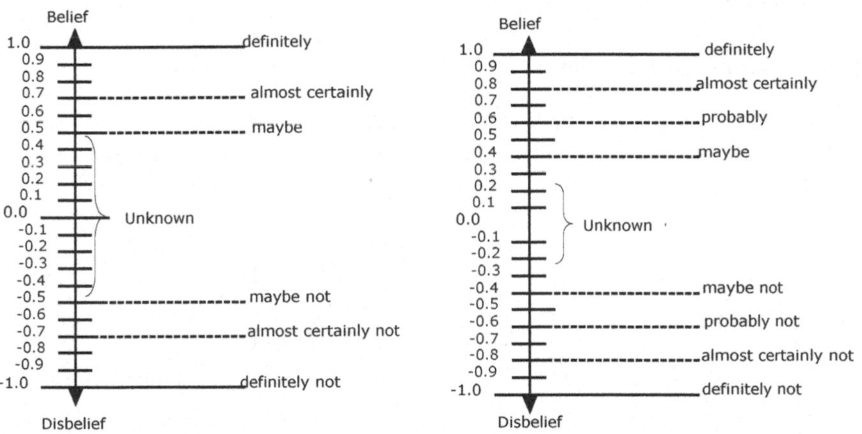

Fig. 8.3 Two examples of uncertain terms and their representations (a general one on the left and a more specific one on the right [15].

8.3.1 Calculating Certainty Factors

Two measures are used to calculate the certainty factors: measure of belief and measure of disbelief.

8.3.1.1 Measure of Belief

For a hypothesis H and an evidence E, the measure of belief – denoted MB(H, E) – represents the degree to which belief in the hypothesis H is supported by observing evidence E. MB(H, E) takes a value between 0 and 1.

MB(H, E) is given by:

$$MB(H,E)=\begin{cases}1, & \textit{if } P(H)=1\\ \dfrac{P(H\mid E)-P(H)}{1-P(H)}, & \textit{if } P(H)<1\end{cases}$$

where:

P(H) is the prior probability of hypothesis H being true;
P(H|E) is the probability that hypothesis H is true given the evidence E.

The above formula is not theoretically derived and it is intended to capture the degree to which the evidence increases probability: p(H|E)-p(H) in proportion to the maximum possible increase in probability: 1-p(H).

To avoid negative values, the following modification is used:

$$MB(H,\,E)=\begin{cases}1, & \textit{if } P(H)=1\\ \dfrac{\max\{P(H\mid E),P(H)\}-P(H)}{1-P(H)}, & \textit{if } P(H)<1\end{cases}$$

8.3.1.2 Measure of Disbelief

For a hypothesis H and an evidence E, the measure of disbelief – denoted MD(H, E) – represents the degree to which disbelief in the hypothesis H is supported by observing evidence E. MD(H, E) takes a value between 0 and 1.

MD(H, E) is given by:

$$MD(H,E)=\begin{cases}1, & \textit{if } P(H)=0\\ \dfrac{P(H)-P(H\mid E)}{P(H)}, & \textit{if } P(H)>0\end{cases}$$

where:

P(H) is the prior probability of hypothesis H being true;
P(H|E) is the probability that hypothesis H is true given the evidence E.

To avoid negative values the following modification can be used:

$$MD(H,E) = \begin{cases} 1, & if\ P(H) = 0 \\ \dfrac{\min\{P(H \mid E), P(H)\} - P(H)}{0 - P(H)}, & if\ P(H) > 0 \end{cases}$$

Certainty factor is calculated in terms of the difference between MB and MD:

$$CF(H,E) = \frac{MB(H,E) - MD(H,E)}{1 - \min\{MB(H,E),\ MD(H,E)\}}$$

The range of certainty factors values is [-1, 1].

8.3.2 Combining Certainty Factors

8.3.2.1 Multiple Rules Providing Evidence for the Same Conclusion

There are situations when multiple sources of evidence produce CFs for the same fact.

For instance, two (or more) rules may provide evidence for the same conclusion:

IF E1
THEN H {CF=0.5}

IF E2
THEN H {CF=0.6}

In such situations we need to combine the CFs. If two rules both support the same hypothesis, then that should increase our belief in the hypothesis.

The combination of the CFs is given by the formula:

$$CF(H, E1 \wedge E2) = \begin{cases} CF(E1) + CF(E2)(1 - CF(E1)), & if\ \ CF(E1), CF(E2) > 0 \\ CF(E1) + CF(E2)(1 + CF(E1)), & if\ \ CF(E1), CF(E2) < 0 \\ \dfrac{CF(E1) + CF(E2)}{1 - \min\{|CF(E1)|, |CF(E2)|}, & if\ \ sign(CF(E1)) \neq sign(CF(E2)) \end{cases}$$

Example 1

Suppose we have the following two rules:

IF E1
THEN H {CF=0.6}

IF E2
THEN H {CF= -0.3}

Then:

$$CF(H, E1 \wedge E2) = \frac{0.6 + (-0.3)}{1 - 0.3} = \frac{0.3}{0.7} = 0.42$$

8.3.2.2 Multiple Rules with Uncertain Evidence for the Same Conclusion

In the previous case, we saw that if the evidence E is observed, then we can conclude H with a CF. However, there are situations where the evidence E itself is uncertain.

For instance, in the rule:

IF E
THEN H {CF=0.5}

evidence E also has a certainty factor associated, say 0.9 (we are not 100% sure about this evidence).

Evidence may also be uncertain when it itself is gained from applying a rule:

Rule 1:
IF A
THEN B {CF=0.4}

Rule 2:
IF B
THEN C {CF=0.3}

If we know absolutely that A is true, then the fact B is estimated with a CF of 0.4.

So, when we go to apply the second rule, we need to take into account that the premise is not certain.

8.3.2.2.1 Rule with Uncertain Evidence: One Premise

When a rule has a single premise, the certainty of the conclusion is the product of the certainty of the premise multiplied by the certainty of the rule:

Rule 1:
IF A
THEN B {CF=0.4}

Rule 2:
IF B
THEN C {CF=0.3}
CF(C) = CF(B) * CF(Rule 1)

If the CF of A is true is 0.9 then:

CF(B) = CF(A)*CF(Rule 1) = 0.9*0.4 = 0.36

and

CF(C) = CF(B)*CF(Rule 2) = 0.36*0.3 = 0.108

8.3.2.2.2 Rule with Uncertain Evidence: Negative Evidence

A rule is only applicable if one believes the premise to be true. If the CF of the premises is negative (one does not believe them) then the rule does not apply.

IF E
THEN H {CF=0.6}

But, if CF(E)=-0.2, then we cannot say anything about E being true.
Thus:

$$CF(H) = \begin{cases} CF(E) \cdot CF(Rule), & if\ CF(E) > 0 \\ 0, & otherwise \end{cases}$$

A value of 0 for CF indicates that we know nothing as the result of applying the rule (we neither believe nor disbelieve). Thus, our knowledge does not change.

8.3.2.2.3 Rule with Uncertain Evidence: Multiple Premises

If the rule has multiple premises joined by AND:

IF E1
AND E2
⋮
AND En
THEN H {CF}

then CF(H) is calculated as:

$$CF(H) = \begin{cases} \min\{CF(E1\},CF(E2),\ldots,CF(En)\} * CF(Rule), & if\ CF(Ei) > 0,\ i = 1,2,\ldots,n \\ 0, & otherwise \end{cases}$$

If the CF of any one premise is ≤ 0 then the CF of the set is ≤ 0 and the rule does not apply. Thus, when evaluating the premises of a rule, one can stop processing if a premise has CF ≤ 0.

Example 1

Consider the following rule with the CFs for each of the premises and the CF of the rule:

IF E1 {CF = 0.8}
AND E2 {CF = 0.7}
AND E3 {CF = 0.5}
AND E4 {CF = 0.3}
AND E5 {CF = 0.9}
THEN H {CF = 0.65}

Then the CF of the conclusion is given by (see also Figure 8.4):

CF(H) = min{CF(E1), CF(E2), CF(E3), CF(E4), CF(E5)} * CF(Rule) = 0.3*0.65 = 0.195.

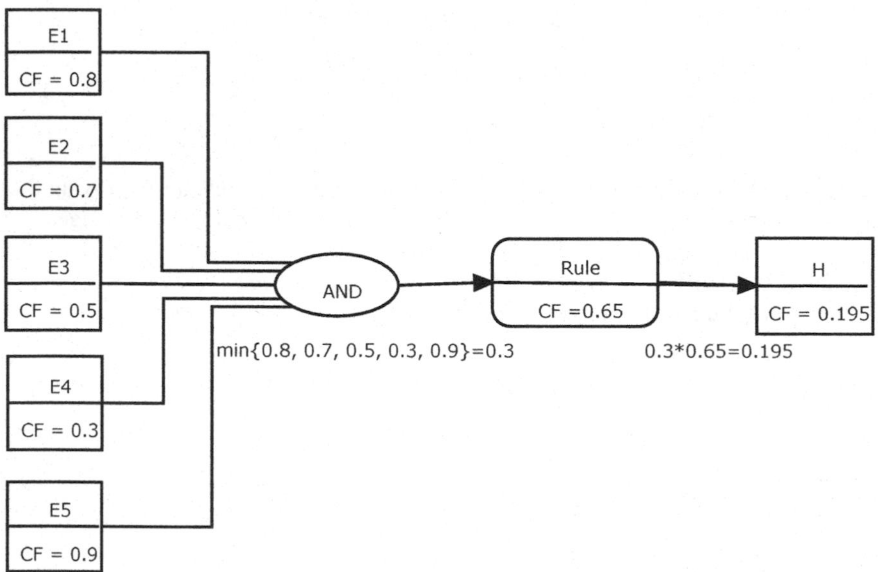

Fig. 8.4 Certainty factor calculation for rules with multiple premises joined by AND.

If the rule has multiple premises joined by OR:

IF E1
OR E2
⋮
OR En
THEN H {CF}

then CF(H) is calculated as:

CF(H) = max{CF(E1), CF(E2), …, CF(En)} * CF(Rule)

Example 2

Consider the same example as above: with the following rules and the CFs for each of the premises and the CF of the rule:

IF E1 {CF = 0.8}
OR E2 {CF = 0.7}
OR E3 {CF = 0.5}
OR E4 {CF = 0.3}
OR E5 {CF = 0.9}
THEN H {CF = 0.65}

Then the CF of the conclusion is given by (see also Figure 8.5):

CF(H) = max{CF(E1), CF(E2), CF(E3), CF(E4), CF(E5)} * CF(Rule) = 0.9*0.65 = 0.585.

As we can see, firing a rule involves the use of two different CFs:

- the CF associated to the antecedent of the rule (premises);
- the CF associated to the rule.

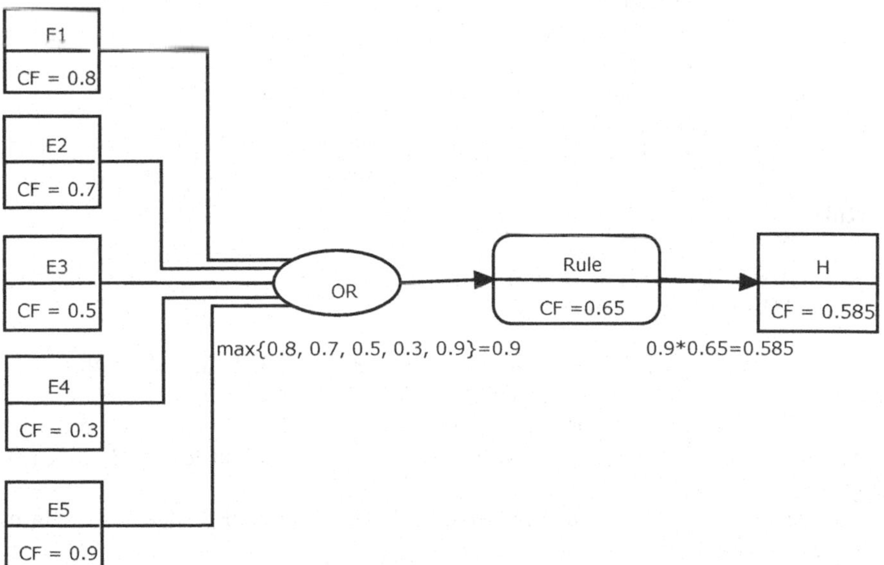

Fig. 8.5 Certainty factor calculation for rules with multiple premises joined by OR.

Example 3
Let us consider the following expert system for diagnosing a cold. The database consists of the following facts: the patient's fever is 37.4, is coughing since less than 24 hours, is sneezing, is having headache with a CF = 0.4 and is having nasal congestion with CF = 0.5

The rule base consists of the following rules:

Rule 1
IF fever < 37.5
THEN Cold symptoms = true {CF = 0.5}

Rule 2
IF fever > 37.5
THEN Cold symptoms = true {CF = 0.9}

Rule 3
IF cough for more than 24 hours
THEN soar troth = true {CF = 0.5}

Rule 4
IF cough for more than 48 hours
THEN soar troth = true {CF = 1}

Rule 5
IF Cold symptoms
AND not sneezing
THEN having cold {CF = -0.2}

Rule 6
IF soar troth
THEN having cold {CF = 0.5}

Rule 7
IF headache
AND nasal congestion
THEN having cold {CF = 0.7}

In order to find the CF of the patient having a cold, we will first draw the inference tree (see Figure 8.6) and the associated CFs for each rule.

Let us now see how the certainty factors are calculated.

The patient has fever less than 37.5, thus the CF of the fact fever <37.5 is 1.0 and the CF of the fact fever >37.5 is -1.0.

The patient is coughing since less that 24 hours, thus both facts cough > 24 h and cough > 48 h have the Cf = -1.0.

The CF of cold symptom as conclusion of Rule 1 is calculated as CF of Rule's 1 premise (which is 1.0) multiplied with the CF of the rule. Thus, we obtain: 1.0*0.5 = 0.5.

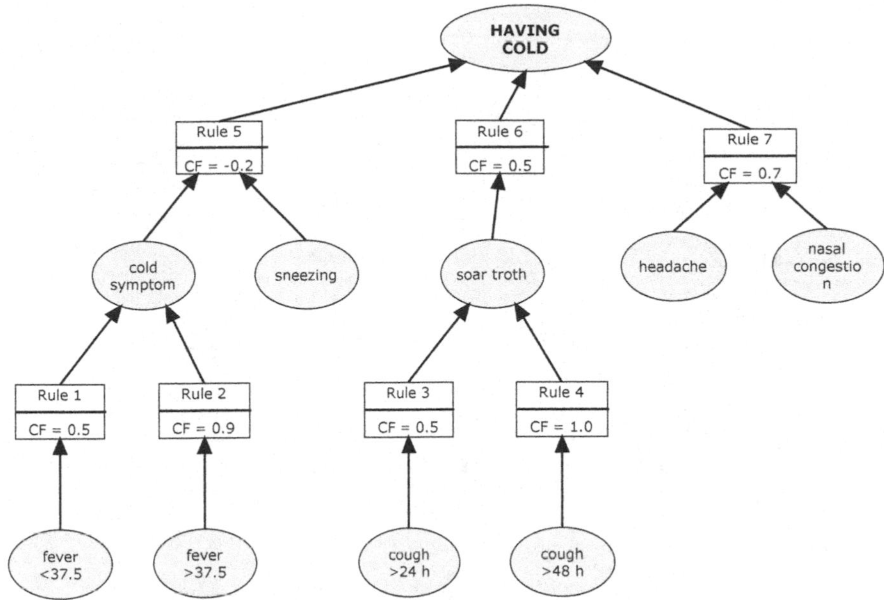

Fig. 8.6 The inference tree for Example 3.

Since the premise of Rule 2 is negative, Rule 2 is not contributing to the CF of cold symptom fact.

Similarly, since the premise of Rule 3 and the premise of Rule 4 are both negative, the CF of soar troth is 0.

The CF of having cold as conclusion of Rule 5 is calculated using the formula for a rule having two evidences joined by AND. Thus, minimum among CF of cold symptom (which is 0.5) and sneezing CF (which is 1.0) is multiplied with Rule's 5 CF;

CF1 = min{0.5, 1.0}*(-0.2) = 0.5*(-0.2)= -0.1

A similar situation is encountered in the case of Rule 7 which has two premises.

The CF of having cold implied by Rule 7 is calculated as minimum among headache's CF and nasal congestion's CF multiplied with Rule 7 CF;

CF2 = min{0.4, 0.5}*0.7 =).4*0.7 = 0.28.

Rules 5, 6 and 7 all affect the same hypothesis: having cold. Among them, only Rule 5 and Rule 6 have a CF different from 0. Thus, using the formula for calculating the CF of a consequent obtained as results of execution of two rules, we obtain (see Figure 8.7):

$$CF = \frac{CF1 + CF2}{1 - \min\{|CF1|, |CF2|\}} = \frac{-0.1 + 0.28}{1 - \min\{|-0.1|, |0.28|\}} = \frac{0.18}{1 - 0.1} = \frac{0.18}{0.9} = 0.2 \cdot$$

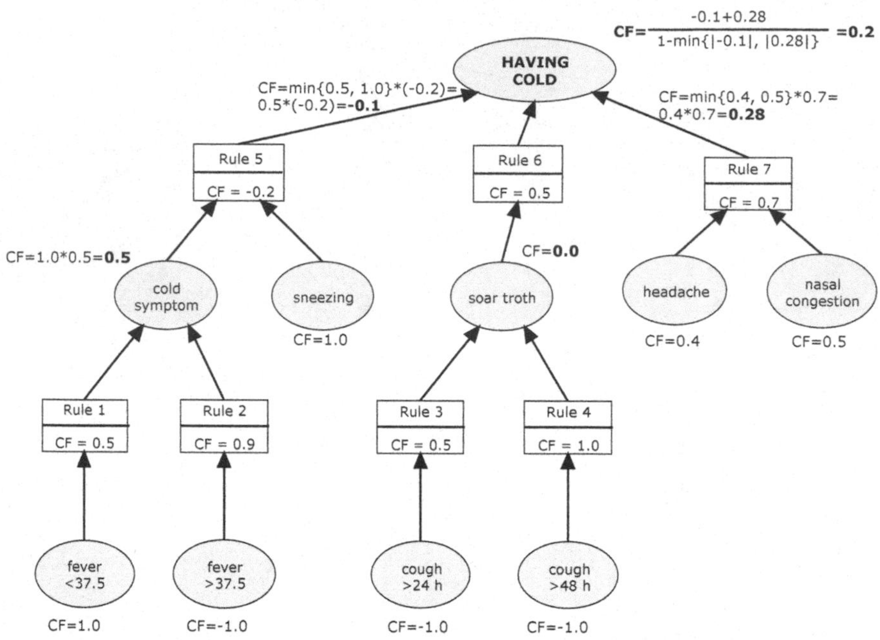

Fig. 8.7 The inference tree with the corresponding CFs for Example 3.

Summaries

This chapter presents various alternatives for the expert systems to deal with uncertain information or in uncertain situations.

Numerous uncertainty theories have been developed. Two main theories are presented in this chapter: Bayesian theory and certainty factors theory.

In Bayesian theory, the probability is interpreted as a *degree of belief*. Beliefs are always subjective, and therefore all the probabilities appearing in Bayesian probability theory are conditional. In particular, under the belief interpretation probability is not an objective property of some physical setting, but is conditional to the prior assumptions and experience of the learning system.

Bayesian formulas are complex enough and not adequate to human's brain reasoning. Reliable statistical information is not available or the independence of evidences cannot be assumed.

Although conditional independences are used to simplify probabilistic inference, exact inference in an arbitrary Bayesian network for discrete variables is NP-hard [20].

Certainty factors theory is an alternative to Bayesian reasoning and introduces a certainty factors calculus based on the human expert heuristics [13].

CF is more flexible and intuitive for the experts than probability.

A certainty factor is used to express how accurate, truthful, or reliable you judge a predicate to be. It is your judgment of how good your evidence is. The issue is how to combine various judgments.

Certainty factors are guesses by an expert about the relevance of evidence; they are *ad hoc* and are tuned by trial and error.

Certainty Factors do adhere to the rules of Bayesian statistics, but it can represent tractable knowledge systems:

- individual rules contribute belief in a hypothesis - basically a conditional probability;
- the formulae for combination of evidence / hypotheses basically assume that all rules are independent ruling out the need for joint probabilities.

References

1. Castillo, E., Alvarez, E.: Expert systems: uncertainty and learning. Elsevier Science publishing Company, New York (1991)
2. Giarratano, J.C., Riley, G.D.: Expert Systems: Principles and Programming, 3rd edn. PWS Publishing Company, Boston (1998)
3. Gonzalez, A.J., Dankel, D.D.: The Engineering of Knowledge-based Systems: Theory and Practice. Prentice-Hall, Englewood Cliffs (1993)
4. Hajek, P., Havranek, T., Jirousek, R.: Uncertain Information Processing in Expert Systems. CRC Press, Boca Raton (1992)
5. Motro, A.: Management of Uncertainty in Database Systems. In: Kim, W. (ed.) Modern Database Systems: the Object Model, Interoperability and Beyond, pp. 457–476. Addison-Wesley/ACM Press (1994)
6. Motro, A.: Sources of Uncertainty, Imprecision and Inconsistency in Information Systems. In: Motro, A., Smets, P. (eds.) Uncertainty Management in Information Systems: From Needs to Solutions, pp. 9–34. Kluwer Academic Publishers, Dordrecht (1996)
7. http://www.onlinefreeebooks.net/free-ebooks-computer-programming-technology/artificial-intelligence/artificial-intelligence-course-material-pdf.html (accessed on February 10, 2011)
8. http://www.cs.ubc.ca/~murphyk/Bayes/bnintro.htm (accessed on February 10, 2011)
9. http://www.stat.ucla.edu/~yuille/ (accessed on February 10, 2011)
10. Heckerman, D.: A tutorial on learning with Bayesian networks, Technical Report MSR-TR-95-06 Microsoft Research Advanced Technology Division, Redmond, WA (1996), http://research.microsoft.com/pubs/69588/tr-95-06.pdf (accessed on February 10, 2011)
11. Cooper, G.: Computational complexity of probabilistic inference using Bayesian belief networks. Artificial Intelligence 42, 393–405 (1990)
12. Nikolopoulos, C.: Expert systems: introduction to first and second generation and hybrid knowledge based systems. CRC Press, New York (1997)
13. Roventa, E., Spircu, T.: Management of Knowledge Imperfection in Building Intelligent Systems. Springer Verlag Series: Studies in Fuzziness and Soft Computing, vol. 227 (2009)

14. http://arantxa.ii.uam.es/~modonnel/IC/04_UncertaintyIV.pdf (accessed on February 10, 2011)
15. Negnevitsky, M.: Artificial Intelligence: a guide to intelligent systems. Addison Wesley, Essex (1992)
16. Kass, R., Raftery, A.: Bayes factors. Journal of the American Statistical Association 90, 773–795 (1995)
17. Jensen, F.: An introduction to Bayesian networks. Springer, Heidelberg (1996)
18. Heckerman, D., Geiger, D., Chickering, D.: Learning in Bayesian networks: the combination of knowledge and statistical data. Machine Learning 20, 197–243 (1995)
19. Finetti, B.: Theory of probability. Wiley and Sons, New York (1970)
20. Dagum, P., Luby, M.: Approximating probabilistic inference in Bayesian belief networks is NP-hard. Artificial Intelligence 60, 141–153 (1993)
21. Bernardo, J., Smith, A.: Bayesian Theory. Wiley and Sons, New York (1994)

Verification Questions

1. What is uncertainty? What are the main sources of uncertainty in an expert system?
2. Write Bayes' rule and explain the terms.
3. What is Bayesian reasoning?
4. What does a Bayesian net represent?
5. Explain the variable ordering in a Bayesian inference.
6. What is a prior and a posterior probability?
7. How does a rule based expert system propagate uncertainties using Bayesian reasoning?
8. What is the role of certainty factors?
9. How you define measure of belief and measure of disbelief?
10. How do you compare certainty factors and conditional probabilities?
11. Explain how does an expert system deals in the following situations:
    a. Multiple rules provide evidence for the same conclusion
    b. Multiple rules with uncertain evidence in the cases:
        i. One premise
        ii. One negative premise
        iii. Multiple premises joint by AND
        iv. Multiple premises joint by OR.

Exercises

1. Consider the following example [14]:

- Premises (or evidences) and their CFs:
    - A, B, C, D, E, F, G;
    - CF(A)= 0.3, CF(B)= 0.0, CF(C)= 0.0, CF(D)= 0.4, CF(E)= 0.0, CF(F)= -0.5, CF(G)= 0.8.
- Rules:

IF A
THEN C {CF = 0.2}

IF B
THEN C {CF = 1.0}

IF D
THEN E {CF = 0.6}

IF E
THEN C {CF = 1.0}

IF F
THEN B {CF = 0.9}

IF G
THEN E {CF = -0.7}

The problem is to find the final CFs for C and B.

2. (Taken from [15]). Knowing that the weather today is rain, the rainfall today is low (with a CF of 0.8), the temperature today is cold (with a Cf of 0.9), the expert system has to predict the weather tomorrow.

The knowledge base consists of the following rules:

Rule 1:
IF today is rain
THEN tomorrow is rain {CF=0.5}

Rule 2:
IF today is dry
THEN tomorrow is dry {CF=0.5}

Rule 3:
IF today is rain
AND rainfall is low
THEN tomorrow is dry {CF=0.6}

Rule 4:
IF today is rain
AND rainfall is low
AND temperature is cold
THEN tomorrow is dry {CF=0.7}

Rule 5:
IF today is dry
AND temperature is warm
THEN tomorrow is rain {CF=0.65}

Rule 6:

IF today is rain
AND temperature is warm
AND sky is overcast
THEN tomorrow is rain {CF=0.55}

3. Given the following set of rules:

Rule 1: IF A
OR B
THEN F {CF = 0.3}

Rule 2: IF not C
THEN E {CF = 0.6}

Rule 3: IF D
THEN G {CF = 0.75}

Rule 4: IF A
OR (F AND G)
THEN H {CF = 0.9}

(i) Draw an inference net from the above set of rules.

(ii) Given CF(A) = 0.5, CF(B) = 0.8, CF(C) = 0.5, CF(D) = -0.3, CF(E) = 0, CF(F) = 0, CF(G) = 0, and CF(H) = 0.2, what is certainty of H after updating?

4. (From [7]). Consider the following probability distribution over 6 variables A,B,C,D,E, and F for which the factorization as stated below holds. Find and draw a Bayesian network that for which this factorization is true, but for which no additional factorizations or any fewer factorizations are true.

$$P(A, B, C, D, E, F) = P(A)\, P(B)\, P(C|A, B)\, P(D|B)\, P(E|C, D)\, P(F|E).$$

5. (From [7]). Consider a situation in which we want to reason about the relationship between smoking and lung cancer. We'll use 5 Boolean random variables representing "has lung cancer" (C), "smokes" (S), "has a reduced life expectancy" (RLE), "exposed to secondhand smoke" (SHS), and "at least one parent smokes" (PS). Intuitively, we know that whether or not a person has cancer is directly influenced by whether he or she is exposed to second-hand smoke and whether he or she smokes. Both of these things are affected by whether the parents smoke. Cancer reduces a person's life expectancy.

1) Draw the network (nodes and arcs only)
2) How many independent values are required to specify all the conditional probability tables (CPTs) for your network?
3) How many independent values are in the full joint probability distribution for this problem domain?

6. (From [7]). Let A, B, C, D be Boolean random variables. Given that:

- A and B are (absolutely) independent.
- C is independent of B given A.
- D is independent of C given A and B.
- $P(A=T) = 0.3$
- $P(B=T) = 0.6$
- $P(C=T|A=T) = 0.8$
- $P(C=T|A=F) = 0.4$
- $P(D=T|A=T,B=T) = 0.7$
- $P(D=T|A=T,B=F) = 0.8$
- $P(D=T|A=F,B=T) = 0.1$
- $P(D=T|A=F,B=F) = 0.2$

Compute the following quantities:

1) $P(D=T)$
2) $P(D=F,C=T)$
3) $P(A=T|C=T)$
4) $P(A=T|D=F)$
5) $P(A=T,D=T|B=F)$.

Chapter 9
Fuzzy Expert Systems

9.1 Introduction

One of the imprecision types of information encountered in an expert system is due to the (natural) language used to express information. If knowledge is not expressed in some formal language, the meaning cannot be interpreted exactly. Since there is no such universal scheme for formal representation language, a particular knowledge representation scheme must be chosen to adequately capture the information about the domain. Many a times the scheme chosen will not provide an exact match with the expert's knowledge. Thus, imprecision will occur.

We saw in the previous chapter that there are several theories which can help the expert system deal with imprecision. Most of them model imprecision using probabilities. But there are situations which do not lead to cases easily modeled by probabilities. Like for instance, when we tell *the speed of the car is about 65 miles*. Similar *fuzzy* words are the ones used in the following examples too:

John is *tall* or
Weather is *warm*.

Both *tall* and *warm* are fuzzy terms.

Such statements are difficult to translate into more precise language without losing some of their semantic value.

The expert system has to reason with such imprecise information and this will lead to *fuzzy reasoning*. The expert systems using fuzzy knowledge and fuzzy reasoning are known as fuzzy expert systems[1][10][11][12][17][20]. While some of the decisions and calculations could be done using traditional logic, fuzzy systems affords a broader, richer field of data and the manipulation of that data than do more traditional methods.

The theory of logic is one of the oldest ones and lasts thousands of years ago when Aristotle and the philosophers who preceded him were trying to devise a concise theory of logic: "Laws of Thought". The "Law of the Excluded Middle," states that every proposition must either be true or false. Even when Parminedes proposed the first version of this law (around 400 B.C.) there were strong and

C. Grosan and A. Abraham: Intelligent Systems, ISRL 17, pp. 219–260.
springerlink.com

immediate objections: for example, Heraclitus proposed that things could be simultaneously true and not true [2].

Plato indicated that there was a third region beyond true and false and Lukasiewicz was the one who first proposed a systematic alternative to the bivalued logic of Aristotle [3].

In the early 90's, Lukasiewicz described a three-valued or trivalent logic; the third value proposed can be translated as the term "possible," and he assigned it a numeric value between true and false. The three-valued logic does not assume the law of the excluded middle; three truth values are possible: true, false, or undecided. As of [5], there are 3072 such logics now. Later, he explored four-valued logics, five-valued logics, and then declared that in principle there was nothing to prevent the derivation of an infinite-valued logic. From an algebraic point of view, the number of truth values is unlimited. One can build logics with four, five, etc. truth values. From the point of view of logical interpretation, such formalisms present serious deficiencies, even in the three valued case.

Lukasiewicz felt that three and infinite valued logics were the most intriguing, but he ultimately settled on a four-valued logic because it seemed to be the most easily adaptable to Aristotelian logic [4] [2]. The logic of probabilities developed by the German logician H. Reichenbach replaces the true-false pair by a continuous scale of values interpreted as probabilities. It was not until relatively recently that the notion of an infinite-valued logic took hold.

In 1965, Lotfi A. Zadeh described the mathematics of fuzzy set theory, and by extension fuzzy logic [18][19]. This theory proposed making the membership function (or the values False and True) operate over the range of real numbers [0.0, 1.0]. New operations for the calculus of logic were proposed, and showed to be in principle at least a generalization of classic logic. In order to better understand how fuzzy expert systems work, we introduce the concept of fuzzy logic, fuzzy rules and fuzzy reasoning.

9.2 Fuzzy Sets

Let us consider X a set and $x \in X$ an element of X.

In classical logic, x either belongs to X ($x \in X$) or does not belongs to X ($x \notin X$).

A fuzzy set $A \subset X$ is characterized by a *membership function* (or *characteristic function*) $f_A(x) : X \rightarrow [0, 1]$ which associates each point in X a real number in the interval [0, 1] [6]. X is called the universe and A is a fuzzy subset of X.

$f_A(x)$ represents the *grade of membership* of x in A.

The value 0 represents false (or non- membership), and the value 1 represents true (or membership). The closer the value of $f_A(x)$ to 1, the higher the grade of membership of x in A.

In classical logic (or ordinary logic), the membership function $f_A(x)$ can only take the values 0 and 1:

$$f_A(x) : X \rightarrow \{0, 1\}$$

$f_A(x) = 1$ corresponds to the case that x belongs to A, and $f_A(x) = 0$ corresponds to the situation that x does not belong to A, respectively.

$$f_A(x) = \begin{cases} 0, \ if \ x \notin A \\ 1, \ if \ x \in A \end{cases}.$$

In the case of fuzzy logic, a fuzzy set A is defined by a membership function $\mu_A(x)$,

$\mu_A(x)$: X →[0, 1]

defined by:

$$\mu_A(x) = \begin{cases} 1, & \text{if } x \text{ is totaly in } A \\ 0, & \text{if } x \text{ is not in } A \\ \in (0,1), & \text{if } x \text{ is partialy in } A \end{cases}.$$

Example 1

Let us consider the example of old people. Consider the age varying between 30 and 80 years old. Table 9.1 presents the degree of membership in both classical and fuzzy logic and the graphical representation is given in Figure 9.1.

Table 9.1 Degree of membership in fuzzy logic and ordinary logic for Example 1.

Age	**Degree of membership**	
	Fuzzy logic	**Ordinary logic**
30	0.0	0.0
35	0.1	0.0
40	0.2	0.0
45	0.3	0.0
50	0.4	1.0
55	0.5	1.0
60	0.6	1.0
65	0.7	1.0
70	0.8	1.0
75	0.9	1.0
80	1.0	1.0

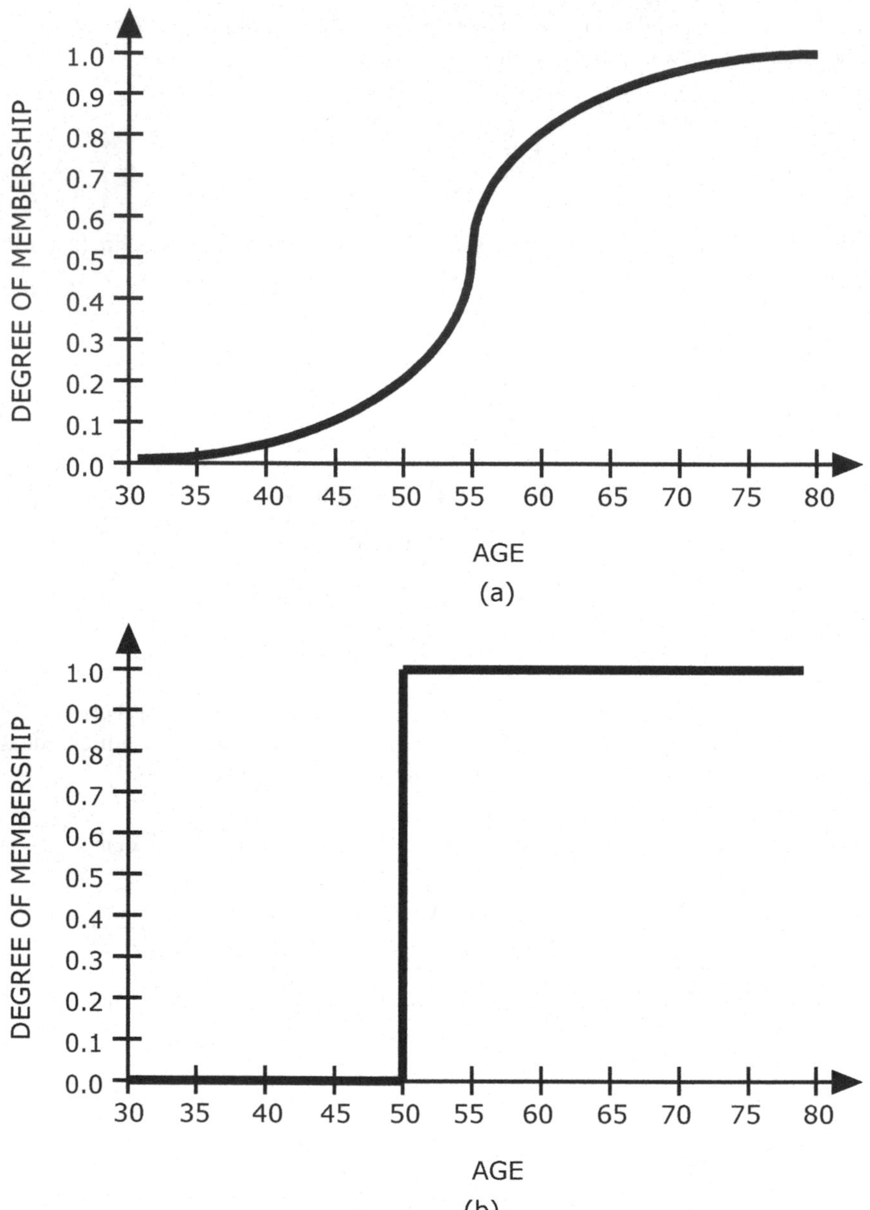

Fig. 9.1 Sets of old people in fuzzy logic (a) and ordinary logic (b) for Example 1.

In ordinary logic, the degree of membership is either 0 or one, the elements belong or not to that set. In fuzzy logic, the degree of membership is a real number between 0 and 1 with the interpretation that a person is partially young or partially old. A person's membership degree of 0.7 means that the person is 0.7 old (or

70% old). There is also a semantic difference between the two types of logic (ordinary and fuzzy): the first logic supposes that a person is or is not old (still caught in the Law of the Excluded Middle). By contrast, fuzzy terminology supposes that a person is "more or less" old, or some other term (corresponding to the value of 0.70 for instance).

9.2.1 Representing Fuzzy Sets

Venn diagrams used for representing ordinary sets cannot be used (are not appropriate) for representing fuzzy sets. Ordinary sets use clear cut on the boundaries and fuzzy sets use grades. The difference is shown in Figure 9.2.

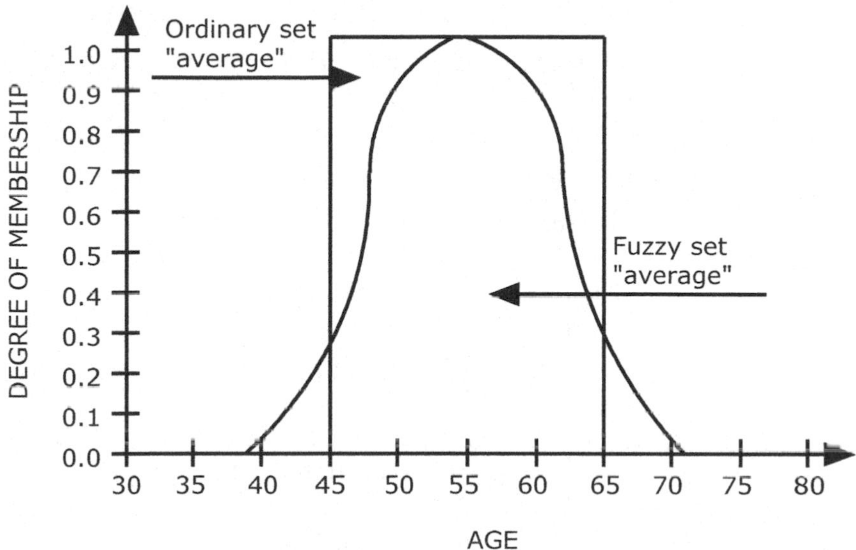

Fig. 9.2 Fuzzy subset and ordinary subset for average age.

As we can see from Figure 9.2, in the ordinary logic, an age of 44 is young and an age of 46 is average, while in fuzzy logic the age 44 is average with a membership degree of 0.2 and the age 46 is average with a membership degree of 0.3.

An important step in solving a problem using a fuzzy system is representing the problem in fuzzy terms (known as *conceptualization in fuzzy terms*). Linguistic variables such as *low*, *below average*, *average*, *above average*, *high*, are fuzzy concepts which may be represented as fuzzy sets.

The process of representing a linguistic variable into a set of linguistic values is called *fuzzy quantization*. The linguistic variables can be quantized into some linguistic labels that can be (graphically) represented by standard functional representations.

There exist several types of standard membership functions used by fuzzy expert systems. Some of the most used are [7]:

- *Singleton* (or single valued):
  $x = a$, where a is a scalar.
- *Triangular*:

$$\mu(x) = 1 - \frac{|x-a|}{|a-b|}.$$

If the triangular membership functions μ_1, μ_2, …μ_n, representing a fuzzy variable are uniformly distributed over the universe of discourse X, then the following propriety holds:

$$\sum_{i=1}^{n} \mu_i(x) = 1, \text{ for all } x \in X$$

- *Trapezoidal*
- *Sigmoid function* (*S-function*)

This membership function is defined as:

$$S(x) = \begin{cases} 0, & \text{if } x \le a \\ 2\left(\dfrac{x-a}{c-a}\right)^2, & \text{if } a < x \le c \\ 1 - 2\left(\dfrac{x-a}{c-a}\right)^2, & \text{if } b < x \le c \\ 1, & \text{if } x > c \end{cases}$$

- Z-function:

  $Z(x) = 1-S(x)$

- *Bell function* (Π - function)

$$\Pi(x) = \begin{cases} S(x), & \text{if } x \le c \\ Z(x), & \text{if } x > c \end{cases}$$

Figure 9.3 presents 4 types of such functions: triangular (Figure 9.3 (a)), trapezoidal (Figure 9.3 (b)), Gaussian (Figure 9.3 (c)), and generalized bell (Figure 9.3 (d)).

For the quantization process, two parameters must be defined:

- the number of fuzzy labels
- the form of the membership function for each fuzzy label.

A fuzzy subset A of a finite reference superset X can be expressed as:

$$A = \{x_1, \mu_A(x_1)\}, \{x_2, \mu_A(x_2)\}, \ldots, \{x_n, \mu_A(x_n)\}$$

or as:

$$A = \{x_1/\mu_A(x_1)\}, \{x_2/\mu_A(x_2)\}, \ldots, \{x_n/\mu_A(x_n)\}.$$

An example of fuzzy sets and ordinary set for our age example is given in Figure 9.4. Three sets are considered: young, average and old.

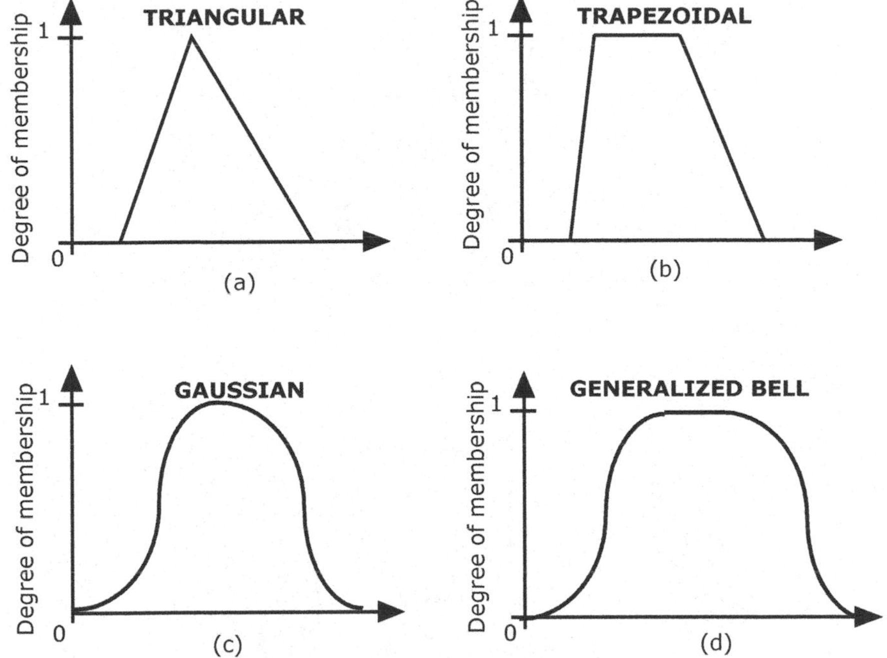

Fig. 9.3 Examples of membership functions: triangular (a), trapezoidal (b), Gaussian (c) and generalized bell (d).

Definition
The *support* of a fuzzy A set can be defined as the subset of the universe who's all elements have a membership degree to A different from 0 (see Figure 9.5):

$$\text{Supp}(A)=\{x \mid x \in X,\ \mu_A(x)>0\}.$$

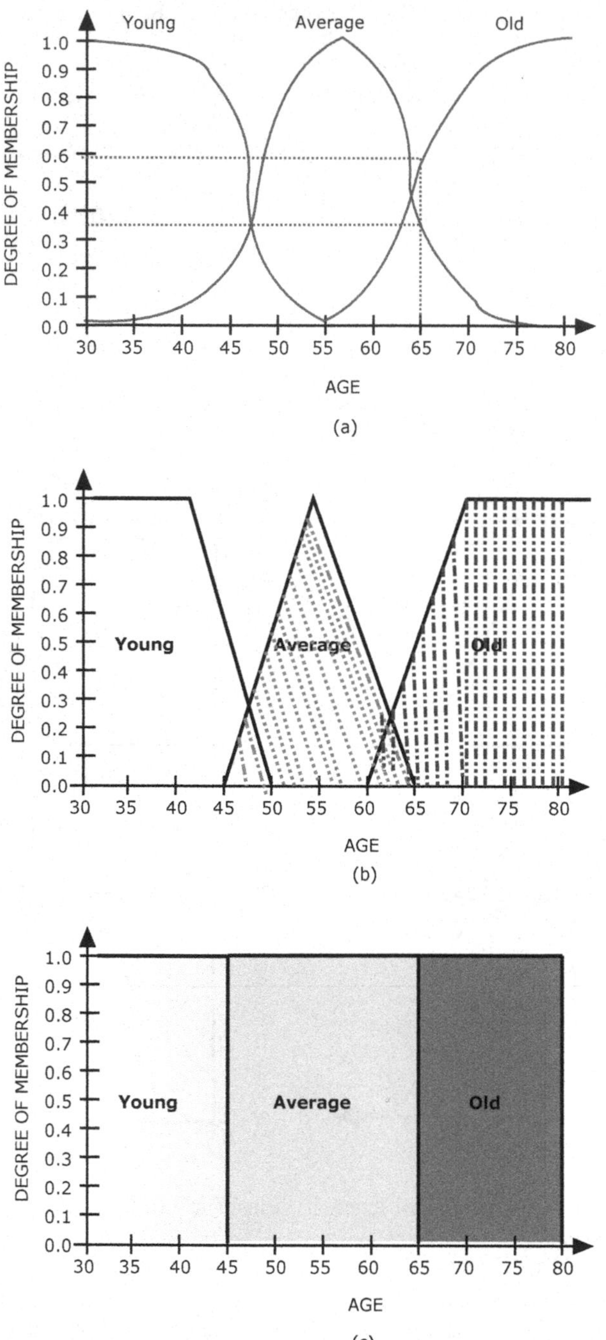

Fig. 9.4 Example of fuzzy sets using curves (a), straight lines (b) and ordinary sets.

In our age example, the support of the fuzzy set average is the interval [39, 71].

Definition
Cardinality of a fuzzy set A is defined by [7]:

$$Cardinality(A) = \sum_{x \in X} \mu_A(x)$$

For ordinary sets, cardinality is the number of elements in the set.

Definition
The *power set* of a fuzzy set A consists of all fuzzy subsets of A.

Definition
A fuzzy set A is a normal fuzzy set if its membership function has a grade 1 for at least one element of the universe.

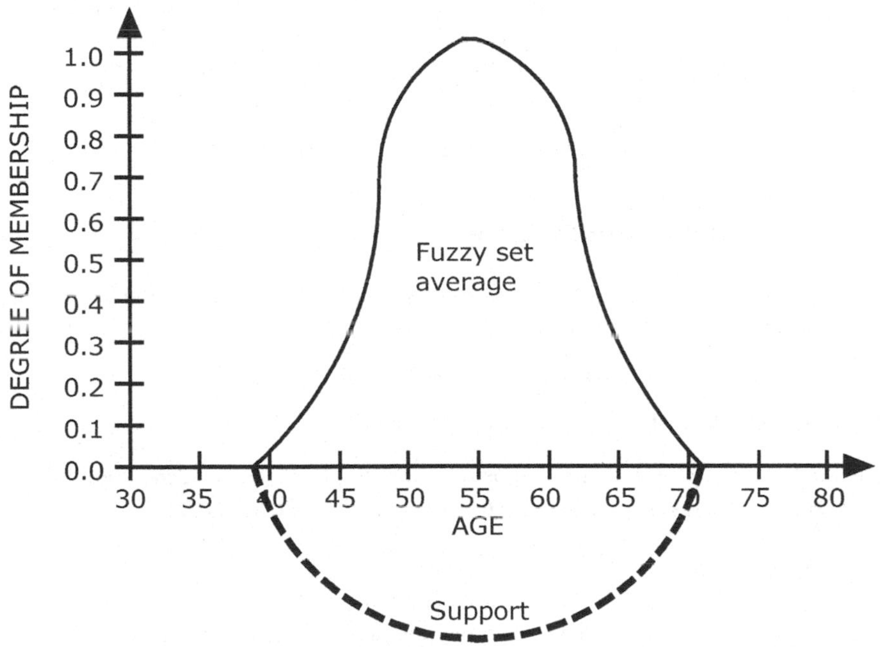

Fig. 9.5 Support of the fuzzy set average.

Definition
The *x-cut* of a fuzzy set *A* is a subset A_a of the universe which consists of values that belong to the fuzzy set A with a membership degree greater (weak cut) or greater or equal (strong cut) than a given value $x \in [0, 1]$.

Every fuzzy set can be represented by its *x*-cut. An example is given in Figure 9.6 [7].

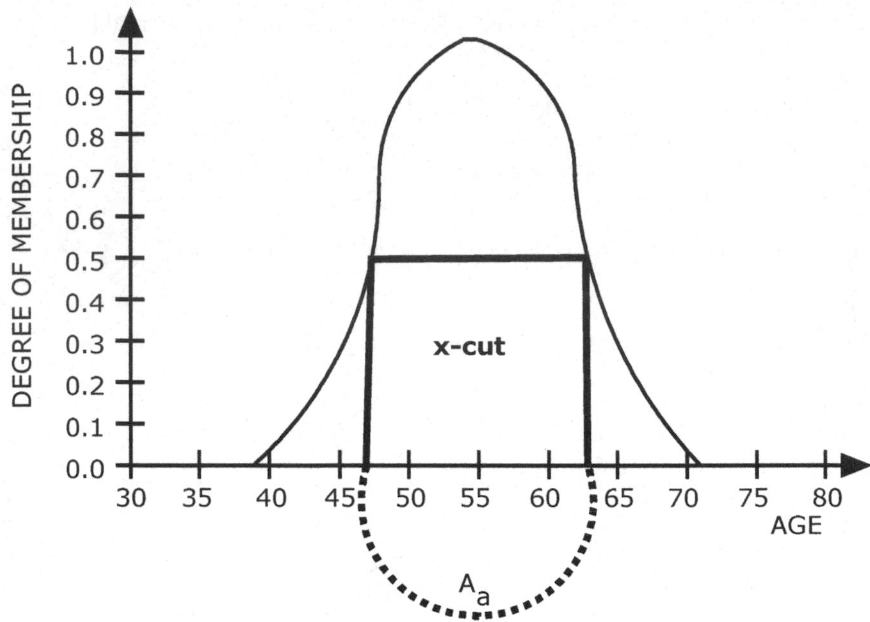

Fig. 9.6 Example of x-cut for a fuzzy set A.

9.2.2 *Operations with Fuzzy Sets*

Since fuzzy logic can be seen as an extension of ordinary logic, the operations used for dealing with ordinary sets can be employed for dealing with fuzzy sets too.

9.2.2.1 Complement

The complement of a set is the opposite of this set. For the set A, the complement is the set Not A (¬A). For a fuzzy set A, the complement ¬A is defined by:

$$\mu_{\neg A}(x) = 1\text{-}\mu_A(x)$$

For example, if we have the set old people, then we can obtain the set of not old people as follows:

Old people = {0, 30}, {0.2, 40}, {0.4, 50}, {0.6, 60}, {0.8, 70}, {1, 80}
Not Old people = {1, 30}, {0.8, 40}, {0.6, 50}, {0.4, 60}, {0.2, 70}, {0, 80}

A graphical representation of a fuzzy set A and the corresponding fuzzy set ¬A is depicted in Figure 9.7.

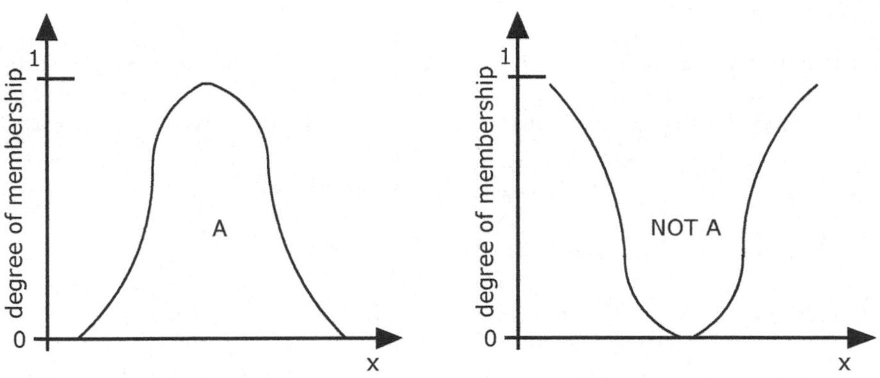

Fig. 9.7 A fuzzy set A and its fuzzy complement ¬A.

9.2.2.2 Containment

A fuzzy set A is a subset of a fuzzy set B if:

$\mu_A(x) \leq \mu_B(x)$, for all $x \in X$.

For example, the set of very old people is included in the set of old people as shown below:

Old people = {0.6, 60}, {0.7, 65}, {0.8, 70}, {0.9, 75}, {1, 80}
Very Old people = {0.6, 60}, {0.67, 65}, {0.7, 70}, {0.8, 75}, {0.95, 80}

The graphical representation of two sets inclusion A and B ($B \subseteq A$) is presented in Figure 9.8.

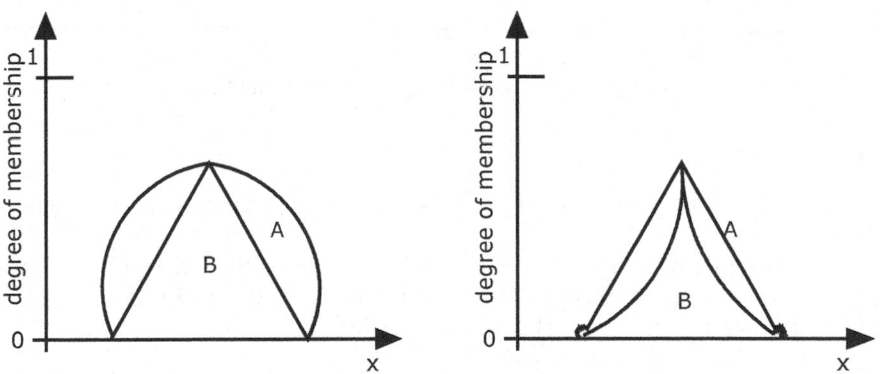

Fig. 9.8 Example of two fuzzy sets inclusion ($B \subseteq A$).

9.2.2.3 Intersection

In the ordinary logic, the intersection of two sets contains the elements shared by these sets. In the fuzzy logic, an element partially belongs to the two sets with different membership grades. Thus, the intersection of two fuzzy sets is given by:

$$\mu_{A\cap B}(x) = \min\{\mu_A(x), \mu_B(x)\} = \mu_A(x) \cap \mu_B(x), \text{ for all } x \in X.$$

If we have the sets of old people and average people as below:

Old people = {0, 30}, {0.1, 40}, {0.2, 50}, {0.6, 60}, {0.7, 65}, {0.8, 70}, {0.9, 75}, {1, 80}
Average people= {0.1, 30}, {0.2, 40}, {0.6, 50}, {0.5, 60}, {0.2, 65}, {0.1, 70}, {0, 75}, {0, 80}

then the intersection Old people ∩ Average people is given by:

Old people ∩ Average = {0, 30}, {0.1, 40}, {0.2, 50}, {0.5, 60}, {0.2, 65}, {0.1, 70}, {0.75}, {0, 80}.

The intersection of two sets A and B is represented graphically in Figure 9.9.

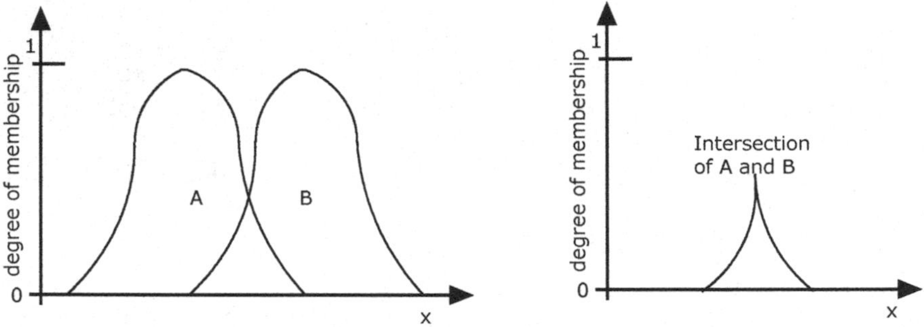

Fig. 9.9 Intersection of two fuzzy sets A and B.

9.2.2.4 Union

In the ordinary logic, the union of two sets contains the elements that fall into either set.

In fuzzy logic, the union of two fuzzy sets *A* and *B* is the largest membership value of the element in either set. The union is the opposite of intersection.

The union of two fuzzy sets *A* and *B* is given by:

$$\mu_{A\cup B}(x) = \max\{\mu_A(x), \mu_B(x)\} = \mu_A(x) \cup \mu_B(x), \text{ for all } x \in X.$$

If we have the sets of old people and average people as below:

Old people = {0, 30}, {0.1, 40}, {0.2, 50}, {0.6, 60}, {0.7, 65}, {0.8, 70}, {0.9, 75}, {1, 80}

Average people= {0.1, 30}, {0.2, 40}, {0.6, 50}, {0.5, 60}, {0.2, 65}, {0.1, 70}, {0, 75}, {0, 80}

Then the union Old people ∪ Average people is given by:

Old people ∪ Average = {0.1, 30}, {0.2, 40}, {0.6, 50}, {0.6, 60}, {0.7, 65}, {0.8, 70}, {0.9, 75}, {1, 80}.

The union of two sets A and B is represented graphically in Figure 9.10.

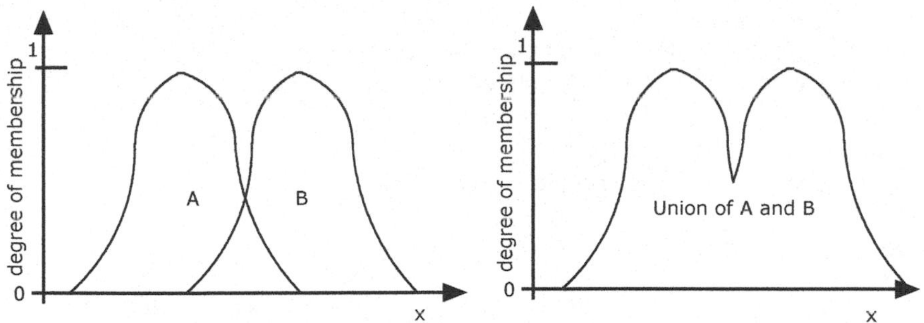

Fig. 9.10 Union of two fuzzy sets A and B.

9.2.2.5 Equality

Two fuzzy sets *A* and *B* are equal if:

$\mu_A(x) = \mu_B(x)$, for all $x \in X$.

9.2.2.6 Algebraic Product

The algebraic product of two fuzzy sets *A* and *B* is given by:

$\mu_{AB}(x) = \mu_A(x) \cdot \mu_B(x)$, for all $x \in X$.

9.2.2.6 Algebraic Sum

The algebraic sum of two fuzzy sets *A* and *B* is given by:

$\mu_{A+B}(x) = \mu_A(x) + \mu_B(x)$, for all $x \in X$.

9.2.3 Proprieties of Fuzzy Sets

The operations over the fuzzy sets have same proprieties as in the case of ordinary sets (associative, commutative, distributive, transitive, etc).

9.2.3.1 Associativity

The associativity propriety of three fuzzy sets A, B and C using AND and OR logical operators is given by:

$$A \cup (B \cup C) = (A \cup B) \cup C$$
$$A \cap (B \cap C) = (A \cap B) \cap C$$

Example

A = set of young people
B = set of average people
C = set of old people.

young people ∪ (average people ∪ old people) =
(young people ∪ average people)∪ old people

young people ∩ (average people ∩ old people) =
(young people ∩ average people) ∩ old people

9.2.3.2 Distributivity

The distributivity propriety of three fuzzy sets A, B and C using AND (AND distributivity) and OR (OR distributivity) logical operators is given by:

$$A \cup (B \cap C) = (A \cup B) \cap (A \cup C)$$
$$A \cap (B \cup C) = (A \cap B) \cup (A \cap C)$$

Example

A = set of young people
B = set of average people
C = set of old people.

young people ∪ (average people ∩ old people) =
(young people ∪ average people) ∩(young people ∪ old people)

young people ∩ (average people ∪ old people) =
(young people ∩ average people) ∪(young people ∩ old people)

9.2.3.3 Commutativity

The commutativity propriety of two fuzzy sets A and B using logical operators AND and OR is given by:

$$A \cup B = B \cup A$$
$$A \cap B = B \cap A$$

Example

A = set of young people
B = set of old people.

young people $\cup$ old people = old people $\cup$ young people
young people $\cap$ old people = old people $\cap$ young people

9.2.3.4 Transitivity

The transitivity propriety of three fuzzy sets A, B and C can be written as:

IF ($A \subset B$) AND ($B \subset C$) THEN ($A \subset C$)

Example

A = set of old people
B = set of very old people
C = set of extremely old people.

IF (old people $\subset$ very old people) AND (very old people $\subset$ extremely old people)
THEN (old people $\subset$ extremely old people)

9.2.3.5 Idempotency

The idempotency propriety of a fuzzy set A with respect to logical operators AND and OR is given by:

$A \cup A = A$
$A \cap A = A$

Example

A = set of young people
young people $\cup$ young people = young people
young people $\cap$ young people = young people

9.2.3.6 Identity

The identity propriety of a fuzzy set A with respect to logical operators AND and OR and given the empty (or null) set $\varnothing$ - having all degrees of membership equal to 0 and the general set X - having all degrees of membership equal to 1, is defined as:

$A \cup \varnothing = A$
$A \cap \varnothing = \varnothing$

$A \cup X = X$
$A \cap X = A$

Example

A = set of young people
∅ = empty set
X = general set

young people ∪ empty set = young people
young people ∩ empty set = empty set

young people ∪ general set = general set
young people ∩ general set = young people

9.2.3.7 Involution

The involution propriety represents the negation of negation and for a fuzzy set A is given by:

$\neg(\neg A) = A$

Example

A = set of young people

NOT (NOT young people) = young people

9.2.3.7 De Morgan's Laws

De Morgan's laws also apply in the case of fuzzy sets. Thus, given two fuzzy sets A and B, De Morgan's laws state as:

$\neg(A \cap B) = \neg A \cup \neg B$
$\neg(A \cup B) = \neg A \cap \neg B$

Example

A = set of young people
B = set of tall people

¬(young people ∩ tall people) = ¬young people ∪ ¬tall people
¬(young people ∪ tall people) = ¬young people ∩ ¬tall people

9.2.4 Hedges

Hedges are modifiers, adjectives, or adverbs, which change the truth values. Hedges modify the shape of a fuzzy set and include terms such as very, more or less, somewhat, about, nearly, etc.

One type of hedges apply to fuzzy numbers and the second type to truth values. Hedges can operate on membership functions as well as on fuzzy rules. We will presents in what follows some examples from the following categories:

- *Hedges which reduce the original truths value (produce a concentration)*
  - ***"very"*** *reduces* the truth value of the term it is applied for. It produces a concentration effect and thus reduces the degree of membership of the fuzzy element it is applied for. The original truth value is raised to square power:

    $\mu_{A}\_very(x) = \mu_A(x)^2$

    If a person has a 0.6 degree of membership to the set of old people, then the same person will have a degree of membership of 0.36 in the set of very old people.
  - ***"extremely"*** *reduces* the truth value. It has a similar influence as *very* but with a greater extend. The original truth value is raised to cube power:

    $\mu_{A}\_extremely(x) = \mu_A(x)^3$

    A person with a 0.6 degree of membership to the set of old people will have 0.21 degree of membership in the set of extremely old people.
  - ***"very very"****reduces* the truth value. It is similar to very and it raises very to square:

    $\mu_{A}\_veryvery(x) = \mu_{A}\_very(x)^2 = \mu_A(x)^4$

    A person with a 0.6 degree of membership to the set of old people will have 0.12 degree of membership in the set of very very old people.
- *Hedges which increase the original truth value (produce a dilatation)*
  - ***"somewhat"*** *increases* the truth value. The new value will be the square root of the original truth value:

    $\mu_{A}\_somewhat(x) = \sqrt{\mu_A(x)}$

    A person having a 0.6 degree of membership in the set of old people will have a degree of membership of 0.77 in the set of somewhat old people.
  - ***"slightly"*** *increases* the truth value. The new value will be the cube root of the original truth value:

    $\mu_{A}\_slightly(x) = \sqrt[3]{\mu_A(x)}$

A person having a 0.6 degree of membership in the set of old people will have a degree of membership of 0.84 in the set of slightly old people.

- *Hedges which intensify the original truth value*
  - ***"indeed"*** has the effect of intensifying the meaning of the sentence. If the actual value of the degree of membership is greater than 0.5 then it will be increased and if the actual value is less than 0.5 it will be decreased. It is given by:

$$\mu_A\_indeed(x) = \begin{cases} 2 \cdot (\mu_A(x))^2, & if\ 0 \le \mu_A(x) \le 0.5 \\ 1 - 2 \cdot (1 - \mu_A(x))^2, & if\ 0.5 < \mu_A(x) \le 1 \end{cases}$$

Thus, a person having a degree of membership 0.6 to the set of old people will have a degree of membership 0.68 to the set of indeed old people.

A person having a degree of membership 0.3 to the set of old people will have a degree of membership 0.18 to the set of indeed old people.

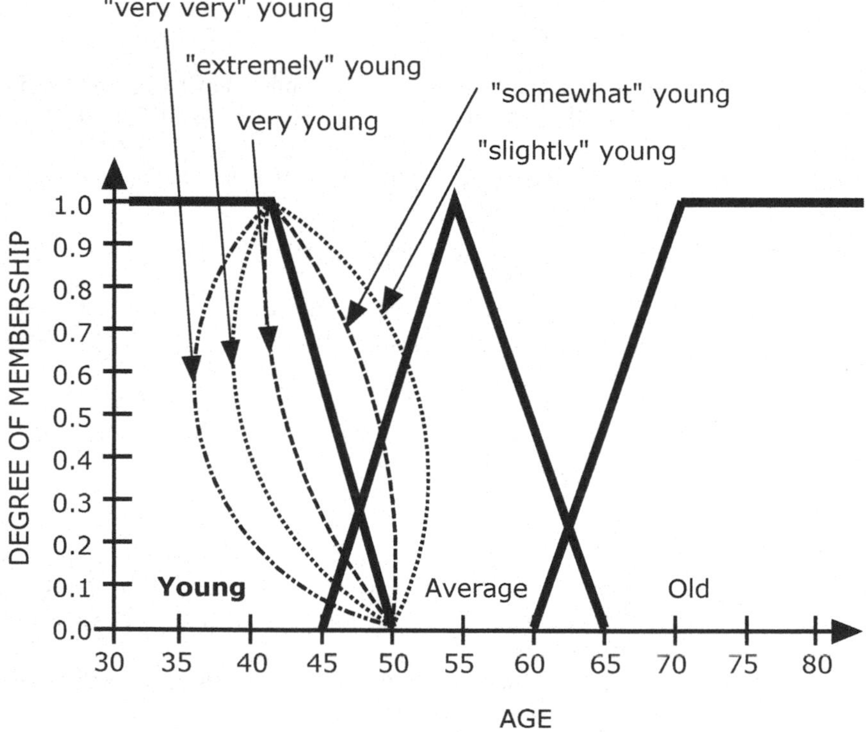

Fig. 9.11 Example of membership function modified by hedges.

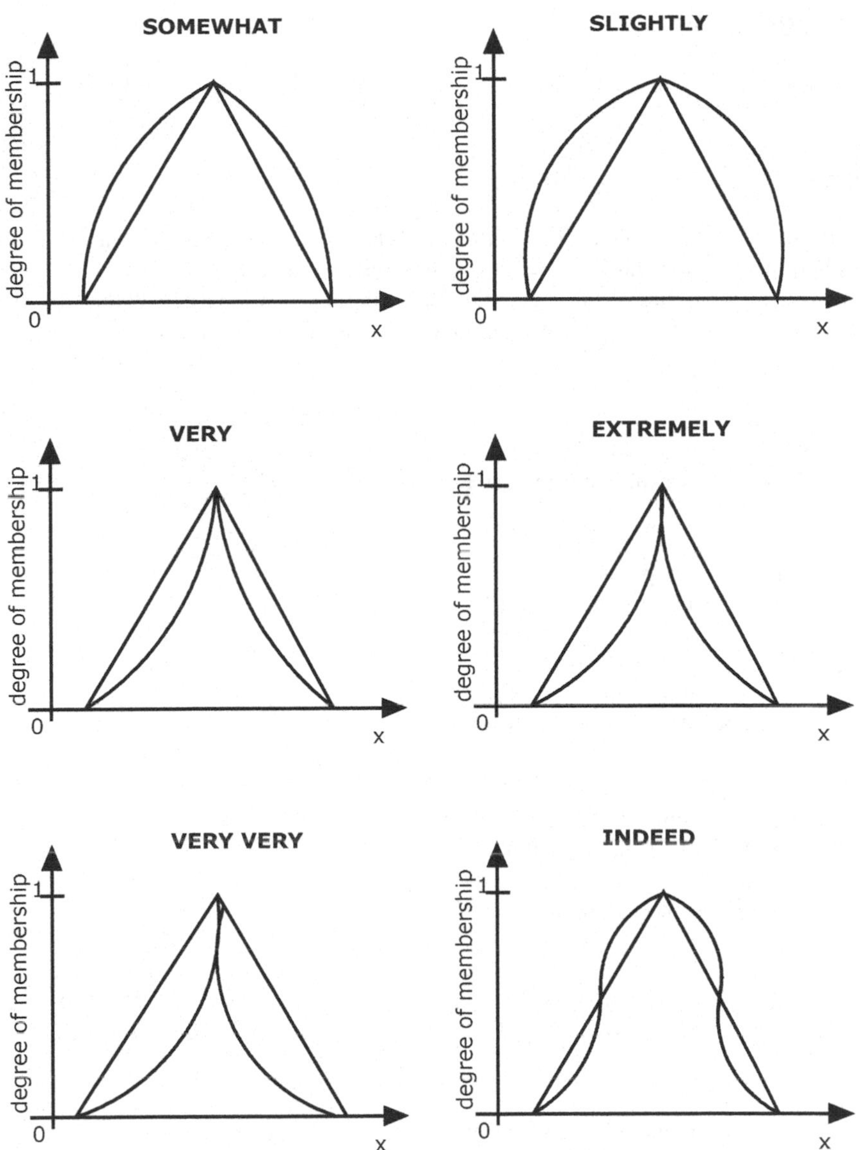

Fig. 9.12 Examples of different hedges in fuzzy logic.

Usually, the original truth value is raised to a power greater than 1 for terms that reduce truth values and less than 1 for terms that increase truth values. Figures 9.11 and 9.12 give a sample membership function modified by hedges.

9.3 Fuzzy Rules

Fuzzy rules are linguistic IF-THEN constructions that have the general form:

IF A
THEN B

where A and B are (collections of) propositions containing linguistic variables. A is called the *premise* and B is the *consequence* of the rule.

In the case of fuzzy rules, A and B are linguistic values determined by the fuzzy sets on two universes of discourses, X and Y, corresponding to A and B respectively.

Example

The following set of rules in ordinary logic:

Rule 1:

IF temperature is -5
THEN the weather is cold

Rule 2:

IF temperature is 15
THEN the weather is warm

Rule 3:

IF temperature is 35
THEN the weather is hot

may be written in fuzzy logic as:

Rule 1:

IF temperature is low
THEN the weather is cold

Rule 2:

IF temperature is average
THEN the weather is warm

Rule 3:

IF temperature is high
THEN the weather is hot

A nice example of graphical representation of fuzzy rules is given in [8]. Using a similar model, our three rules above are depicted in Figure 9.13.

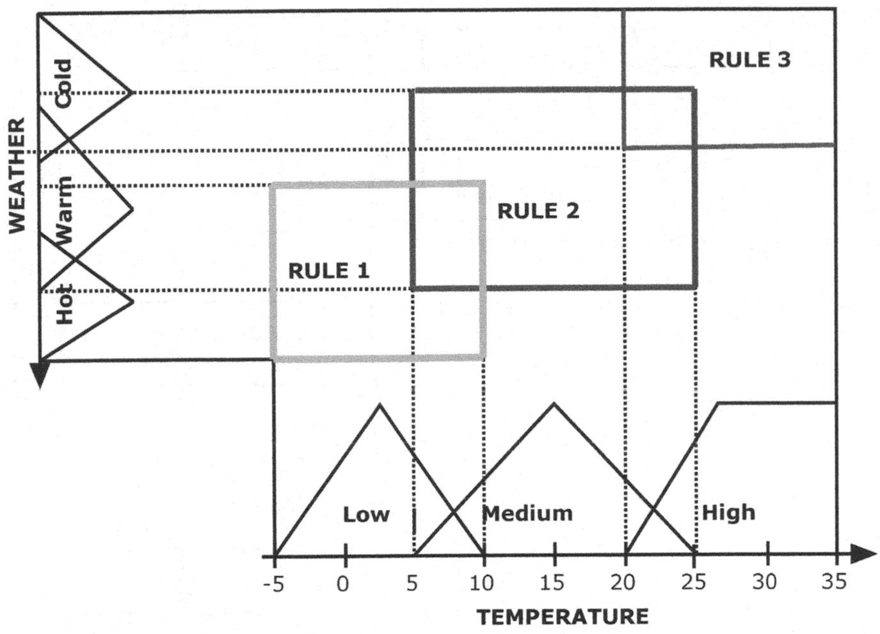

Fig. 9.13 Example of fuzzy rules.

9.4 Fuzzy Inference

The process of fuzzy reasoning is incorporated into what is called a Fuzzy Inferencing System (FIS). It is comprised of several steps (see Figure 9.14):

Step 1 - Define Fuzzy Sets
Step 2 - Relate Observations to Fuzzy Sets
Step 3 - Define Fuzzy Rules
Step 4 - Evaluate Each Case for all Fuzzy Rules
Step 5 - Combine Information from Rules
Step 6 - Defuzzify Results

Three steps are of interest at this stage:

1) Fuzzification
2) Rule Evaluation
3) Defuzzification.

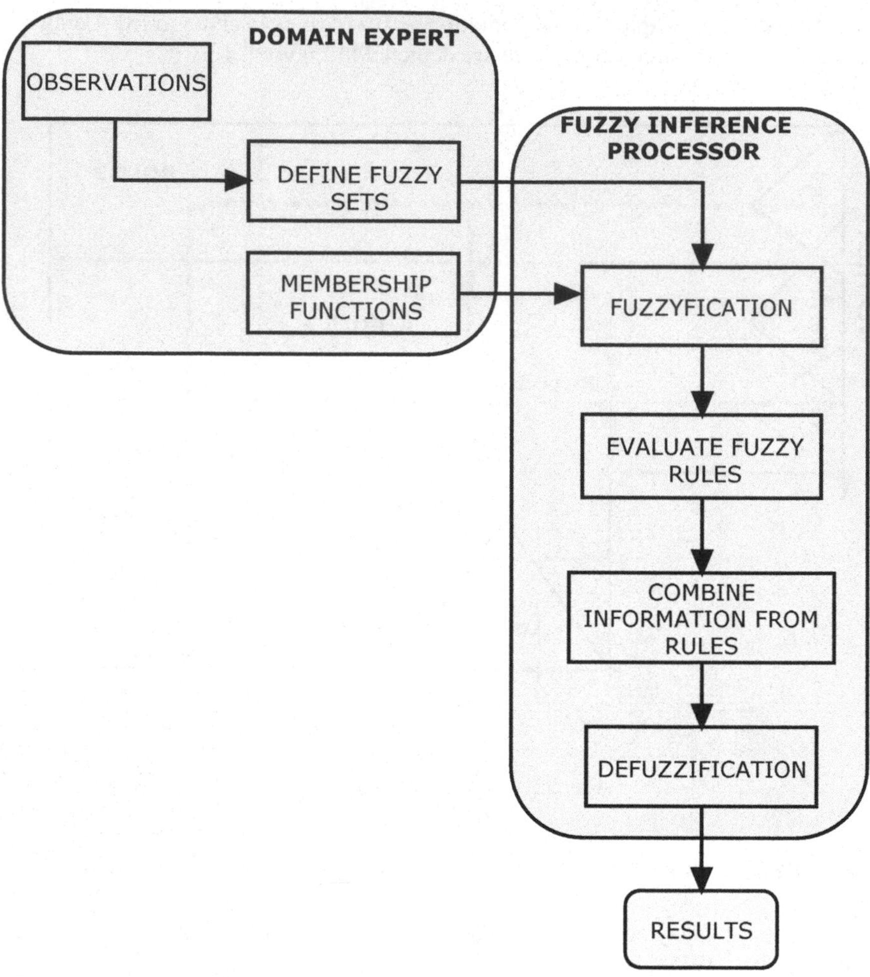

Fig. 9.14 The scheme of a fuzzy inference system.

9.4.1 Fuzzyfication

The first step in the fuzzy inference process is ***fuzzification***. During this step, the standard (ordinary inputs) are transformed into fuzzy inputs. Each ordinary (crisp) input has its own group of membership functions or sets to which they are transformed. This group of membership functions exists within a universe of discourse that holds all relevant values that the crisp input can possess. Figure 9.15 shows an example of membership functions within a universe of discourse for an ordinary input. The universe of discourse is divided into five fuzzy sets and the associated membership function (triangular or trapezoidal form) are depicted.

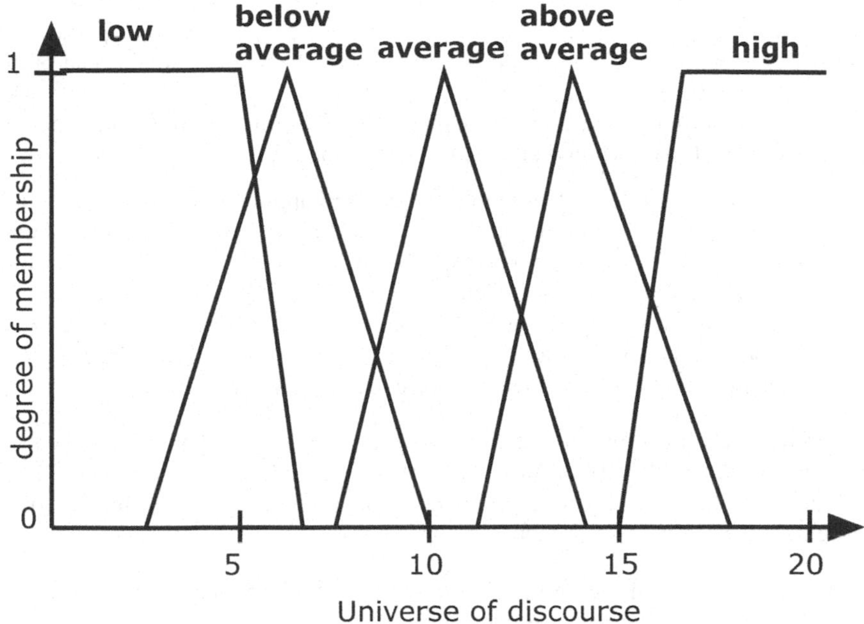

Fig. 9.15 Example of membership functions.

When designing the number of membership functions for an input variable, labels must initially be determined for the membership functions. The number of labels correspond to the number of regions that the universe should be divided, such that each label describes a region of behavior. A scope must be assigned to each membership function that numerically identifies the range of input values that correspond to a label. Choosing the number of labels is an important point for a fuzzy system. If the fuzzy labels are correctly chosen, then the fuzzy discretization does not lead to loose of information.

Fuzzy quantization is possible not only for numerical variables but for qualitative variables such as truth values. In this case, fuzzy membership functions will be represented on a scale of truthfulness. In order to represent a single real value, or a function or a set by a corresponding fuzzy membership function, the following relation is used: if we have a function $f: X \rightarrow Y$ between two crisp sets X and Y and if we know the membership function μ_A of a subset $A \subseteq X$, then the fuzzy representation of $f(A)$ in Y is given by [7]:

$$\mu_{f(A)}(f(x)) = \mu_A(x).$$

The *scope* or domain of a membership function represents the width of the membership function, the range of concepts, usually numbers, over which a membership function is mapped. The shape of the membership function should be representative of the variable. However this shape is also restricted by the computing resources available. Complicated shapes require more complex descriptive equations or large lookup tables. Choosing a standard type of membership function resembles choosing the Gaussian probability distribution for the conditional probabilities in the Bayesian theory [7].

When considering the number of membership functions within the universe of discourse, the following should be taken into account [2]:

i) too few membership functions for a given application will cause the response of the system to be too slow and fail to provide sufficient output control in time to recover from a small input change. This may also cause oscillation in the system.

ii) too many membership functions may cause rapid firing of different rule consequents for small changes in input, resulting in large output changes, which may cause instability in the system.

The membership functions should also be overlapped. No overlap reduces the system to a system based on Boolean logic.

Marsh [9] noted some interesting points which should be taken into account while defining the domain of membership functions:

- Every point in the universe of discourse should belong to the domain of at least one membership function.
- Two membership functions can not have the same point of maximal meaningfulness (1).
- When two membership functions overlap, the sum of membership grades for any point in the overlap should be less than or equal to 1.
- When two membership functions overlap, the overlap should not cross the point of maximal meaningfulness of either membership function.

Marsh has proposed two indices to describe the overlap of membership functions quantitatively. These are *overlap ratio* and *overlap robustness* as illustrated (together with their meaning) in Figure 9.16.

$$Overlap\,ratio = \frac{overlap\ scope}{adjacent\ membership\ function\ scope}$$

$$Overlap\,robustness = \frac{area\ of\ summed\ overlap}{maximum\ area\ of\ summed\ overlap} = \frac{\int_L^U (\mu_1 + \mu_2)dx}{2 \cdot (U - L)}$$

The fuzzification process maps each crisp input on the universe of discourse, and its intersection with each membership function is transposed onto the μ (degree of membership) axis. The μ values are the degrees of truth for each crisp input and are associated with each label as fuzzy inputs. These fuzzy inputs are then passed on to the next step, Rule Evaluation.

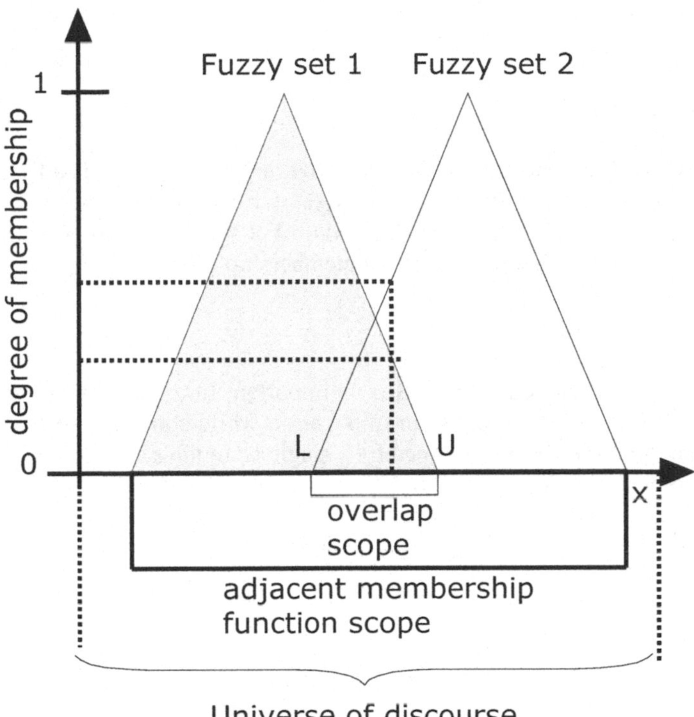

Fig. 9.16 Example of overlap indices.

9.4.2 Rule Evaluation and Inference

A fuzzy rule is expressed as:

IF antecedent
THEN consequent

The antecedent consists of input variable label and is equal to its associated fuzzy input or truth value $\mu(x)$. The consequent consists of output variable label.

As in the case of rule based expert systems, a fuzzy rule can have multiple antecedents joined by logical operators AND or OR.

The logical operator AND represents the intersection or *minimum* between the two sets, expressed as:

$$\mu_{A\cap B}(x) = \min\{\mu_A(x), \mu_B(x)\}$$

The logical operator OR represents the union or *maximum* between the two sets, expressed as:

$$\mu_{A\cup B}(x) = \max\{\mu_A(x), \mu_B(x)\}$$

Negation, or the logical operator NOT represents the opposite of the set, expressed as:

$$\mu_{\neg A}(x) = 1 - \mu_A(x)$$

Fuzzy *relations* link two fuzzy sets in a predefined manner. If A is a fuzzy set defined over the universe X and B is a fuzzy set defined over the universe Y, then a fuzzy relation R(A, B) is a fuzzy set defined on $X \times Y = \{(x, y) / x \in X, y \in Y\}$.

A fuzzy relation is characterized by a membership function [7]:

$$\mu_{R(x, y)} : X \times Y \rightarrow [0, 1].$$

Fuzzy *implication* denoted as A → B is an important fuzzy relation. In fuzzy logic there are different ways to define an implication while compared to propositional logic where the implication is defined by a single truth table.

For example, the rule:

IF x is A
THEN y is B

can be described as a relation by:

$$R(x, y) = \sum_{x_i, y_i} \frac{\mu(x_i, y_i)}{(x_i, y_i)}$$

$$R(x, y) = \int_{x_i, y_i} \frac{\mu(x_i, y_i)}{(x_i, y_i)}$$

where $\mu(x,y)$ is the relation we want to discover.

There are over 40 implication relations reported in the literature.

There are two ways of interpreting the implication P → Q:

(i) P is coupled to Q and the implication is a T-norm operator

- Examples:
  - Mamdani

$$R(x_i, y_i) = \sum_{x_i, y_i} \mu_A(x_i) \wedge \mu_B(y_i) / (x_i, y_i)$$

  - Larson

$$R(x_i, y_i) = \sum_{x_i, y_i} \mu_A(x_i) \times \mu_B(y_i) / (x_i, y_i)$$

- Bounded Difference

$$R(x_i, y_i) = \sum_{x_i, y_i} 0 \vee (\mu_A(x_i) + \mu_B(y_i) - 1) / (x_i, y_i)$$

(ii) P entails Q and implications are generalizations of the material implications in two-valued logic as in:

- $a \rightarrow b = \neg a \vee b$

$$R(x_i, y_i) = \sum_{x_i, y_i} (1 - \mu_A(x_i)) \vee \mu_B(y_i) / (x_i, y_i)$$

- $a \rightarrow b = \neg a \vee (a \wedge b)$

$$R(x_i, y_i) = \sum_{x_i, y_i} (1 - \mu_A(x_i)) \vee (\mu_A(x_i) \wedge \mu_B(y_i)) / (x_i, y_i)$$

- Examples:
  - Goguen

$$R(x, y) = \begin{cases} 1, & \text{if } \mu_A(x) \leq \mu_B(x) \\ \dfrac{\mu_A(x)}{\mu_B(x)}, & \text{if } \mu_A(x) > \mu_B(x) \end{cases}$$

  - Kurt Godel

$$R(x, y) = \begin{cases} 1, & \text{if } \mu_A(x) \leq \mu_B(x) \\ \mu_B(x), & \text{if } \mu_A(x) > \mu_B(x) \end{cases}$$

Fuzzy inference refers to computational procedures used for evaluating fuzzy rules of the form:

IF x is A
THEN y is B

There are two important inferencing procedures

- Generalized modus ponens (GMP) – mode that affirms
- Generalized modus tollens (GMT) – mode that denies

There are a few well known inference systems (we will give some examples later in this section):

- Mamdani Fuzzy models
- Sugeno Fuzzy models
- Tsukamoto Fuzzy models

9.4.3 Defuzzyfication

During the deffuzification process, the fuzzy output of the inference engine is converted to crisp values using membership functions analogous to the ones used by the fuzzifier.

In the case of crisp inputs and outputs, a fuzzy inference system implements a nonlinear mapping from its input space to output space.

There are five commonly used defuzzifying methods (the meaning of A and Z in the formulas below are considered as given in Figure 9.17):

- Centroid of area (COA):

$$COA = \frac{\int_Z \mu_A(z) \cdot z \, dz}{\int_Z \mu_A(z) \, dz}$$

If Z is limited by *a* and *b*, then COA can be written as:

$$COA = \frac{\int_a^b \mu_A(z) \cdot z \, dz}{\int_a^b \mu_A(z) \, dz}$$

which can be further expressed as:

$$COA = \frac{\sum_{z=a}^{b} \mu_A(z) \cdot z}{\sum_{z=a}^{b} \mu_A(z)}$$

- Bisector of area (BOA)

$$BOA = \int_{\alpha}^{BOA} \mu_A(z) \, dz = \int_{BOA}^{\beta} \mu_A(z) \, dz$$

- Mean of maximum (MOM)

$$MOM = \frac{\int_{Z'} z\, dz}{\int_{Z'} dz}$$

where $Z' = \{z \,/\, \mu_A(z) = \mu^*)$

- Smallest of maximum (SOM)
- Largest of maximum (LOM)

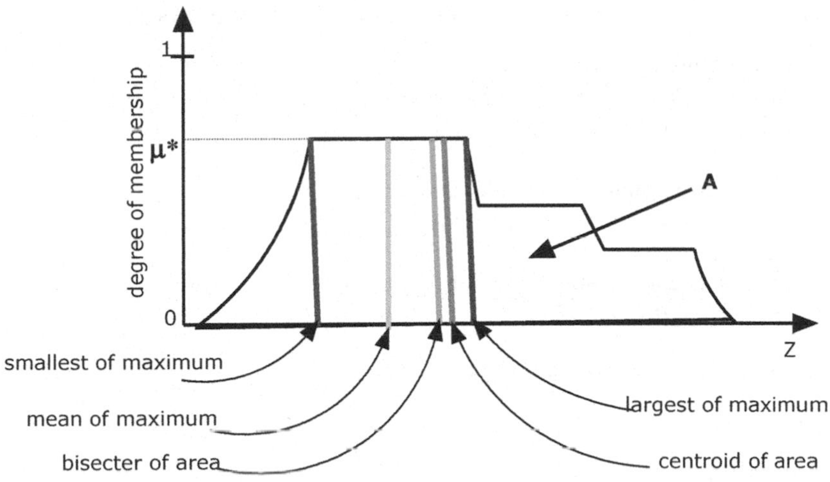

Fig. 9.17 Deffuzification examples.

9.4.4 Mamdani Fuzzy Model

Mamdani fuzzy model has been proposed in 1975 by E. Mamdani [14]. It is one of the first fuzzy systems and one of the most used ones.

In order to see how Mamdani method works, we consider a simple example: a system consisting of the three rules below:

Rule1:

IF	temperature is low
AND	wind blowing is strongly
THEN	weather is cold

Rule2:

IF	temperature is medium
OR	wind blowing is gentle
THEN	weather is average

Rule3:

IF	temperature is high
OR	wind blowing is gentle
THEN	weather is hot

Mamdani fuzzy inference process consists on the four standard steps: fuzzyfication of input variables, rule evaluation, aggregation of the results and defuzyfication.

We have three linguistic variables: temperature, wind and weather. The linguistic values determined by the fuzzy sets of the variable temperature are low, medium and high. Similarly, the linguistic values determined by the fuzzy sets of the variable wind are strongly and gentle and for the variable weather are cold, average and hot.

In order to simplify the representation let us use the following notations:

- temperature : x
  fuzzy sets low, medium, high: A1, A2, A3
  universe of discourse: X
- wind: y
  fuzzy sets strongly and gentle: B1, B2
  universe of discourse: Y
- weather: z
  fuzzy sets cold, average and hot: C1, C2, C3
  universe of discourse: Z

Thus, the rules above can be simply rewritten as:

Rule1:

IF	x is A1
AND	y is B1
THEN	z is C1

Rule2:

IF	x is A2
OR	y is B2
THEN	z is C2

Rule3:

IF	x is A3
OR	y is B2
THEN	z is C3

The input data are: temperature value is 25 degrees and the wind speed is 35 km/h.

Step 1. Fuzzyfication

The graphical representation of the fuzzy sets is given in Figure 9.18.

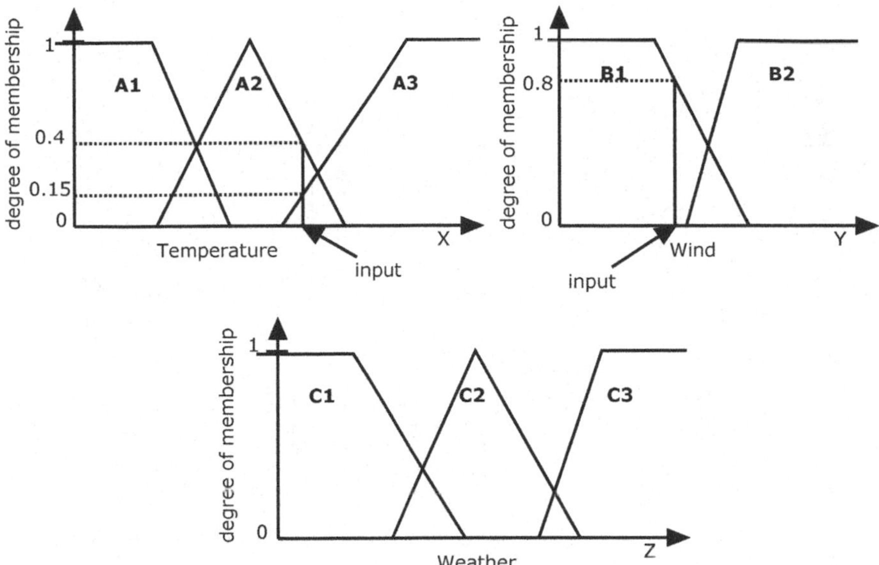

Fig. 9.18 Fuzzy sets for the Mamdani example.

For the input variables, we obtain the following membership values:
$\mu_{A1}(x)=0$;
$\mu_{A2}(x)=0.4$;
$\mu_{A3}(x)=0.15$;

$\mu_{B1}(y)=0.8$;
$\mu_{B2}(y)=0$;

Step 2. Rules evaluation

We have three rules; all of them will be fired. The order is Rule 1, then Rule 2 and in the end Rule 3. Rule evaluation and the results obtained are depicted in Figure 9.19.

Step 3. Results aggregation

We have obtained the following results:

$\mu_{C1}(z)=0$;
$\mu_{C2}(z)=0.4$;
$\mu_{C3}(x)=0.15$.

By aggregating these we obtain the results given in Figure 9.20.

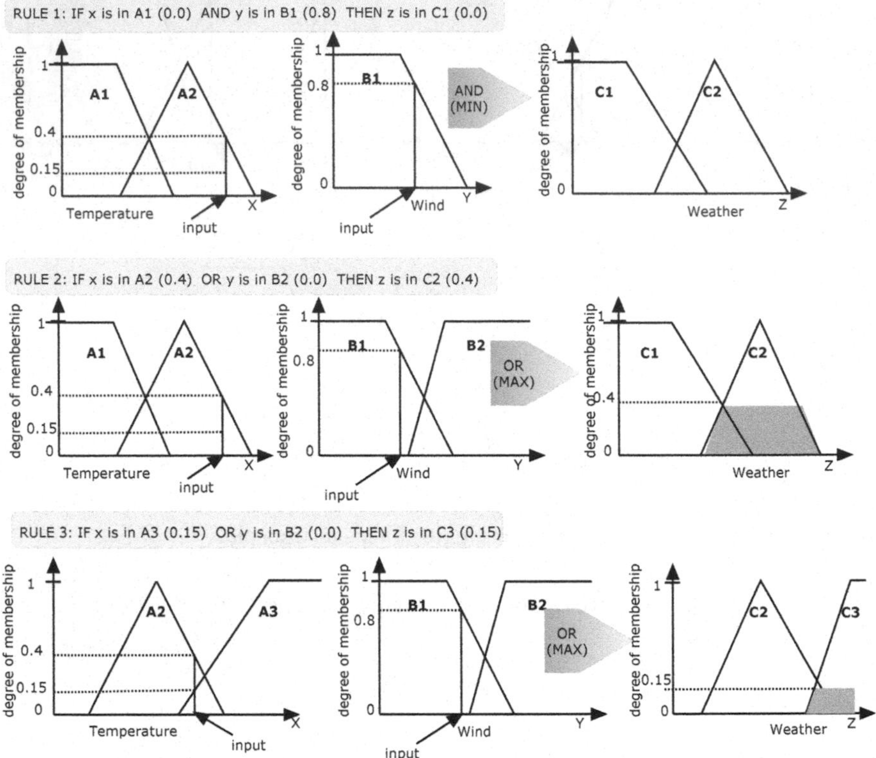

Fig. 9.19 Rule evaluation following Mandani fuzzy model.

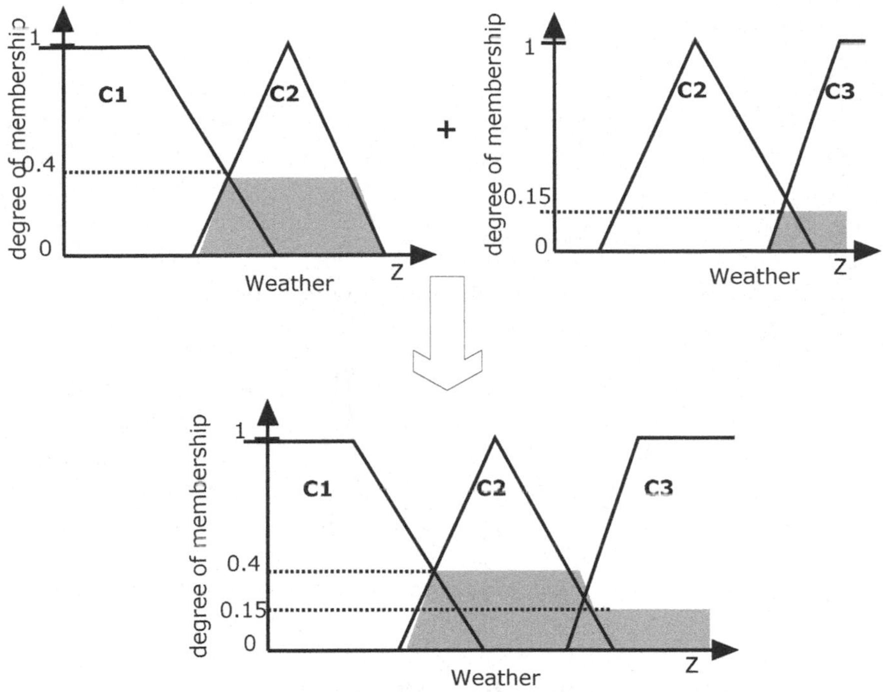

Fig. 9.20 Results aggregation.

Step 4. Defuzzyfication

We have now obtained the fuzzy results and have to defuzzyfy them to get a crisp output. Using the COG formula, the crisp output is calculated as (see Figure 9.21):

$$COG = \frac{(0+10+20+30)\cdot 0.0+(40+50+60+70)\cdot 0.4+(80+90+100)*0.15}{0+0+0+0+0.4+0.4+0.4+0.4+0.15+0.15+0.15}$$
$$= \frac{88+40.5}{2.05} = \frac{128.5}{2.05} = 62.68$$

The interpretation of the output may be that the weather is hot with a 62.68%.

9.4.5 Sugeno Fuzzy Model

Sugeno (or Takagi-Sugeno-Kang) method of fuzzy inference has been introduced in 1985 [15][16][21][22][25] and it is similar to the Mamdani method in many respects. The first two parts of the fuzzy inference process, fuzzifying the inputs and applying the fuzzy operator, are exactly the same. The main difference between Mamdani and Sugeno is that the Sugeno output membership functions are either linear or constant.

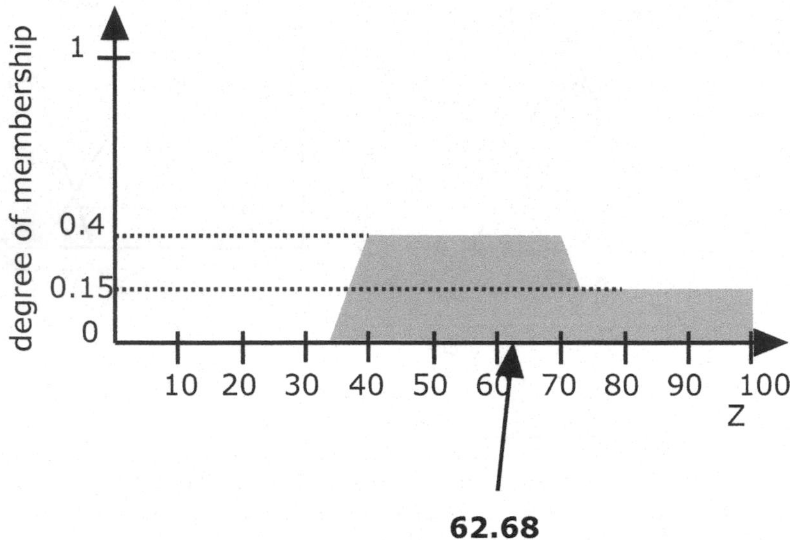

Fig. 9.21 Defuzzyfication.

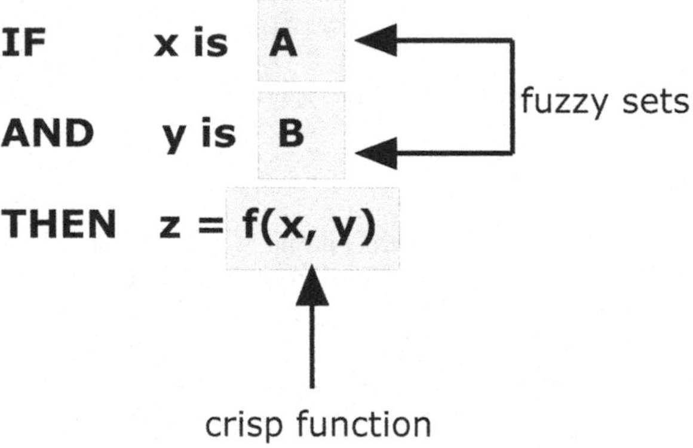

Fig. 9.22 A rule in a Sugeno fuzzy model.

A typical rule in a Sugeno fuzzy model has the form given in Figure 9.22, where the crisp function f(x, y) is a polynomial function of x and y.

If f is of the form:

$ax+by+c$

then we have a first order Sugeno model.

If the output z is a constant then we have a zero-order Sugeno model. In this case, all the membership functions corresponding to the consequent are represented by singletons. Mandani and Sugeno fuzzy models are very similar, with just a small difference regarding the consequent. The reasoning process is also very similar for both methods.

In order to show how Sugeno model works, we consider the same example as in the case of Mandani style.

Step 1 (Fuzzyfication) and Step 2 (Rule evaluation) of the Mandani model are same for Sugeno model, with a different style for output representation as it can be seen from Figure 9.23). The output representation also induces differences in the Rule aggregation (Step 3, depicted in Figure 9.24) and Defuzzyfication (Step 4) steps.

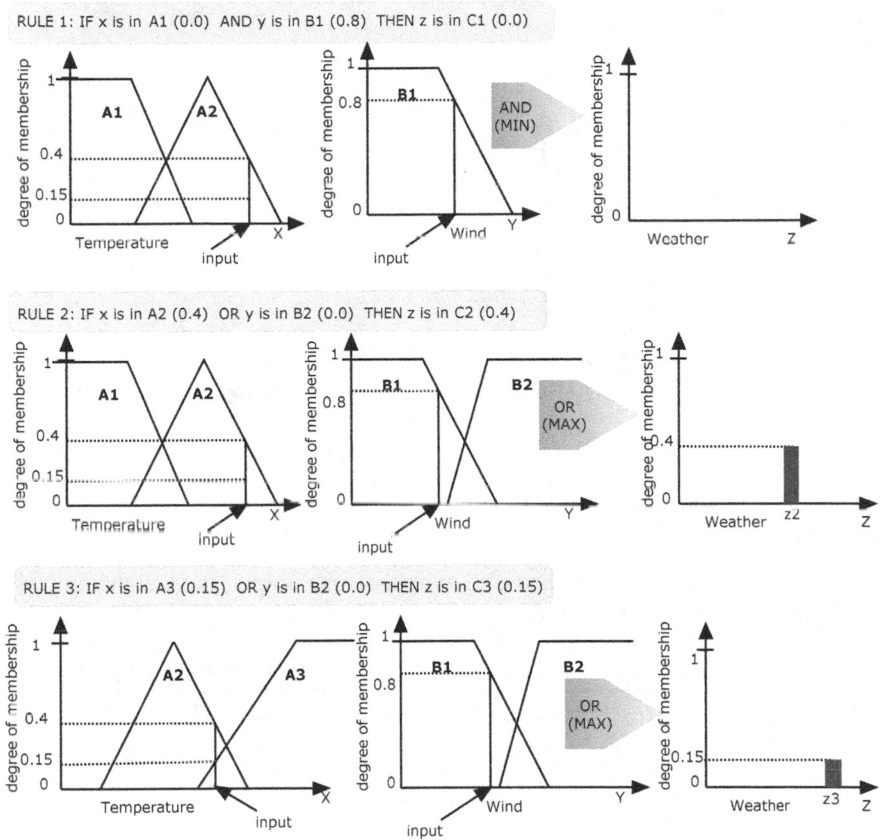

Fig. 9.23 Rule evaluation in Sugeno fuzzy model.

Rule aggregation consists in simply summing z1 (which is 0 for this example), z2 and z3 as it can be seen in Figure 9.24.

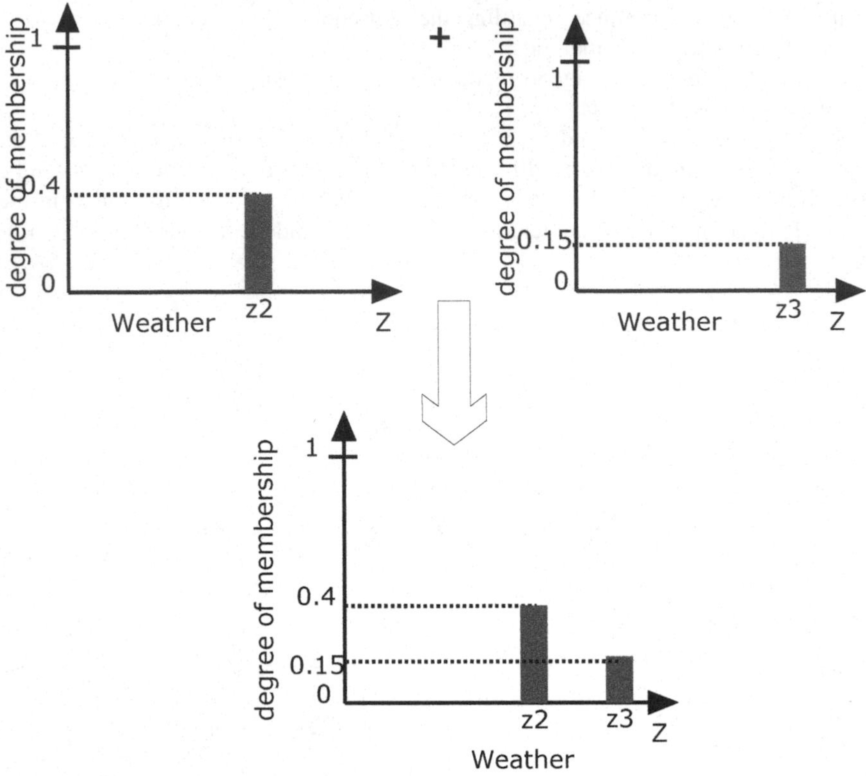

Fig. 9.24 Agregation of the results for Sugeno model.

For defuzzyfying the results and obtaining the crisp output, the weighted average of the singletons is considered:

$$Crisp\,output = \frac{\mu(z1)\cdot z1 + \mu(z2)\cdot z2 + \mu(z3)\cdot z3}{\mu(z1)+\mu(z2)+\mu(z3)} = \frac{0+0.4\cdot 0.55+0.15\cdot 0.75}{0+0.4+0.15}$$

$$= \frac{0.22+0.11}{0.55} = \frac{0.33}{0.55} = 0.6$$

9.4.6 Tsukamoto Fuzzy Model

In Tsukamoto fuzzy model, the consequent of each fuzzy IF-THEN rule is represented by a fuzzy set with a monotonic membership function fuzzy set.

As a result, the inferred output is a crisp value induced by the rule's firing strength. The overall (final) output is obtained by taking the weighted average of each rule's output.

This fuzzy model spends less time for defuzzyfication (same like Sugeno model) but it is less transparent compared to Mandani or Sugeno models and also not so popular [23] [24].

We consider the same example as in the case of Mandani and Sugeno models and show how Tsukamoto model works.

Fuzzyfication and defuzzyfication steps are similar to Sugeno model. Rules evaluation and aggregation of the results are presented in Figure 9.25 and Figure 9.26 respectively.

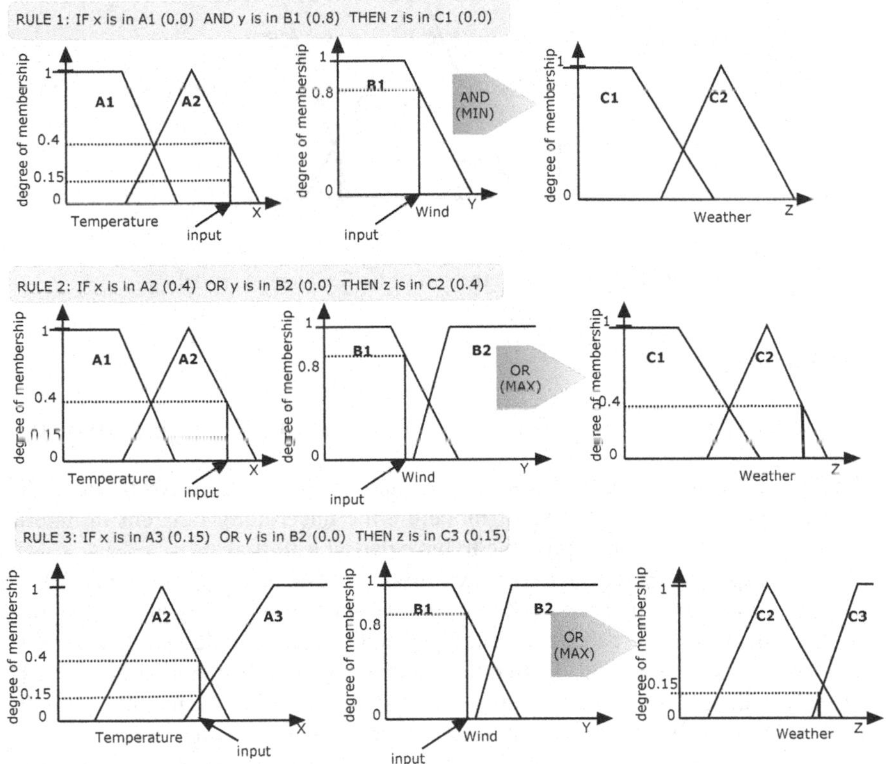

Fig. 9.25 Rule evaluation in Tsukamoto fuzzy model.

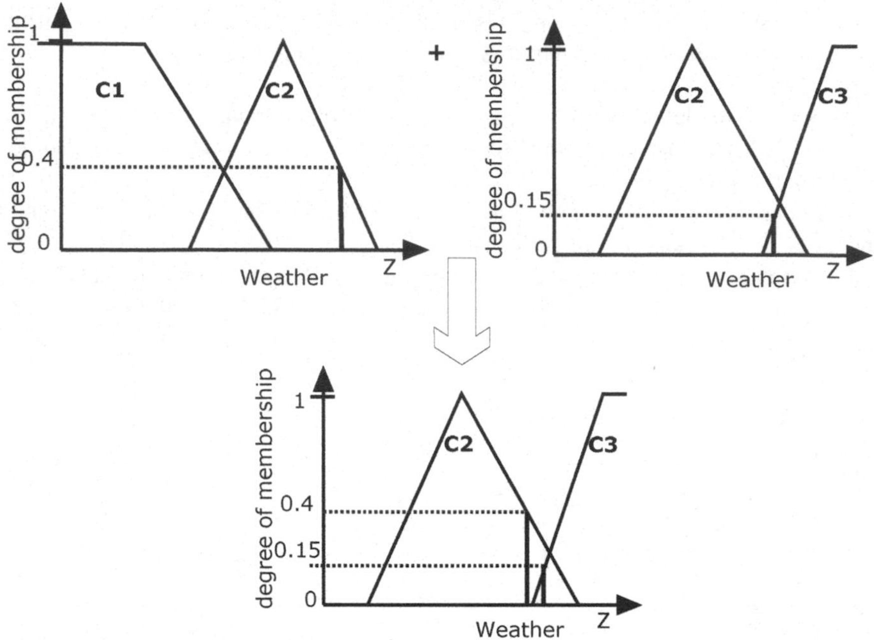

Fig. 9.26 Agregation of the results for Tsukamoto model.

Summaries

This chapter introduces fuzzy logic and all its elements and then fuzzy expert systems.

Fuzzy logic is a superset of Boolean logic and has been introduced by Lotfi Zadeh in the 1960s for the purpose of modeling the uncertainty inherent in natural language. In fuzzy logic, it is possible to have partial truth-values.

A fuzzy expert system is an expert system that uses fuzzy logic instead of Boolean logic. Fuzzy logic is used to define rules of inference, and membership functions that allow a expert system to draw conclusions. In other words, a fuzzy expert system is a collection of membership functions and rules that are used to reason about data. Unlike conventional expert systems, which are mainly symbolic reasoning engines, fuzzy expert systems are oriented toward numerical processing.

Even thought proposed in 1965, fuzzy sets were for a long time not accepted by the AI community. Now they have become highly evolved and their techniques are well established. Inference process in fuzzy expert systems has four steps: fuzzyfication, rule evaluation, aggregation of the results and defuzzyfication.

Several methods of inferencing in fuzzy systems have been proposed. In this chapter three of them are presented: Mamdani model, Sugeno model and Tsukamoto model.

In fuzzy inference systems, rules can be viewed as a set of fuzzy points, which as a whole approximate the compatibility relation.

Zero-order polynomial function in the case of Sugeno model can be viewed either as a special case of Mamdani model in which each rule's consequent is specified by a fuzzy singleton or a special case of Tsukamoto model, in which each rule's consequent, is specified by a membership function of a step function crossing at the constant. Experts rely on common sense when they solve problems. Fuzzy logic reflects how people think. It attempts to model our decision making, and our common sense and thus it leads to new, more human, intelligent systems.

Fuzzy expert systems are the most common use of fuzzy logic. They are used in several wide-ranging fields, including:

- Linear and nonlinear control;
- Pattern recognition;
- Financial systems;
- Operations research;
- Data analysis.

There are several practical applications of fuzzy control systems, among them:

- Fuzzy car;
- Fuzzy logic chips and fuzzy computers;
- Fuzzy washing machine;
- Fuzzy vacuum cleaner;
- Fuzzy air conditioner;
- Fuzzy camcorder;
- Fuzzy Automatic Train Operation systems;
- Fuzzy automatic container crane operations;
- Etc.

Some of the advantages of using fuzzy systems are:

- easy to develop and debug
- easy to understand
- easy and cheap to maintain

References

1. Siler, W., Buckley, J.J.: Fuzzy Expert Systems and Fuzzy Reasoning. Wiley and Sons, Chichester (2005)
2. `http://www.onlinefreeebooks.net/free-ebooks-computer-programming-technology/artificial-intelligence/artificial-intelligence-course-material-pdf.html` (accessed on February 10, 2011)
3. Łukasiewicz, J.: Aristotle's Syllogistic from the Standpoint of Modern Formal Logic. Clarendon Press, Oxford (1951)
4. Lukasiewicz, J.: Elements of Mathematical Logic, 2nd edn. Macmillan, New York (1964)
5. `http://mathworld.wolfram.com/` (accessed on February 10, 2011)

6. Zadeh, L.A.: Fuzzy sets. Information and Control 8, 338–353 (1965)
7. Kasabov, N.K.: Foundations of Neural Networks. Fuzzy Systems and Knowledge Engineering. MIT Press, Cambridge (1996)
8. Kosko, B.: Fuzzy Thinking: The New Science of Fuzzy Logic. Hyperion, New York (1993)
9. Marsh, S., Huang, W.Y., Sibigtroth, J.: Center for Emerging Computer Technologies, Motorola, Inc.: "Fuzzy Logic Program 2.0" (1994)
10. Altrock, C.: Fuzzy Logic and Neuro Fuzzy Applications Explained. Prentice-Hall, Englewood Cliffs (1995)
11. Cox, E.: The fuzzy systems handbook: a practitioner's guide to building and maintaining fuzzy systems. Elsevier Science & Technology Books, Amsterdam (1994)
12. Li, H., Gupta, M.: Fuzzy logic and intelligent systems. Kluwer Academic Publishers, Boston (1995)
13. Kandel, A.: Fuzzy expert systems. CRC Press, Boca Raton (1992)
14. Mamdani, E.H., Assilian, S.: An experiment in linguistic synthesis with fuzzy logic controller. International Journal of Man-Machine Studies 7(1), 1–13 (1975)
15. Sugeno, M.: Industrial Applications of fuzzy control. Elsevier Science Ltd., Amsterdam (1985)
16. Terano, T., Asai, K., Sugeno, M.: Fuzzy Systems Theory and Its Applications. Academic Press, London (1992)
17. Driankov, D., Hellendoorn, H., Reinfrank, M.: An Introduction to Fuzzy Control. Springer, Heidelberg (1993)
18. Zadeh, L.A.: Fuzzy Logic. Computer 1(4), 83–93 (1988)
19. Zadeh, L.A., Knowledge, L.A.: representation in fuzzy logic, . IEEE Transactions on Knowledge and Data Engineering 1, 89–100 (1989)
20. Dubois, D.: H. Prade, . In: Fuzzy Sets and Systems: Theory and Applications. Academic Press, New York (1980)
21. Sugeno, M., Kang, G.T.: Structure identification of fuzzy model. Fuzzy Sets and Systems 28, 15–33 (1988)
22. Takagi, T., Sugeno, M.: Fuzzy identification of systems and its application to modeling and control. IEEE Transactions on Systems, Man and Cybernetics 15(1), 116–132 (1985)
23. Jang, J.-S.R., Sun, C.T., Mizutani, E.: Neuro-Fuzzy and Soft Computing: A Computational Approach to Learning and Machine Intelligence. MATLAB Curriculum Series. Prentice Hall, Upper Saddle River (1997)
24. Tsukamoto, Y.: An approach to fuzzy reasoning method. In: Advances in Fuzzy Set Theory and Applications, pp. 137–149. North-Holland, Amsterdam (1979)
25. Sugeno, M.: Fuzzy measures and fuzzy integrals: A survey. In: Gupta, M., Saridis, G.N., Gaines, B.R. (eds.) Fuzzy Automata and Decision Processes, pp. 89–102. North-Holland, New York (1977)

Verification Questions

1. How does fuzzy logic differ from classical logic?
2. How to represent fuzzy sets?
3. What are the proprieties of fuzzy logic operators?
4. What are the main operations which can be performed with fuzzy sets?

5. What do hedges represent? Give some examples.
6. What are the main steps of fuzzy inference process?
7. Give examples of fuzzyfication methods.
8. Give examples of defuzzyfication methods.
9. Explain Mandani, Sugeno and Tsukamoto inference models and specify the main differences between them.
10. Give examples of practical applications of fuzzy systems.
11. What are the advantages and disadvantages of fuzzy expert systems.

Exercises

1. Given the variable weight (of a person), create 3 fuzzy sets – underweighted, normal, overweighted – corresponding to it. Then, given a person's weight, calculate the degree of membership to each of the fuzzy sets.

2. Given the variables weight and height, define three fuzzy sets for each of them.
Then, knowing that a person is overweight to a degree of 0.6 and tall to a degree of 0.3, calculate the degrees that the person is:

1) overweight and tall
2) overweight or tall
3) normal and short
4) normal and tall
5) normal or tall
6) not overweight
7) not tall
8) not (overweight and tall)

3. Fuzzy controller for setting a thermostat value:
Given the linguistic variables "outside temperature, amount of time spent at home, thermostat value":

1) Determine by fuzzy sets on universe of discourse for each variable their linguistic values
2) Use a Mamdani-style fuzzy inference (you may consider some crisp inputs as of your choice)
    a. Mention all the steps required and the operations performed at each step
    b. Use diagrams to show the rules inferences.
3) Explain the results

4. Fuzzy controller for setting the speed of a car:

Given the linguistic variables "weather condition, car type, car speed, road quality":

1) Determine by fuzzy sets on universe of discourse for each variable their linguistic values

2) Use a Sugeno-style fuzzy inference (you may consider some crisp inputs as of your choice)
    a. Mention all the steps required and the operations performed at each step
    b. Use diagrams to show the rules inferences.
3) Explain the results.

5. Fuzzy controller for setting the final grade of a student giving admission to a college based on his overall high school results and on the results of two given exam:

Given the linguistic variables "overall high school results, exam1 result, exam 2 result, final grade"

1) Determine by fuzzy sets on universe of discourse for each variable their linguistic values
2) Use a Tsukamoto-style fuzzy inference (you may consider some crisp inputs as of your choice)
    a. Mention all the steps required and the operations performed at each step
    b. Use diagrams to show the rules inferences.
3) Explain the results.

6. Fuzzy controlled for defining the number of professors required for dealing with students for their high school graduation exam:

Given the linguistic variables "number of students, number of different exams, professional quality (of the professor, in general), number of professors"

1. Determine by fuzzy sets on universe of discourse for each variable their linguistic values
2. Use a Sugeno-style fuzzy inference and a Mandani –style fuzzy inference (you may consider some crisp inputs as of your choice)
    2.1 Mention all the steps required and the operations performed at each step
    2.2 Use diagrams to show the rules inferences.
    2.3 Use at lest two deffuzyfication methods for each of the Mamdani and Sugeno models.
3. Compare and explain the results.

Chapter 10
Machine Learning

10.1 Introduction

Machine Learning[6][8][12] is concerned with the study of building computer programs that automatically improve and/or adapt their performance through experience. Machine learning can be thought of as "programming by example" [11]. Machine learning has many common things with other domains such as statistics and probability theory (understanding the phenomena that have generated the data), data mining (finding patterns in the data that are understandable by people) and cognitive sciences (human learning aspire to understand the mechanisms underlying the various learning behaviors exhibited by people such as concept learning, skill acquisition, strategy change, etc.) [1].

The goal of machine learning is to devise learning algorithms that do the learning automatically without human intervention or assistance. Rather than program the computer to solve the task directly, in machine learning, we seek methods by which the computer will come up with its own program based on examples that we provide [11].

Dietterich [1] mentioned 4 situations in which it is not easy for software engineers to design the software for solving a problem, but there are more similar situations:

- problems for which there exist no human experts. As an example, the need to predict machine failures before they occur in modern automated manufacturing facilities. This can be performed by analyzing sensor readings. There are no human experts who can be interviewed by a programmer to provide the knowledge necessary to build a computer system. A machine learning system can study recorded data and subsequent machine failures and learn prediction rules.
- problems where human experts exist, but where they are unable to explain their expertise. This in domains such as speech recognition, handwriting recognition, and natural language understanding. Experts cannot describe the detailed steps that they follow as they perform them. Humans can provide machines with examples of the inputs and correct

C. Grosan and A. Abraham: Intelligent Systems, ISRL 17, pp. 261–268.
springerlink.com

outputs for these tasks, so machine learning algorithms can learn to map the inputs to the outputs.

- dynamic problems where phenomena are changing rapidly. Many a times, people would like to be able predict the future behavior of certain phenomena such as the stock market, exchange rates or even weather forecast. These behaviors change frequently, so that even if a programmer could construct a good predictive computer program, it would need to be rewritten frequently. A learning program can relieve the programmer of this burden by constantly modifying and tuning a set of learned prediction rules.
- applications that need to be customized for each computer user separately. For instance, a program to filter unwanted electronic mail messages. Different users will need different filters. A machine learning system can learn which mail messages the user rejects and maintain the filtering rules automatically.

Some examples of machine learning problems are [3][4][5][9][10][11]:

- character (including digit) recognition
- handwriting recognition
- face detection
- spam filtering
- sound recognition
- spoken language understanding
- stock market prediction
- weather prediction
- medical diagnosis
- fraud detection
- fingerprint matching
- etc.

A computer program is said to *learn* from experience *E* with respect to some class of tasks *T* and performance measure *P*, if its performance at tasks in *T*, as measured by *P*, improves with experience *E* [2].

A learning system is characterized by the following elements:

- task (or tasks) T
- experience E
- performance measure P.

For example, a learning system for playing tic-tac-toe game (or nuggets and crosses) will have the following corresponding elements:

- T: Play tic-tac-toe;
- P: Percentage of games won (and eventually drawn);
- E: Playing against itself (can also be playing against others).

Thus, a generic learning system can be defined by the following components [2] (see Figure 10.1):

- *Goal:* Defined with respect to the task to be performed by the system;
- *Model:* A mathematical function which maps perception to actions;
- *Learning rules:* Used to update the model parameters with new experience in a way which optimizes the performance measures with respect to the goals. Learning rules help the algorithm search for the best model;
- *Experience:* A set of perception (and possibly the corresponding actions).

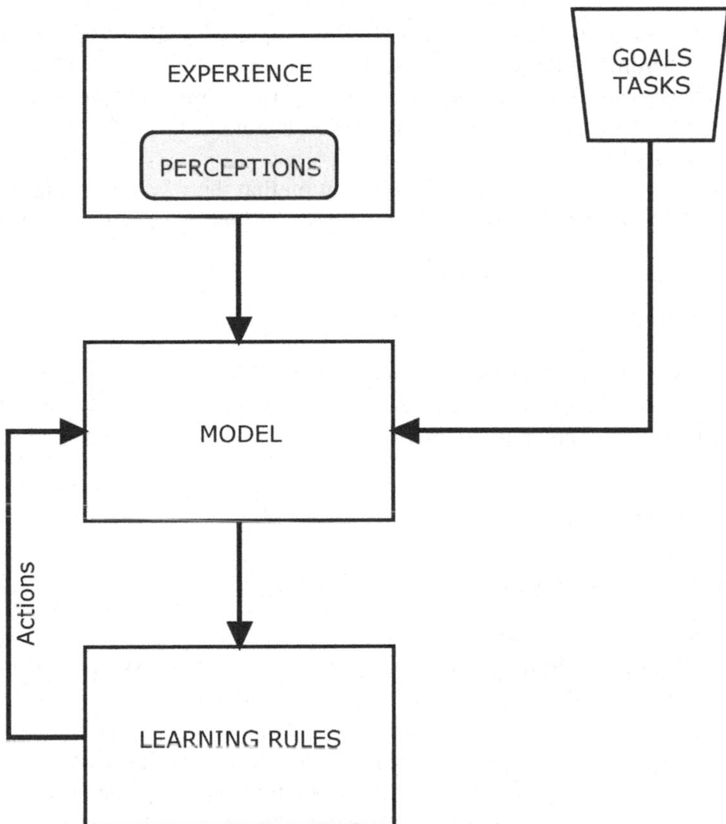

Fig. 10.1 Generic scheme of a learning system

10.2 Terminology

Before getting into the core of a learning system, we first need to define the basic notions [11].

An *example* (sometimes also called an *instance*) is the object that is being classified. For instance, if we consider the cancer tumor patients classification, then the patients are the examples.

An example is described by a set of *attributes*, also known as *features* or *variables*.

For instance, the cancer tumor patients classification, a patient might be described by attributes such as gender, age, weight, tumor size, tumor shape, blood pressure, etc.

The *label* is the category that we are trying to predict. For instance, in the cancer patient classification, the labels can be "avascular", "vascular" and "angiogenesis".

During training, the learning algorithm is supplied with *labeled examples*, while during testing, only *unlabeled examples* are provided.

In certain situation we can assume that only two labels are possible that we might as well call 0 and 1 (for instance, cancer or not cancer). This will make the things much simple.

We will also make the simplifying assumption that there is a mapping from examples to labels. This mapping is called a *concept*. Thus, a concept is a function of the form

$$c : X \rightarrow \{0, 1\}$$

where X is the space of all possible examples called the *domain* or *instance space*. A collection of concepts is called a *concept class*. We will often assume that the examples have been labeled by an unknown concept from a *known* concept class.

10.3 Learning Steps

The main steps in a learning process are as follows:

- *data and assumptions*
  Data refers to the data available for the learning task and assumptions represent what we can assume about the problem.
- *representation*
  We should define how to represent the examples to be classified. There are many representation methods for the same data. The choice of representation may determine whether the learning task is very easy or very difficult.
- *method and estimation*
  Method takes into account what are the possible hypotheses and estimation helps to adjust the predictions based on the feedback (such as updating the parameters when there is a mistake, etc).
- *evaluation*
  This evaluates how well the method is working (for instance, considering the ratio of wrong classified data and the whole dataset).

- *model selection*

  It tells us whether we can rethink the approach to do even better or to make it more flexible, or, whether we should choose an entirely different model that would be more suitable.

10.4 Learning Systems Classification

Learning systems can be classified based on their components.

10.4.1 Classification Based on Goal, Tasks, Target Function

Based on the goals of a learning system, we can have the following classification [2]:

- *Prediction*: the system predicts the desired output for a given input based on previous input/output pairs.

  Example: prediction of a stock value given other (input) parameters values like market index, interest rates, currency conversion, etc.

- *Regression*: the system estimates a function of many variables (multivariate) or single variable (univariate) from scattered data.

  Example: a simple univariate regression problem is $x^4+x^3+x^2+x+1$

- *Classification (categorization)*: the system classifies an object into one of several categories (or classes) based on features of the object.

  Example: A diagnosis system which has to classify a patient's cancer tumor into one of the three categories: avascular, vascular, angiogenesis.

- *Clustering*: the system task is to organize a group of objects into homogeneous segments.

  Example: a satellite image analysis system which groups land areas into forest, urban and water body, for better utilization of natural resources.

- *Planning*: the system has to generate an optimal sequence of actions to solve a particular problem.

  Example: robot path planning (to perform a certain task or to move from one place to another, etc).

Learning tasks can be classified (among others) in [1]:

- empirical learning and
- analytical learning.

Empirical learning is learning that relies on some form of external experience (the program cannot infer the rules of the game analytically - it must interact with a teacher to learn them), while analytical learning requires no external inputs (the program is able to improve its performance just by analyzing the problem).

10.4.2 Classification Based on the Model

The model is actually the algorithm used and there are several learning models:

- Decision trees
- Linear separators (perceptron model)
- Neural networks
- Genetic programming
- Evolutionary algorithms
- Graphical models
- Support vector machines
- Hidden Markov models

10.4.3 Classification Based on the Learning Rules

Learning rules are usually related with the model of learning used. Some common rules are:

- gradient descent
- least square error
- expectation maximization
- margin maximization.

10.4.4 Classification Based on Experience

The nature of experiences available varies with applications. Some common situations are [2] [7].

- *Supervised learning:*
  In supervised learning, the machine is given the desired outputs and its goal is to learn to produce the correct output given a new input.
  In supervised learning a teacher or oracle is available which provides the desired action corresponding to a perception. A set of perception action pair provides what is called a training set. Examples include an automated vehicle where a set of vision inputs and the corresponding steering actions are available to the learner.

- *Unsupervised learning*:
  In unsupervised learning the goal of the machine is to build a model of input that can be used for reasoning, decision making, predicting things, and communicating.
  In unsupervised learning no teacher is available. The learner only discovers persistent patterns in the data consisting of a collection of perceptions. This is also called exploratory learning. Finding out malicious network attacks from a sequence of anomalous data packets is an example of unsupervised learning.

- *Active learning:*
  Here not only a teacher is available, the learner has the freedom to ask the teacher for suitable perception-action example pairs which will help the learner to improve its performance. Consider a news recommender system which tries to learn a user's preference and categorize news articles as interesting or uninteresting to the user. The system may present a particular article (of which it is not sure) to the user and ask whether it is interesting or not.

- *Reinforcement learning*:
  In reinforcement learning, the machine can also produce actions which affect the state of the world, and receive rewards (or punishments). The goal is to learn to act in a way that maximizes rewards in the long term.
  In reinforcement learning a teacher is available, but the teacher instead of directly providing the desired action corresponding to a perception, return reward and punishment to the learner for its action corresponding to a perception. Examples include a robot in an unknown terrain where its get a punishment when its hits an obstacle and reward when it moves smoothly.

10.5 Machine Learning Example

Consider the data given in Table 10.1.

	INPUT						OUTPUT
Nr.	x_1	x_2	x_3	x_4	x_5	x_6	Y
1.	1	0	2	0	1	1	0
2.	0	2	0	2	3	0	1
3.	1	1	0	3	0	0	0
4.	0	0	3	1	0	1	0
5.	4	0	0	1	0	0	1
6.	1	1	2	3	0	0	1
7.	3	2	5	0	1	3	1
8.	2	1	0	0	2	3	1
9.	1	1	1	0	2	2	0
10.	1	1	2	1	2	1	0
11.	0	6	1	1	1	3	0
12.	1	0	0	2	0	0	1
13.	0	0	3	3	3	0	0
14.	0	3	1	1	1	0	1
15.	2	0	1	1	0	1	0
16	1	2	1	1	0	0	?

For this example we have:

- Space of all possible examples: X = (x_1, x_2, x_3, x_4, x_5, x_6)
- Instance space = |{X}| = 15
- Concept learning/regression: we have to find *f* such as $Y=f(X)$
- Prediction: $f(1, 2, 1, 1, 0, 0)$ =?
- Evaluation:

$$\text{Error}(f) = \frac{\left|\{X \,/\, f(X) \neq Y\}\right|}{15}$$

The following chapters will cover various topics of machine learning such as learning decision trees, neural network learning, statistical learning methods, evolutionary computation and reinforcement learning.

References

1. Dietterich, T.G.: Machine Learning. In: Nature Encyclopedia of Cognitive Science. Macmillan, London (2003)
2. http://www.onlinefreeebooks.net/free-ebooks-computer-programming-technology/artificial-intelligence/artificial-intelligence-course-material-pdf.html (accessed on February 10, 2011)
3. Jordan, M.: An introduction to Graphical Models, Center for Biological and Computational Learning, Massachusetts Institute of Technology (1997), http://www.cis.upenn.edu/~mkearns/papers/barbados/jordan-tut.pdf (accessed on February 10, 2011)
4. Duda, R., Hart, P., Stork, D.: Pattern Classification, 2nd edn. Wiley and Sons, New York (2001)
5. Bishop, C.M.: Neural Networks for Pattern Recognition. Oxford University Press, Oxford (1995)
6. Mitchell, T.: Machine Learning. McGraw-Hill, New York (1997)
7. Machine Learning course, University of Cambridge, UK, http://learning.eng.cam.ac.uk/zoubin/ml06/index.html (accessed on February 10, 2011)
8. Kearns, M.J., Vazirani, U.V.: An introduction to computational learning theory. MIT Press, Cambridge (1994)
9. Vapnik, V.N.: The nature of statistical learning theory. Springer, Heidelberg (1995)
10. Vapnik, V.N.: Statistical learning theory. Wiley and Sons, Chichester (1998)
11. Foundations of machine learning theory course, Princeton University, http://www.cs.princeton.edu/courses/archive/spr03/cs511/ (accessed on February 10, 2011)
12. http://www-it.fmi.uni-sofia.bg/markov/courses/ml.html#IntroductiontoMachineLearning (accessed on February 10, 2011)

Chapter 11
Decision Trees

11.1 Introduction

Decision trees are suitable for scientific problems entail labeling data items with one of a given, finite set of classes based on features of the data items. Decision Trees are classifiers that predict class labels for data items [3]. A decision tree learning algorithm approximates a target concept using a tree representation, where each internal node corresponds to an attribute, and every terminal node corresponds to a class[5][6][10].

There are two types of nodes in the tree:

- Internal node: splits into different branches according to the different values the corresponding attribute can take. Example: fever < 37 or fever > 37, cough weak or cough strong in the example below.
- Terminal Node: decides the class assigned.

A decision tree is a branching structure, which consists of nodes and leafs. The root node is at the top and leafs at the bottom. Each node tests the value of some feature of an example, and each leaf assigns a class label to the example. Consider a simple example of classifying a patient's symptoms into two classes – class1 and class2 – corresponding to whether the patient has or has not a cold (see Figure 11.1).

The elements of a decision tree representation have the following meaning:

- each internal node tests an attribute;
- each branch corresponds to an attribute value;
- each leaf node assigns a classification.

One can also use a re-representation as if-then rules: disjunction of conjunctions of constraints on the attribute value instances.

To classify an example X we start at the root of the tree, and check the value of that attribute on X. We follow the branch corresponding to that value and jump to the next node. We continue until we reach a terminal node and take that class as our best prediction [1].

C. Grosan and A. Abraham: Intelligent Systems, ISRL 17, pp. 269–280.
springerlink.com

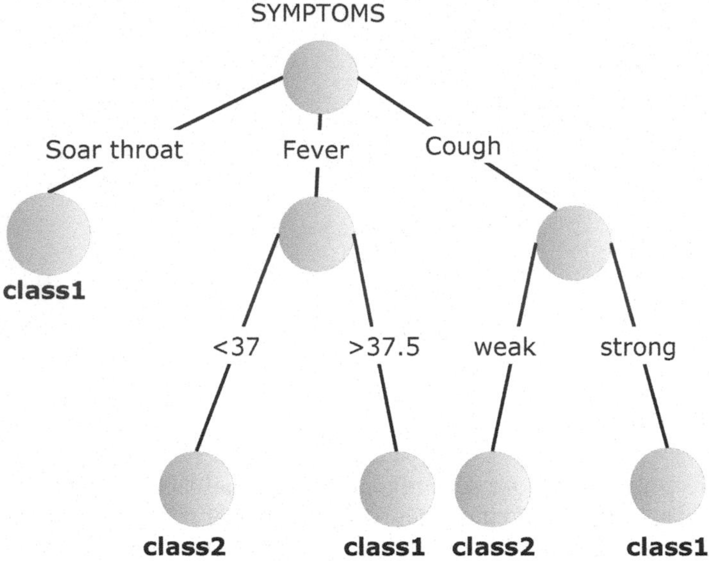

Fig. 11.1 Decision tree for classifying whether a patient has cold (class1) or not (class2).

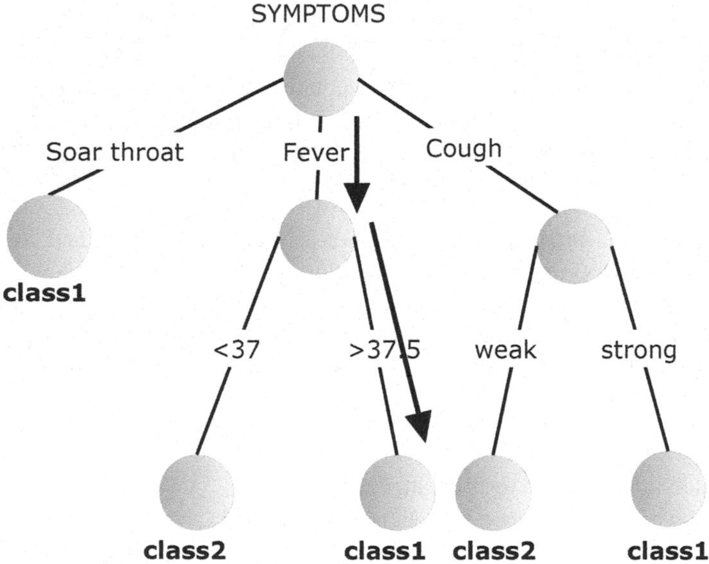

Fig. 11.2 Example of assigning a class to a patient based on the symptoms.

In our example, if we have for instance

X = (Symptoms: fever; fever >37.5)

then the assigned class is represented in Figure 11.2

11.2 Building a Decision Tree

Decision trees are constructed by analyzing a set of training examples for which the class labels are known. They are then applied to classify previously unseen examples.

There are different ways to construct trees from data. In what follows we will present the top-down approach.

11.2.1 Top-Down Induction of Decision Tree

The algorithm is presented below:

```
Step 1. Create a root for the tree
Step 2.
        Step 2.1. If all examples are of the same class
        or the number of examples is below a threshold
        Then return that class
        Step 2.2. If no attributes available return
        majority class
Step 3.
        Step 3.1. Let A be the best attribute for the
        next node
        Step 3.2. Assign A as decision attribute for
        node
Step 4. For each possible value v of A
        Create a new descendant of node. Add a branch
        below A labeled "A = v"
Step 5. Let Sv be the subsets of example where attribute
        A=v
        Recursively apply the algorithm to Sv
Step 6. If training examples are perfectly classified
        Then Stop
        Else iterate over new leaf nodes.
End
```

The algorithm terminates either when all the attributes have been exhausted, or the decision tree perfectly classifies the examples.

Example

Let us illustrate this using the following example (adapted from [1]) for classifying a patient with cold symptoms into one of the two classes: cold or not-cold.

The graphical representation of the initial data is given in Figure 11.3.

We can observe we have two attributes: fever and cough.

Fever has three values:

- < 37;
- >37 and < 38;
- >38.

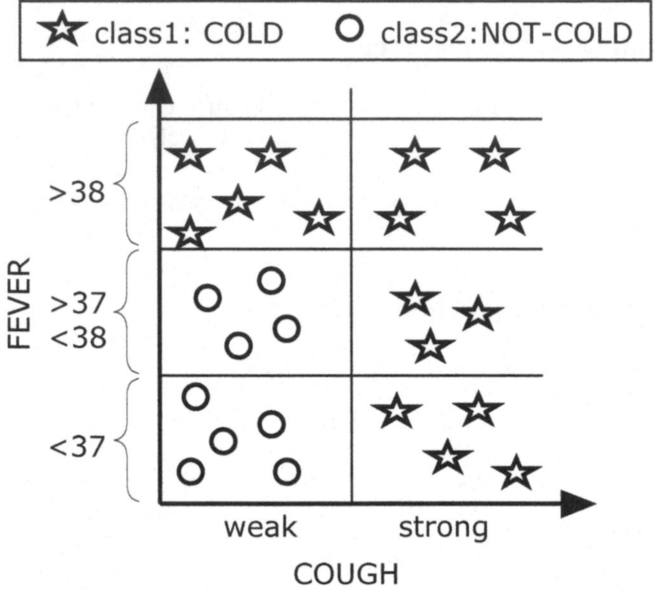

Fig. 11.3 Graphical representation of the initial data.

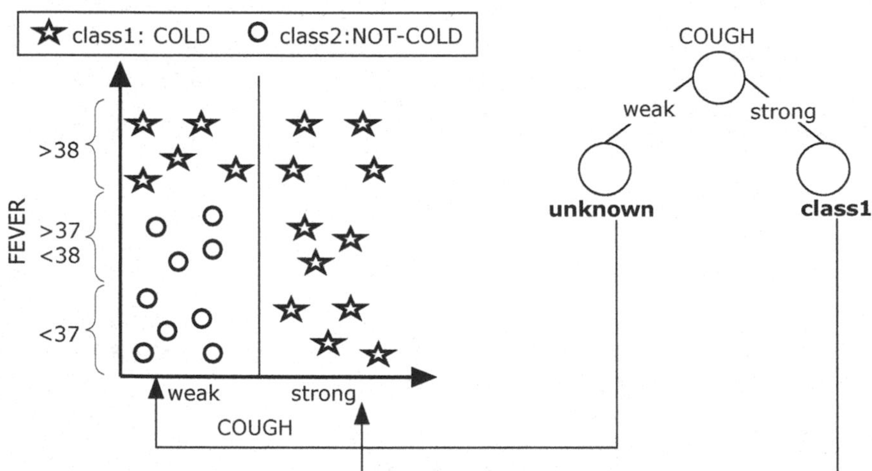

Fig. 11.4 Decision tree construction after considering the first attribute: *cough*.

Cough attribute has two values:

- weak;
- strong.

Suppose we have cough as best attribute. The decision tree obtained until now is depicted in Figure 11.4.

By considering the first attribute – cough – we can now have the root of the tree and two branches corresponding to the two values – weak and strong.

The right branch – strong – leads to a class (which is class1 - cold), but we still don't have a final classification on the left branch. We should now consider the next best attribute – which is the only remaining one: fever. The decision tree obtained is depicted in Figure 11.5.

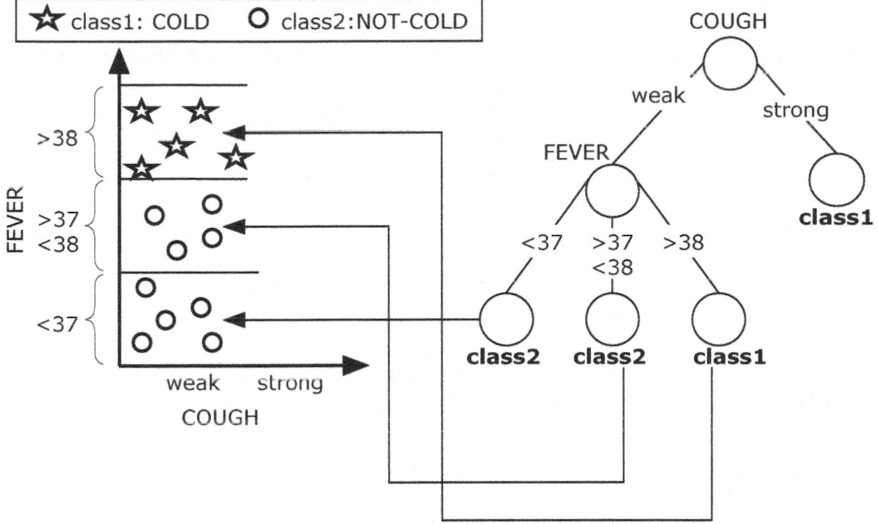

Fig. 11.5 Constructed decision tree.

11.2.2 How to Chose the Best Attribute?

Putting together a decision tree is all a matter of choosing which attribute to test at each node in the tree. We shall define a measure called *information gain* which will be used to decide which attribute to test at each node. Information gain is itself calculated using a measure called *entropy*, which we first define for the case of a binary decision problem and then define it for the general case.

In the example above we considered *cough* as the best attribute. But how to chose this or how to define this in a consequent manner?

In order to choose an attribute to best split the data we will use entropy based splitting function.

Let us use the following notations:

- S – sample of training data
- p_+– the proportion of positive examples in S
- p_-– the proportion of negative examples in S

Entropy measures the impurity in S and it is given by:

$$\textit{Entropy}\ (S) = -\ p_+\log_2 p_+ -\ p_-\log_2 p_-$$

Entropy (S) is the expected number of bits needed to encode the class (+ or -) of randomly drawn member of S under the optimal, shortest length code [2].

As known from information theory, the optimal length code assigns - $\log_2 p$ bits to message having probability p.

Thus, the expected number of bits to encode + or – random member of S is:

$$\sum_{p\in\{p_+,\, p_-\}} p(-\log_2 p)$$

And thus:

$$\textit{Entropy}(S) = \sum_{p\in\{p_+,\, p_-\}} p(-\log_2 p)$$

Coming back to our example, the attribute *cough* divides the sample set into two subsets S1 and S2 (as shown in Figure 11.6):

- S1 = {5 +, 9 -}
- S2 = {11 +, 0 -}.

Then we have:

$$\textit{Entropy}(S1) = -\frac{5}{14}\log_2\left(\frac{5}{14}\right) - \frac{9}{14}\log_2\left(\frac{9}{14}\right)$$

$$\textit{Entropy}(S2) = 0$$

The attribute *fever* divides the sample set into three subsets S1, S2 and S3 (as shown in Figure 11.8):

- S1 = {9 +, 0 -}
- S2 = {3 +, 4 -}.
- S3 = {4+, 5-}

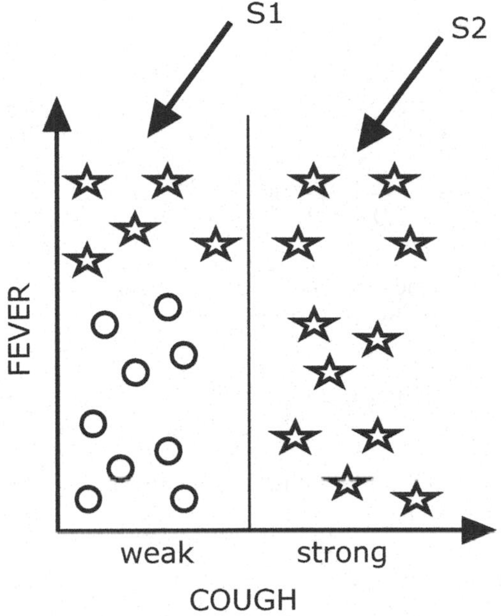

Fig. 11.6 The split of data into two subsets S1 and S2 using the attribute *cough*.

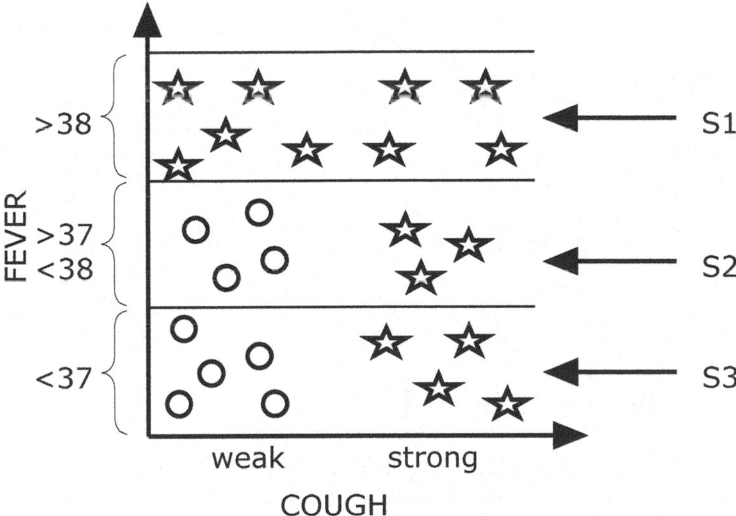

Fig. 11.7 The split of data into three subsets S1, S2 and S3 using the attribute *fever*.

In this case we have:

$$Entropy(S1) = 0$$

$$Entropy(S2) = -\frac{3}{7}\log_2\left(\frac{3}{7}\right) - \frac{4}{7}\log_2\left(\frac{4}{7}\right)$$

$$Entropy(S3) = -\frac{4}{9}\log_2\left(\frac{4}{9}\right) - \frac{5}{9}\log_2\left(\frac{5}{9}\right)$$

Information gain gives the expected reduction in entropy due to sorting on A. Information gain is given by the formula:

$$gain(S, A) = Entropy(S) - \sum_{v \in values(A)} \frac{|Sv|}{|S|} \cdot Entropy(Sv)$$

where *Entropy(Sv)* is the entropy of one sub-sample after partitioning S based on all possible values of attribute A.

For the example considered in Figure 11.6 we have:

$$Entropy(S1) = -\frac{5}{14}\log_2\left(\frac{5}{14}\right) - \frac{9}{14}\log_2\left(\frac{9}{14}\right)$$

$$Entropy(S2) = 0$$

$$Entropy(S) = -\frac{16}{25}\log_2\left(\frac{16}{25}\right) - \frac{9}{25}\log_2\left(\frac{9}{25}\right)$$

$$\frac{|S1|}{|S|} = \frac{14}{25}$$

$$\frac{|S2|}{|S|} = \frac{11}{25}.$$

11.3 Overfitting in Decision Trees

Overfitting is a common problem in machine learning. Decision trees suffer from this, because they are trained to stop when they have perfectly classified all the training data, i.e., each branch is extended just far enough to correctly categorize the examples relevant to that branch. Many approaches to overcoming overfitting in decision trees have been attempted.

In order to define overfitting consider an hypothesis h and $error_{train}(h)$ the error of hypothesis h over training data and $error_D(h)$ the error of hypothesis h over the entire distribution D of data.

Hypothesis h overfits training data if there is an alternative hypothesis h' such that [2]:

$$(i) \qquad error_{train}(h) < error_{train}(h')$$

and

$$(ii) \qquad error_D(h) > error_D(h').$$

The depth of the tree is related to the generalization capability of the tree. If not carefully chosen it may lead to overfitting. A tree *overfits* the data if we let it grow deep enough so that it begins to capture "aberrations" in the data that harm the predictive power on unseen examples [1].

If we add some noise in our example given in Figure 11.3, the tree will grow deeper to capture this noise (as it can be observed in Figure 11.8).

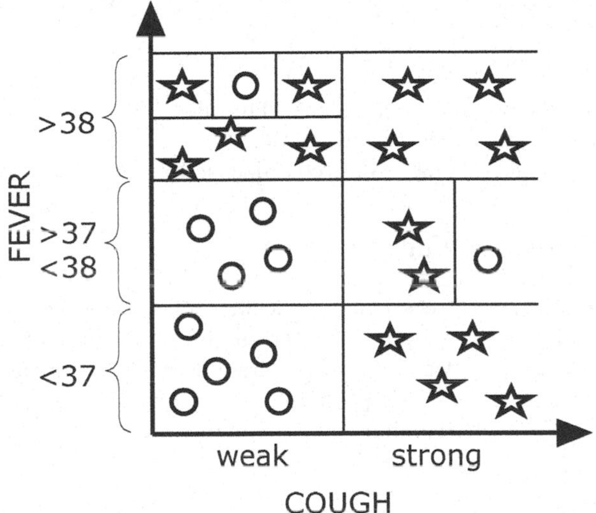

Fig. 11.8 Example from figure 11.3 modified by adding some noise which increases the tree size.

There are two main ways to avoid overfitting [1][2]:

1) Stop growing the tree when data split is not statistically significant.
   This is hard to implement in practice because it is not clear what a good stopping point is.
2) Grow the full tree and then post-prune.

11.3.1 Pruning a Decision Tree

The algorithm below describes the main steps required to prune a decision tree [1][2]:

```
Step 1. Split data into training and validation set.
Step 2. Consider all internal nodes in the tree.
Step 3. Do
        Step 3.1 For each node in the tree evaluate im-
        pact on validation set of pruning it (plus those
        nodes below it) and assign it to the most common
        class.

        Step 3.2 In a Greedy way remove the one (and its
        sub-tree) that most improves validation set accu-
        racy (yields the best performance).
Until further pruning is harmful (or no more improve-
ments are possible):
end
```

Approaches for extracting decision rules from decision trees have also been successful.

Post-pruning of rules follows the steps [2]:

1) Convert the tree to an equivalent set of rules;
2) Prune each rule independently of others;
3) Sort the final rules into a desired sequence for use.

11.4 Decision Trees Variants

Although single decision trees can be excellent classifiers, increased accuracy often can be achieved by combining the results of a collection of decision trees. This forms ensembles of decision trees and are sometimes among the best performing types of classifiers [3] [7] [8] [9].

Two of the strategies for combining decision trees are:

- random forests and
- boosting.

Random forests is a machine learning ensemble classifier in which many different decision trees are grown by a randomized tree-building algorithm. The training set is sampled with replacement to produce a modified training set of equal size to the original but with some training items included more than once. In addition, when choosing the question at each node, only a small, random subset of the features is considered. With these two modifications, each run may result in a slightly different tree. The predictions of the resulting ensemble of decision trees are combined by taking the most common prediction [3].

One of the random forests disadvantages is that does not handle large numbers of irrelevant features as well as ensembles of entropy-reducing decision trees.

Maintaining a collection of good hypotheses, rather than committing to a single tree, reduces the chance that a new example will be misclassified by being assigned the wrong class by many of the trees.

Boosting is a machine-learning method used to combine multiple classifiers into a stronger classifier by repeatedly re-weighting training examples to focus on the most problematic.

In practice, boosting is often applied to combine decision trees [3]. Although it is usually applied to decision tree models, it can be used with any type of model and it is a special case of the model averaging approach.

Alternating decision trees are a generalization of decision trees that result from applying a variant of boosting to combine weak classifiers based on decision stumps, which are decision trees that consist of a single question. In alternating decision trees, the levels of the tree alternate between standard question nodes and nodes that contain weights and have an arbitrary number of children. In contrast to standard decision trees, items can take multiple paths and are assigned classes based on the weights that the paths encounter.

Alternating decision trees can produce smaller and more interpretable classifiers than those obtained from applying boosting directly to standard decision trees [3].

Summaries

There are many different learning algorithms that have been developed for supervised classification and regression. These can be grouped according to the formalism they employ for representing the learned classifier or predictor: decision trees, decision rules, neural networks, linear discriminant functions, Bayesian networks, support vector machines, etc. [4].

A decision tree is a branching structure, which consists of nodes and leafs. The root node is at the top and leafs are at the bottom. Each node tests the value of some feature of an example, and each leaf assigns a class label to the example. This chapter presented a top-down algorithm for learning decision trees.

Decision tree learning provides a practical method for concept learning/learning discrete-valued functions. Decision trees are sometimes more interpretable than other classifiers such as neural networks and support vector machines because they combine simple questions about the data in an understandable way [3].

This algorithm gets into trouble overfitting the data. This occurs with noise and correlations in the training set that are not reflected in the data as a whole. In order to deal with overfitting, one can restrict the splitting, so that it splits only when the split is useful or can allow unrestricted splitting and prune the resulting tree where it makes unwarranted distinctions. One of the advantages of using decision trees is that, if they are not too large, they can be interpreted by humans. This can be useful both for gaining insight into the data and also for validating the reasonableness of the learned tree [4].

We should consider decision trees in the following situations [2]:

- instances can be described by attribute-value pairs;
- attributes are both numeric and nominal.
- target function is discrete valued;
- disjunctive hypothesis may be required;
- possibly noisy training data;
- data may have errors.

References

1. http://www.onlinefreeebooks.net/free-ebooks-computer-programming-technology/artificial-intelligence/artificial-intelligence-course-material-pdf.html (accessed on February 10, 2011)
2. Mitchell, T.M.: Machine learning. McGraw-Hill, New York (1997)
3. Kingsford, C., Salzberg, S.L.: What are decision trees? Nature Biotechnology 26, 1011–1013 (2008)
4. Dietterich, T.G.: Machine Learning. In: Nature Encyclopedia of Cognitive Science. Macmillan, London (2003)
5. Quinlan, J.R.: C4.5: Programs for Machine Learning. Morgan Kaufmann Publishers, San Mateo (1993)
6. Breiman, L., Friedman, J., Olshen, R., Stone, C.: Classification and Regression Trees. Wadsworth International Group, Belmont (1984)
7. Breiman, L.: Random forests. Machine Learning 45, 5–32 (2001)
8. Heath, D., Kasif, S., Salzberg, C.: Committees of decision trees. In: Gorayska, B., Mey, J. (eds.) Cognitive Technology: In Search of a Human Interface, pp. 305–317. Elsevier Science, Amsterdam (1996)
9. Schapire, R.E.: The boosting approach to machine learning: an overview. In: Denison, D.D., Hansen, M.H., Holmes, C.C., Mallick, B., Yu, B. (eds.) Nonlinear Estimation and Classification, pp. 141–171. Springer, New York (2003)
10. Freund, Y., Mason, L.: The alternating decision tree learning algorithm. In: Bratko, I., Džeroski, S. (eds.) Proceedings of the 16th International Conference on Machine Learning, pp. 124–133. Morgan Kaufmann, San Francisco (1999)

Verification Questions

1) What are the steps required to build a decision tree?
2) Explain some ways to choose the best attribute
3) What is overfitting?
4) Enumerate the steps for pruning a decision tree.
5) Nominate some decision trees variants.

Chapter 12
Artificial Neural Networks

12.1 Introduction

Artificial Neural Networks (ANN) are inspired by the way biological neural system works, such as the brain process information. The information processing system is composed of a large number of highly interconnected processing elements (neurons) working together to solve specific problems. ANNs, just like people, learn by example. Similar to learning in biological systems, ANN learning involves adjustments to the synaptic connections that exist between the neurons.

The first artificial neuron was produced in 1943 by the neurophysiologist Warren McCulloch and the logician Walter Pits. But the technology available at that time did not allow them to do too much. There is a significant difference between the ways neural networks solve a problem while compared to a standard algorithm. In a conventional algorithmic approach the computer follows a set of instructions in order to solve a problem. The specific steps that the computer needs to follow are known and without this the computer cannot solve the problem. But computers would be so much more useful if they could do things that we don't exactly know how to do. The way the problem is to be solved must be known and stated in small unambiguous instructions. These instructions are then converted to a high level language program and then into machine code that the computer can understand. These machines are totally predictable; if anything goes wrong is due to a software or hardware fault [3].

Neural networks learn by example. They cannot be programmed to perform a specific task. The examples must be selected carefully otherwise useful time is wasted or even worse the network might be functioning incorrectly. Since the network finds out how to solve the problem by itself, its operation can be unpredictable[5][8][9][12][14].

Neural networks and conventional algorithms do not compete but complement each other. There are tasks ,which are more suited to an algorithmic approach and tasks that are more suitable to neural networks approach. And, more than this, a large number of tasks require systems that combine the two approaches (such as, for example, a conventional computer is used to supervise the neural network) in order to perform at maximum efficiency [3].

C. Grosan and A. Abraham: Intelligent Systems, ISRL 17, pp. 281–323.
springerlink.com

12.2 Similarities between Biological and Artificial Neural Networks

The human brain contains about 100 billions of nervous cells. The brain's billions of neurons connect with one another in complex networks. Total number of neurons in human cerebral cortex is about 10 billion and the total number of synapses in cerebral cortex is about 60 trillion [7]. C. Koch [8] lists the total number of neurons in the cerebral cortex at 20 billion and the total synapses in the cerebral cortex at 240 trillion.

The mass of a large sensory neuron is about 10^{-6}gram [6] and the number of synapses for a typical neuron varies between 1,000 and 10,000.

All physical and mental functioning depends on the establishment and maintenance of neuron networks. Connections the brain finds useful become permanent; those not useful are eliminated as the brain selectively strengthens and prunes connections based on experience. A person's habits and skills - such as nail-biting or playing a musical instrument - become embedded within the brain in frequently activated neuron networks. When a person stops performing an activity, the neural networks for the activity fall into disuse and eventually may disappear [4].

A neuron consists of (see Figure 12.1) a soma (cell body), axon (a long fiber) and dendrites. The axon sends signals and the dendrites receive signals. A synapse connects an axon to a dendrite. Given a signal, a synapse might increase (excite) or decrease (inhibit) electrical potential. A neuron fires when its electrical potential reaches a threshold.

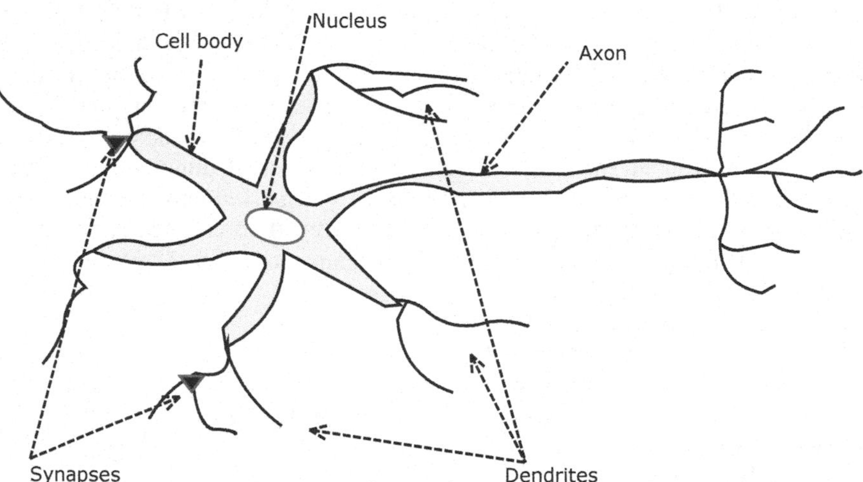

Fig. 12.1 Structure of a biological neuron.

An artificial neural network consists of a number of neurons (units) similar to the biological neurons in the brain (often arranged in layers), a number of connections which are performed by weighted links and whose role is to transmit signals from one neuron to another, and weights. The output signal is transmitted through the neuron's outgoing connection (analogue to the axon in the biological neurons). The outgoing connection splits into a number of branches that transmit the same signal. The outgoing branches terminate at the incoming connections of other neurons in the network.

Inputs and outputs are numeric.

The correspondence between a biological and an artificial neural network is given in Figure 12.2.

Biological Neural Network	Artificial Neural Network
soma (cell body)	neuron
dendrites, axon	connections (input, output)
synapse	weight
potential	weighted sum
signal	activation

Fig. 12.2 Correspondence between biological and artificial neural network.

The *neuron* is the simplest component of an artificial neural network. The diagram of an artificial neuron is depicted in Figure 12.3.

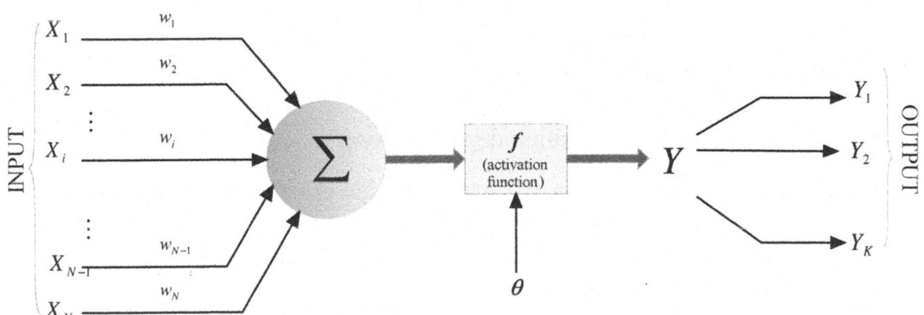

Fig. 12.3 Diagram of an artificial neuron.

The main components of a neural network (or a parallel distributed model) are [20] [21] [22]:

- a set of processing units (called neurons or cells);
- a state of activation Y_i for every unit, which is equivalent to the output of the unit;

- connections between the units; each connection is defined by a weight w_{jk} which determines the effect which the signal of unit j has on unit k. The contribution for positive w_{jk} is considered as an excitation and for negative w_{jk} as inhibition.
- a propagation rule, which determines the effective input X_i of a unit from its external inputs;
- an activation function f, which determines the new level of activation based on the effective input $X_i(t)$ and the current activation $Y_i(t)$;
- an external input (also know as bias, offset) θ_i for each unit;
- a method for information gathering (the learning rule);
- an environment within which the system must operate, providing input signals and / if necessary / error signals.

Te *neurons* or *input units* are of 3 types:

- input units;
- hidden units;
- output units.

The neurons receive input from their neighbors or external sources and use this to compute an output signal which is propagated to other units. Apart from this processing, a second task is the adjustment of the weights. The system is inherently parallel in the sense that many units can carry out their computations at the same time [20].

The aim of a neural network is to train the network to achieve a balance between the ability to respond correctly to the input patterns that are used for training and the ability to give reasonable responses to a new input which is similar but not identical to those used for training.

12.3 Neural Networks Types

The neural networks can be classified depending on:

- the nature of information processing carried out at individual nodes:
  - single layer network (perceptron);
  - multi-layer network;
- the connection geometries:
  - feedforward network;
  - backpropagation network;
- the algorithm for adaptation of link weights.

12.3.1 Layered Feed-Forward Network

A layered feed-forward network is characterized by a collection of input neurons whose role is to supply input signals from the outside world into the rest of the network. Following this can come one or more intermediate layers of neurons and finally an output layer where the output of the computation can be communicated to the outside world. The intermediate – also known as *hidden layers* – have no direct contact with the outside world.

For this class of networks, there are no connections from a neuron-to-neuron (s) in previous layer, other neurons in the same layer or to neurons more than one layer ahead.

Every neuron in a given layer receives inputs from layers below its own and sends output to layers above its own. Thus, given a set of inputs from the neurons in the input layer, the output vector is computed by a succession of forward passes, which compute the intermediate output vectors of each layer in turn using the previously computed signal values in the earlier layers. One of the simplest such networks consists of a single layer and is called a *perceptron*.

12.3.2 The Perceptron

This consists of a single neuron with multiple inputs and a single output. It has restricted information processing capability. The information processing is done through a transfer function, which is either linear or non-linear.

12.3.3 Feedforward Radial Basis Function (RBF) Network

Feed-forward radial basis function network is a feed-forward network with an input layer, output layer and a hidden layer. The hidden layer is based on a radial basis function. The RBF generally used is the Gaussian function. Several RBF's in the hidden layer allow the RBF network to approximate a more complex activation function than a typical feed-forward neural network. RBF networks are used for pattern recognition. They can be trained using genetic, annealing or one of the propagation techniques. Other means must be employed to determine the structure of the RBF's used in the hidden layer.

12.3.4 Recurrent Networks

Some networks allow the output signal of each neuron in a given layer to be connected not only to the layer ahead but also to that same neuron as an input signal. Such networks are called *associative recurrent networks. Backpropagation* networks not only have feed-forward connections but each hidden layer also receives an error feedback connection from each of the neurons above it.

12.3.4.1 Hopfield Neural Network

This network is a simple single layer recurrent neural network. The Hopfield neural network is trained with a special algorithm that teaches it to learn to recognize patterns. The Hopfield network will indicate that the pattern is recognized by echoing it back. Hopfield neural networks are typically used for pattern recognition.

12.3.4.2 Simple Recurrent Network (SRN) Elman Style

This network is a recurrent neural network that has a context layer. The context layer holds the previous output from the hidden layer and then echos that value back to the hidden layer's input. The hidden layer then always receives input from its previous iteration's output. Elman neural networks are generally trained using genetic algorithm, simulated annealing, or one of the propagation techniques. Elman neural networks are typically used for prediction.

12.3.4.3 Simple Recurrent Network (SRN) Jordan Style

This network is a recurrent neural network that has a context layer. The context layer holds the previous output from the output layer and then echos that value back to the hidden layer's input. The hidden layer then always receives input from the previous iteration's output layer. Jordan neural networks are generally trained using genetic algorithm, simulated annealing, or one of the propagation techniques. Jordan neural networks are typically used for prediction.

12.3.5 Self-Organizing Maps

Self Organizing Maps[16][17][18][19] (or Kohonen networks) have a grid topology, with unequal grid weights. The topology of the grid provides a low dimensional visualization of the data distribution. These are thus used in applications which typically involve organization and human browsing of a large volume of data. Learning is performed using a winner take all strategy in an unsupervised mode.

12.4 The Perceptron

The simplest model of a neural network consists of a single neuron[23]. The neuron can be trained to learn different simple tasks by modifying its threshold and input weights. Inputs are presented to the neuron and each input has a desired output. If the neuron doesn't give the desired output, then it has made a mistake. To rectify this, its threshold and/or input weights must be changed. A learning algorithm determines how to change the weights and threshold.

The output of the perceptron is usually constrained to boolean values – which can be interpreted as true and false, 0 and 1 or -1 and 1.

The architecture of a simple perceptron is depicted in Figure 12.4 where:

- X_1, X_2, ..., X_i, …, X_N are inputs. These could be real numbers or boolean values depending on the problem.
- Y is the output and is boolean.
- w_1, w_2, ..., w_i, …, w_N are weights of the edges and are real value.
- θ is the threshold and is a real value.

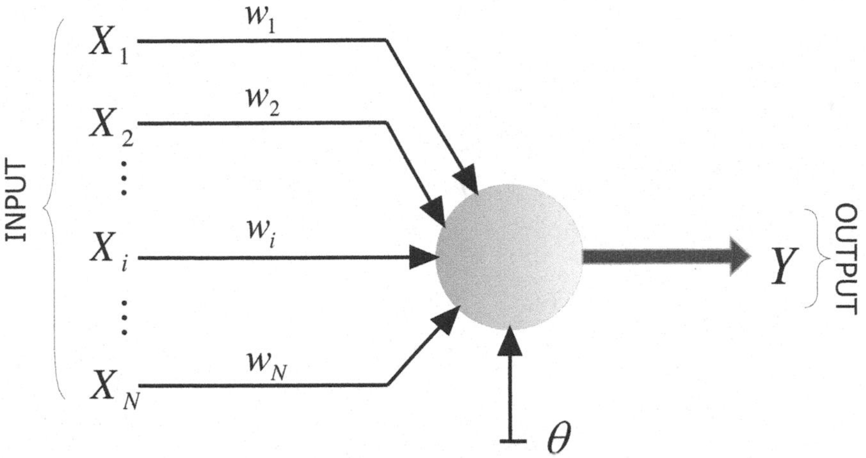

Fig. 12.4 Architecture of a simple perceptron.

The role of the perceptron is to classify a set of stimuli into one of the two available classes. The decision regions are separated by a hyperplane whose equation is given by:

$$\sum_{i=1}^{N} w_i X_i - \theta = 0$$

Thus, the perceptron can be successfully applied only in the particular case of linearly separable data, which means the data are situated on one side and the other of a hyperplane.

The neuron computes the weighted sum of the input signals and compares the result with a *threshold value*, θ. If the net input is less than the threshold, the neuron output is 0. But if the net input is greater than or equal to the threshold, the neuron becomes activated and its output attains a value +1.

In this case, the neuron uses the following *activation function* (sign function):

$$X = \sum_{i=1}^{N} w_i X_i \qquad (1)$$

$$Y = \begin{cases} 0, \ if \ X < \theta \\ 1, \ if \ X \geq \theta \end{cases}$$

12.4.1 Activation Functions

There are several activation functions which can be used for neural networks. Some of them are as follows (see Figure 12.5 where some models are depicted):

- *Sign:*

$$\text{Activation}(X) = \begin{cases} 0, & if\ X < \theta \\ 1, & if\ X \geq \theta \end{cases}$$

- *Step:*

$$\text{Activation}(X) = \begin{cases} -1, & if\ X < \theta \\ 1, & if\ X \geq \theta \end{cases}$$

- *Sigmoid (logistic):*

$$\text{Activation}\ (X) = \frac{1}{1+e^{-X}}$$

- *Identity (linear):*

$$\text{Activation}\ (X) = X$$

- *Than:*

$$\text{Activation}(X) = \frac{1-e^{-2X}}{1+e^{-2X}}$$

- *Arctan:*

$$\text{Activation}(X) = 2 \cdot \frac{\arctan(X)}{\pi}$$

- *Exponential:*

$$\text{Activation}(X) = e^{X}$$

- *Reciprocal*

$$\text{Activation}(X) = \frac{1}{X}$$

- *Gaussian*

$$\text{Activation}(X) = e^{-X^2/\sigma^2}$$

- *Sine*

$$\text{Activation}(X) = \sin(X)$$

- *Cosine*

$$\text{Activation}(X) = \cos(X)$$

- *Elliott*

$$\text{Activation}(X) = \frac{X}{1+|X|}$$

where X is given by the equation (1).

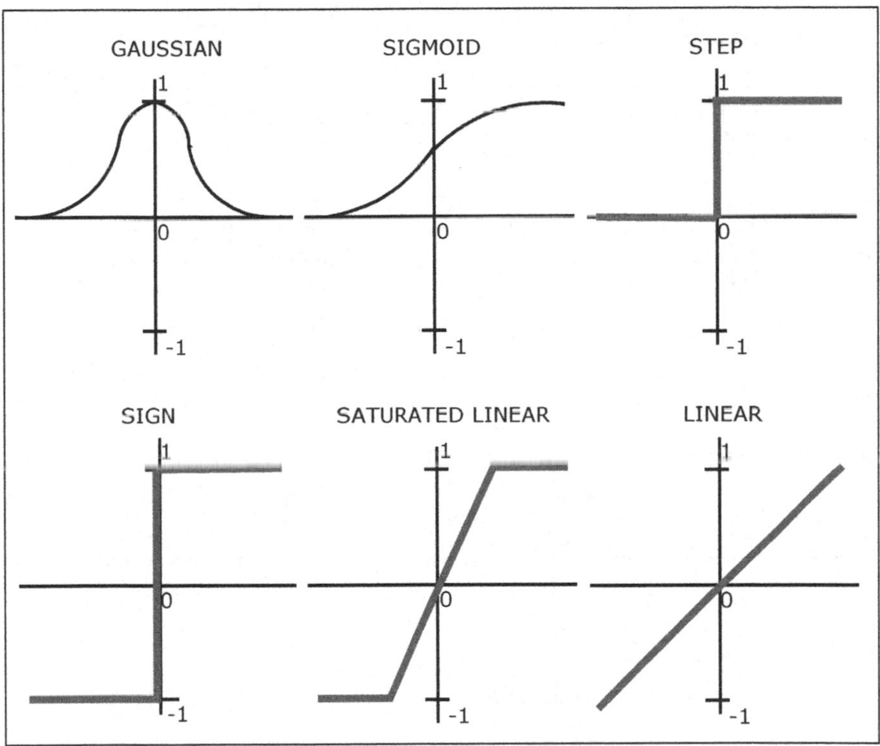

Fig. 12.5 Examples of activation functions.

All activation functions enumerated above use a linear combination of weights and input activations. There is also another category which computes Euclidean distance between weights and input activations (outputs of other neurons). The only activation function in this category is radial Basis and is given by:

$$\text{Activation}(X) = e^{f \cdot \log(altitude) - X}$$

where:

- f is the fan-in of each unit in the layer, that is the number of other units feeding into that unit, and
- altitude is a positive number stored in the neuron (or neural layer or neural network). The default is altitude = 1.0, for that value the activation function reduces to the simple e^{-X}.

12.4.2 How the Perceptron Learns a Task?

The perceptron is a sign (or step) function based on a linear combination of real-valued inputs. If the combination is above a threshold it outputs a 1, otherwise it outputs a 0 (1 and -1 respectively in the case of sign activation function).

As depicted in Figure 12.6, the output is:

$$Output = \begin{cases} 1, & if \ step(w_1 X_1 + w_2 X_2 + \ldots w_N X_N - \theta) \geq 0 \\ 0, & otherwise \end{cases}$$

A perceptron draws a hyperplane as the decision boundary over the (n-dimensional) input space (see Figure 12.7 for 2-dimensional and 3-dimesional case respectively).

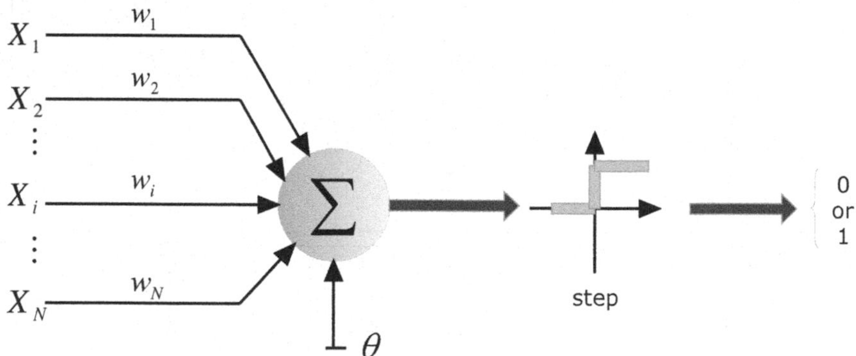

Fig. 12.6 Learning perceptron.

A perceptron can learn only examples that are called linearly separable. These are examples that can be perfectly separated by a hyperplane.

Learning a perceptron means finding the right values for *w*. The hypothesis space of a perceptron is the space of all weight vectors.

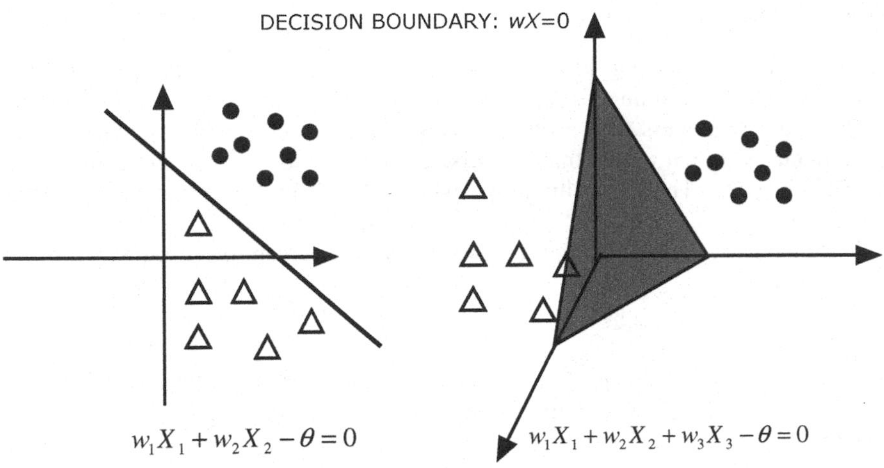

Fig. 12.7 Perceptron separation hyperplane for 2 and 3 dimensions.

In order to train the perceptron, a series of inputs are presented to the perceptron - each of the form (X_1, X_2, ..., X_N). For each such input, there is a desired output - either 0 or 1. The actual output is determined by the net input which is:

$$w_1 X_1 + w_2 X_2 + \ldots + w_n X_N.$$

If the net input is less than threshold then the output is 0, otherwise output is 1. If the perceptron gives a wrong (undesirable) output, then one of two things could have happened:

- The desired output is 0, but the net input is above threshold. So the actual output becomes 1. In such a case we should decrease the weights.
- The desired output is 1, but the net input is below threshold. We should now increase the weights.

There are two popular weight update rules which will be described in what follows.

1) The perceptron rule, and
2) Delta rule

Let us use the following notations:

$X(t) = [X_1(t), X_2(t), \ldots, X_N(t)]^T$ – input vector ;
$w(t) = [w_1(t), w_2(t), \ldots, w_N(t)]^T$ – synaptic weigths vector; $0 \le w_i \le 1$, $i=1, \ldots, N$
$\theta(t)$ – threshold ;
$Y(t)$ – actual output;
$Y_d(t)$ – desired output;
$\alpha(t)$ – learning rate; $0 < \alpha < 1$;

12.4.2.1 The Perceptron Rule

In 1958, Frank Rosenblatt [10] introduced a training algorithm that provided the first procedure for training a simple perceptron.

The algorithm starts with a random hyperplane and incrementally modifies the hyperplane such that points that are misclassified move closer to the correct side of the boundary. The algorithm stops when all learning examples are correctly classified.

With the above notations, the steps of Rosenblatt perceptron learning algorithm are as follows:

Step 1: Set $t = 1$.
Initialize the weights $w_1, w_2,\ldots, w_N$ to random numbers in the range [–0.5, 0.5].
Initialize the threshold θ (with a value between [-0.5, 0.5]).
Repeat
Step 2: Activate the perceptron.
The inputs are $X_1(t), X_2(t),\ldots, X_N(t)$ and desired output $Y_d(t)$.
The actual output is:

$$Y(t) = step\left[\sum_{i=1}^{N} X_i(t)\, w_i(t) - \theta\right]$$

Step 3: Calculate the error:
$e(t) = Y_d(t)\text{-}Y(t)$
Step 4: Update the weights of the perceptron:

$$w_i(t+1) = w_i(t) + \alpha \cdot X_i \cdot e(t)$$

Step 5: $t = t+1$
Go to Step 2.
Until convergence.

Remarks

(i) If we use the notation:

$\Delta w_i = \alpha{\cdot}X_i{\cdot}e(t)$

Then the weight update rule can be simple expressed as:

$w_i(t+1) = w_i(t) + \Delta w_i$.

(ii) Provided the examples are linearly separable and a small value for α is used, the rule is proved to classify all training examples correctly (i.e. is consistent with the training data).

(iii) An *epoch* is the presentation of the entire training set to the neural network. In the case of the AND (OR, XOR) function an epoch consists of four sets of inputs being presented to the network

Perceptron Convergence Theorem

The *Perceptron convergence theorem* states that for any data set which is linearly separable the Perceptron learning rule is guaranteed to find a solution in a finite number of steps.

In other words, the perceptron learning rule is guaranteed to converge to a weight vector that correctly classifies the examples provided the training examples are linearly separable.

A function is said to be linearly separable when its outputs can be discriminated by a function which is a linear combination of features that is we can discriminate its outputs by a line or a hyperplane.

12.4.2.2 Delta Rule

An important generalization of the perceptron training algorithm was presented by Widrow and Hoff[13] as the least mean square (LMS) learning procedure, also known as the delta rule. The main functional difference with the perceptron training rule is the way the output of the system is used in the learning rule:

- the perceptron learning rule uses the output of the threshold function for learning;
- the delta-rule uses the net output without further mapping into output values.

When the data are not linearly separable we try to approximate the real concept using the delta rule. The key idea is to use a *gradient descent search*. We will try to minimize the following error:

$$e = \sum_i \left(Y_{d_i} - Y_i\right)^2$$

where:

- the sum goes over all training examples;
- Y_i is the inner product wX and not sign(wX) as in the case of perceptron rule.

The idea is to find a minimum in the space of weights and the error function e.

The delta rule works as follows:

For a new training example $X = (X_1, X_2,\ldots, X_N)$ the weights are updated according to the rule:

$$w_i = w_i + \Delta w_i$$

where $\Delta w_i = -\alpha \cdot \dfrac{e'(W)}{w_i}$

$$\frac{e'(W)}{w_i} = \sum_i \left(Y_{d_i} - Y_i\right) \cdot \left(-X_i\right)$$

α denotes the learning rate.

Thus, we obtain the following equation [1]:

$$w_i = \alpha \cdot \sum_i \left(Y_{d_i} - Y_i\right) \cdot X_i$$

There are two differences between the perceptron and the delta rule [1]:

1) the perceptron is based on an output from a step function whereas the delta rule uses the linear combination of inputs directly;
2) the perceptron is guaranteed to converge to a consistent hypothesis assuming the data is linearly separable. The delta rule converges in the limit but it does not need the condition of linearly separable data.

There are two main difficulties with the gradient descent method:

1) convergence to a minimum may take a long time;
2) there is no guarantee we will find the global minimum.

These are handled by using momentum terms and random perturbations to the weight vectors.

12.4.3 Example: Perceptron for OR Function

With a function such as OR (with only two inputs) we can easily decide what weights to use to give us the required output from the neuron.

For OR function, the truth table and the graphical representation are given in Figure 12.8.

The initial perceptron (the initial values for weights w_1 and w_2, the threshold θ and the learning rate α are provided in Figure 12.9.

With the initial configuration, the equation of the separator is:

$$w_1X_1 + w_2X_2 - \theta = 0$$

which is:

$$-0.4X_1 + 0.1 \cdot X_2 - 0.2 = 0$$

or:

$$4X_1 - X_2 + 2 = 0$$

X_1	X_2	Y
0	0	0
1	0	1
0	1	1
1	1	1

Fig. 12.8 Truth table and graphical representation for OR function.

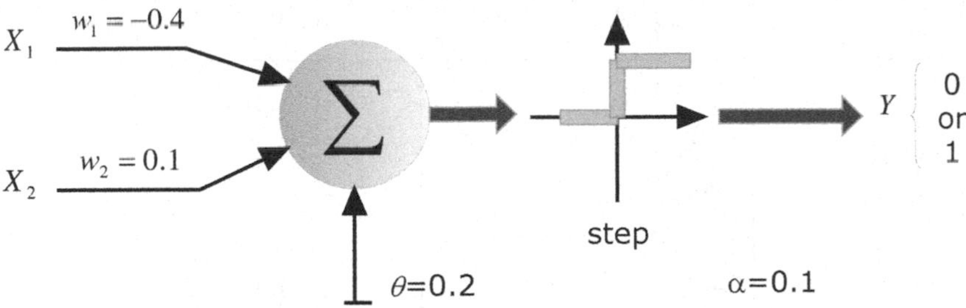

Fig. 12.9 The initial perceptron for OR function.

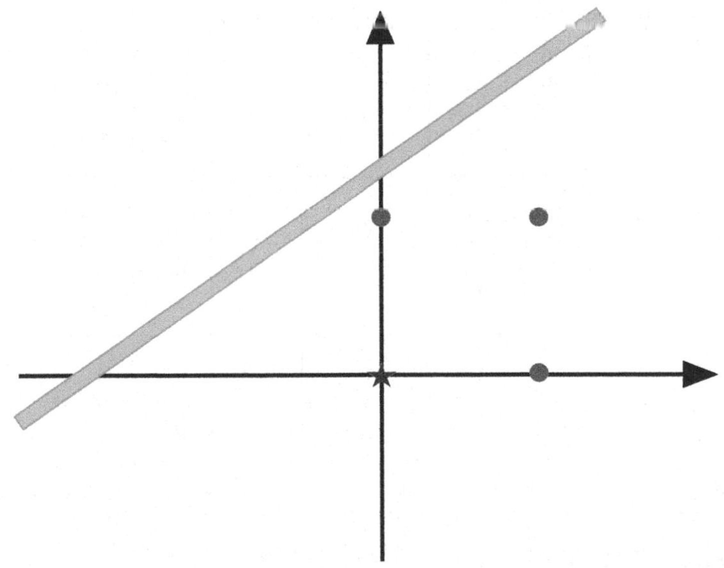

Fig. 12.10 The initial separator given by the equation $4X_1 - X_2 + 2 = 0$.

whose graphical representation is given in Figure 12.10 (we can observe all the points belong to the same plane).

In what follows we will use the perceptron rule to adapt the weights in order to obtain a correct classification (separation).

Table 12.1 presents the percepton training process for OR function during 6 epochs. At the end of the 6th epoch the error is 0 for each of the input data.

Table 12.1 Perceptron learning model for OR function.

Input		Desired output	Initial weights		Obtained output	error	Updated weights		Epoch
X_1	X_2	Y_d	w_1	w_2	Y	e	w_1	w_2	
0	0	0	-0.4	0.1	0	0	-0.4	0.1	1
1	0	1	-0.4	0.1	0	1	-0.3	0.1	
0	1	1	-0.3	0.1	0	1	-0.3	0.2	
1	1	1	-0.3	0.1	0	1	-0.2	0.3	
0	0	0	-0.2	0.3	0	0	-0.2	0.3	2
1	0	1	-0.2	0.3	0	1	-0.1	0.3	
0	1	1	-0.1	0.3	1	0	-0.1	0.3	
1	1	1	-0.1	0.3	1	0	-0.1	0.3	
0	0	0	-0.1	0.3	0	0	-0.1	0.3	3
1	0	1	-0.1	0.3	0	1	0	0.3	
0	1	1	0	0.3	1	0	0	0.3	
1	1	1	0	0.3	1	0	0	0.3	
0	0	0	0	0.3	0	0	0	0.3	4
1	0	1	0	0.3	0	1	0.1	0.3	
0	1	1	0.1	0.3	1	0	0.1	0.3	
1	1	1	0.1	0.3	1	0	0.1	0.3	
0	0	0	0.1	0.3	0	0	0.1	0.3	5
1	0	1	0.1	0.3	0	1	0.2	0.3	
0	1	1	0.2	0.3	1	0	0.2	0.3	
1	1	1	0.2	0.3	1	0	0.2	0.3	
0	0	0	0.2	0.3	0	0	0.2	0.3	6
1	0	1	0.2	0.3	1	0	0.2	0.3	
0	1	1	0.2	0.3	1	0	0.2	0.3	
1	1	1	0.2	0.3	1	0	0.2	0.3	

The error in all the 6 epochs is presented in Figure 12.11.

The separator obtained at the end of each epoch is depicted in Figure 12.12.

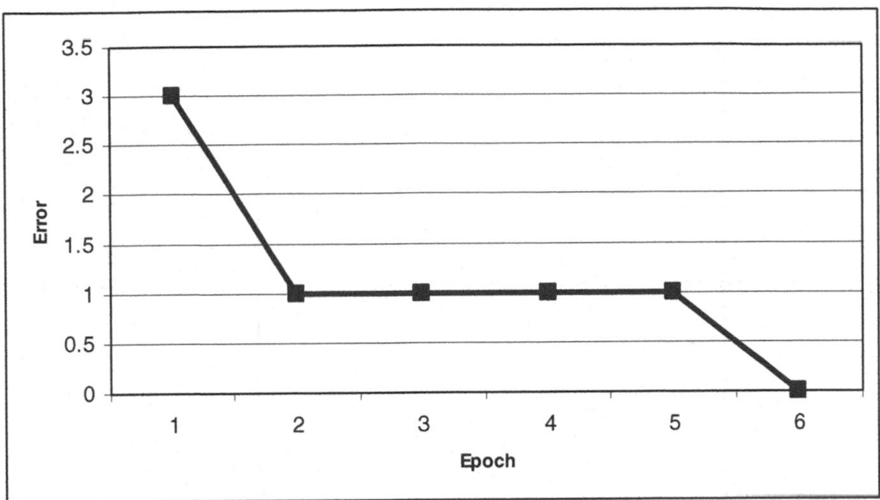

Fig. 12.11 Perceptron error for the 6 epochs in OR example.

Let us shortly describe the process during the first epoch (rest are all similar).

We consider the first training data with values 0 for all X_1, X_2 and Y_d. Initial weights values are -0.4 and 0.1. We activate the perceptron using step activation function. The value of the threshold is 0.2.

The obtained output is:

$$Y = \text{step}[0\cdot(-0.4)+0\cdot 0.1 - 0.2] = \text{step}\ [-0.2] = 0$$

The obtained output is 0 which coincides with the desired output. Thus, the error in this case is 0 and the weights remain unchanged.

Consider the second training data, $X_1 = 1$, $X_2 = 0$ and $Y_d = 1$.

The obtained output is:

$$Y = \text{step}[1\cdot(-0.4)+0\cdot 0.1 - 0.2] = \text{step}\ [-0.6] = 0$$

The obtained output is different from the desired output. In this case the error is 1.

We now have to update the weights:

$$w_1 = w_1 + \alpha X_1 \cdot e$$

The learning rate value is 0.1 in our example.

$$w_1 = -0.4 + 0.1\cdot 1\cdot 1 = -0.3$$
$$w_2 = 0.1 + 0.1\cdot 0\cdot 1 = 0.1$$

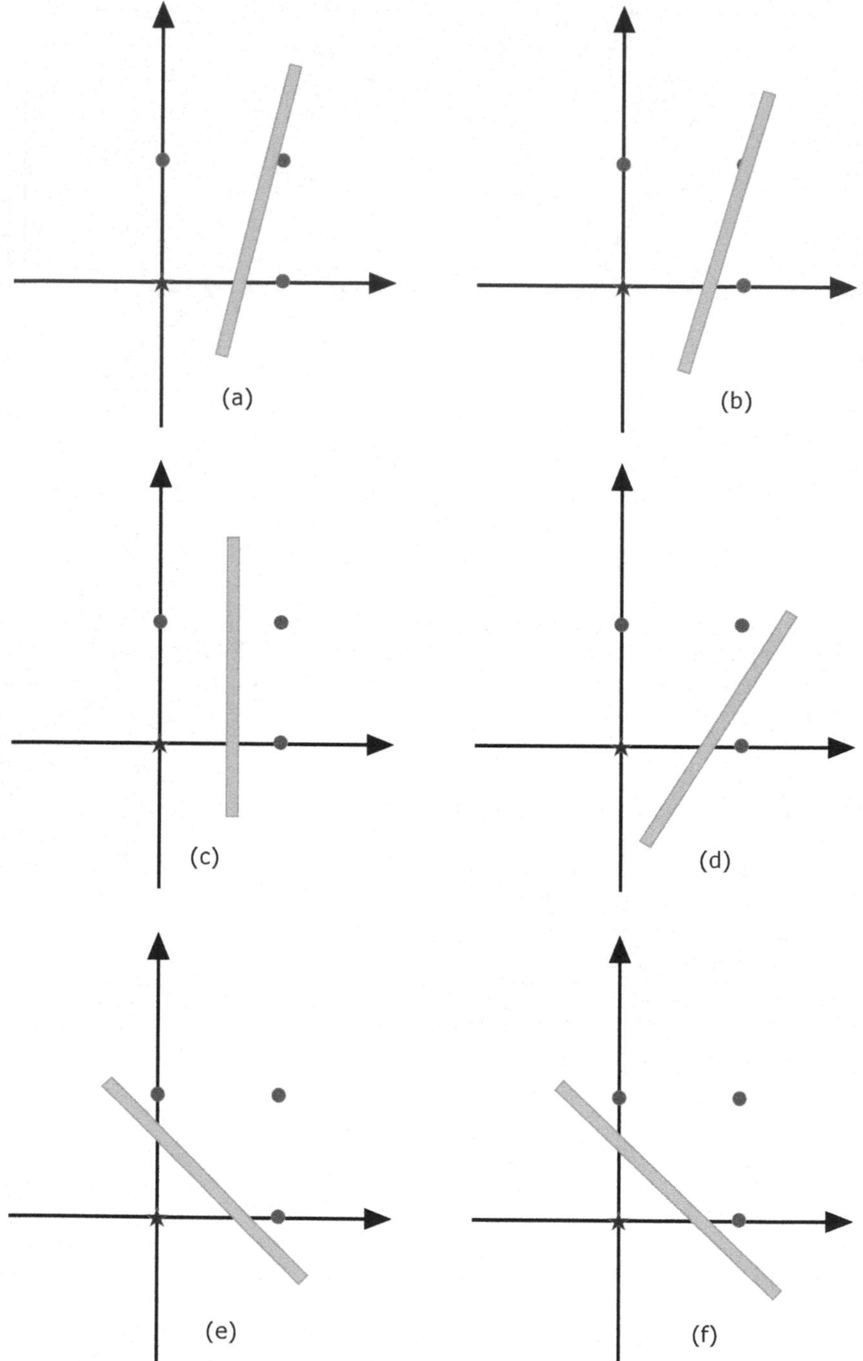

Fig. 12.12 Separator obtained at the end of each of the 6 epochs.

We present the third training data ($X_1 = 0$, $X_2 = 1$ and $Y_d = 1$) to the perceptron with the new updated weights $w_1 = -0.3$ and $w_2 = 0.1$.

The obtained output is:

$$Y = \text{step}[0\cdot(-0.3)+1\cdot0.1 - 0.2] = \text{step}\,[-0.1] = 0$$

Again, we have an error 1 as difference between the desired and the obtained output. We update again the weights and the new weights are $w_1 = -0.3$ and $w_2 = 0.2$.

We present the forth data to the perceptron: $X_1 = 1$, $X_2 = 1$ and $Y_d = 1$.

The obtained output is 0 and the desired one is 1. We have again a difference of 1. By updating the weights we obtain the values: $w_1 = -0.2$ and $w_2 = 0.3$.

With this, the first epoch ends – we have presented all the 4 sets of data to the perceptron. The total error is 3 for all the data which means 3 data were incorrect classified. We have to continue until all the data are correctly classified. The separator obtained at the end of this epoch is depicted in Figure 12.12 (a).

At the end of the 6th epoch all the data are correctly classified (the sum of errors is 0) and the linear separator is depicted in Figure 12.12 (f).

12.4.4 Limitations of the Perceptron

Single perceptron can only model functions whose graphical models are linearly separable. If there is no line (or, in the general sense, no hyperplane) that divides the data, then it impossible for the perceptron to learn to behave with that data.

For instance, the boolean function XOR is not linearly separable, so you this cannot be modeled with only one perceptron. The weight values just keep on shifting, and the perceptron never actually converges to one value.

The perceptron has several limitations which restrict it to certain classes of problems:

- learning is efficient if weights are not very large;
- attributes are weighted independently;
- can only learn lines-hyperplanes (cannot learn exclusive OR for example).

Perceptrons enable a pattern to be broken up into simpler parts that can each be modeled by a separate perceptron in a network. So, even though perceptrons are limited, they can be combined into one powerful network that can model a wide variety of patterns, such as XOR and many complex boolean expressions of more than one variable. These algorithms, however, are more complex in arrangement, and thus the learning function is slightly more complicated.

12.5 Multi-layer Perceptron

As we could see in the previous sections, the perceptron can be successfully used for AND and OR logical functions. But a simple perceptron cannot decide in the case of XOR function (see Figure 12.13).

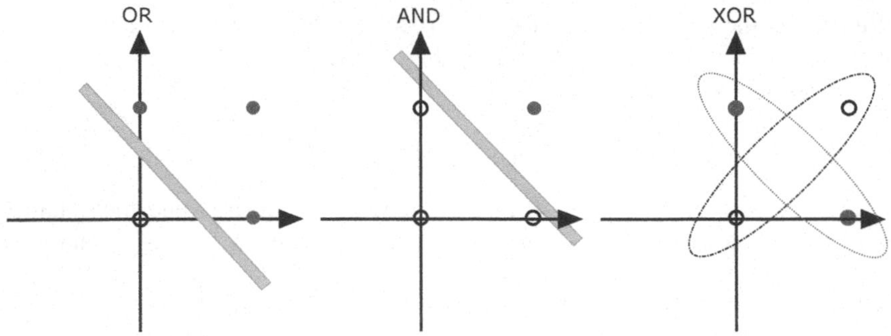

Fig. 12.13 Graphical representation of separation in case of logical AND, OR and XOR.

In all the tree situations, the input space consists of four points. In the XOR case, the solid circles cannot be separated by a straight line from the two empty circles. But this can be overcome and Minsky and Papert [15] show that for binary inputs any transformation can be carried out by adding a layer of predicates which are connected to all inputs.

Thus, the XOR problem can be solved by introducing hidden units which involves extension of the network to a multi-layered perceptron (see Figure 12.14).

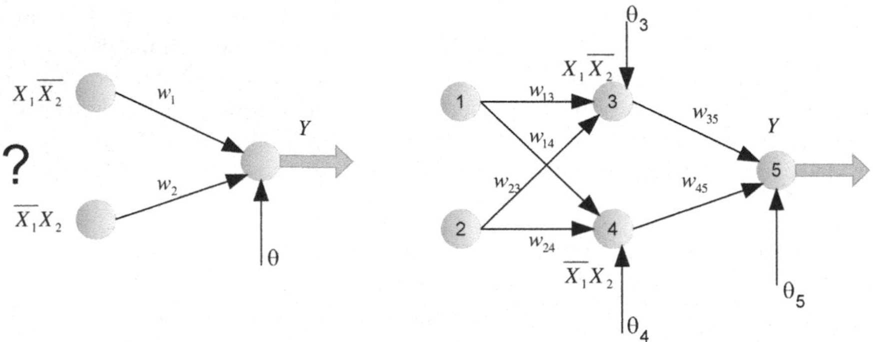

Fig. 12.14 The hidden layer for the Exclusive OR problem.

A feed-forward network has a layered structure. Each layer consists of units which receive their input from units from a layer directly below and send their output to units in a layer directly above the unit. There are no connections within a layer.

No processing takes place in the input neurons. The activation of a hidden neuron is a function of the weighted inputs plus a bias (threshold).

By contrast to perceptrons, multilayer networks can learn not only multiple decision boundaries, but the boundaries may be nonlinear. The typical architecture of a multi-layer perceptron (MLP) is shown in Figure 12.15 (with one hidden layer) and Figure 12.16 (with M hidden layers).

A standard multilayer perceptron contains:

- an input layer;
- one or more hidden layers;
- an output layer.

A feedforward network is an acyclic directed graph of units. The input units provide the input values. All other units compute an output value from their inputs. Hidden units are internal.

If the hidden units compute a linear function, the network is equivalent to one without hidden units. Thus, the output of hidden units is produced by a nonlinear activation function. This is optional for output units.

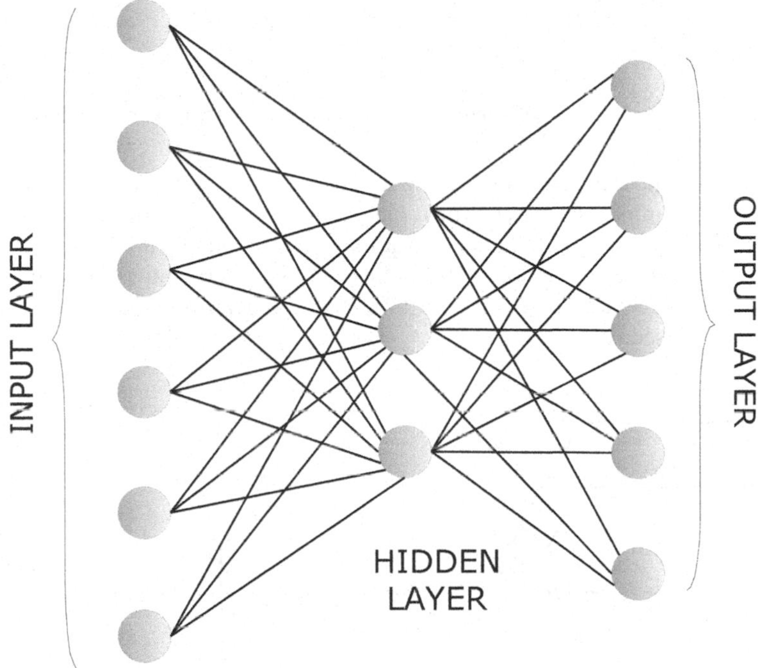

Fig. 12.15 The architecture of a multi layer perceptron containing one hidden layer.

There are three important characteristics of a multilayer perceptron:

1) the neurons in the input, hidden and output layer use, in their mathematical model, activation functions which are nonlinear and which are differentiable at any point;
2) the multilayer perceptron contains one or more hidden layers used for complex tasks;
3) have a high connectivity.

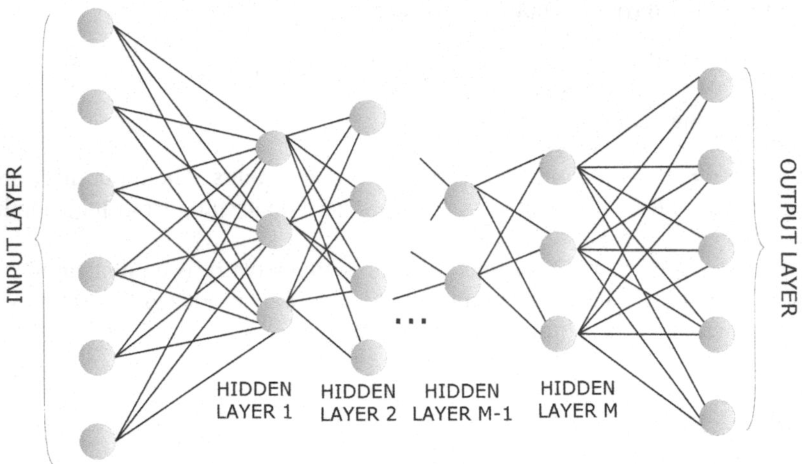

Fig. 12.16 The architecture of a multi layer perceptron containing M hidden layers.

To make nonlinear partitions on the space each unit has to be defined as a nonlinear function (unlike the perceptron). One solution is to use the sigmoid unit. Another reason for using sigmoids are that they are continuous unlike linear thresholds and are thus differentiable at all points. Figure 12.16 shows the activation of a multilayer perceptron by contrast with a single perceptron in the sense of activation function used.

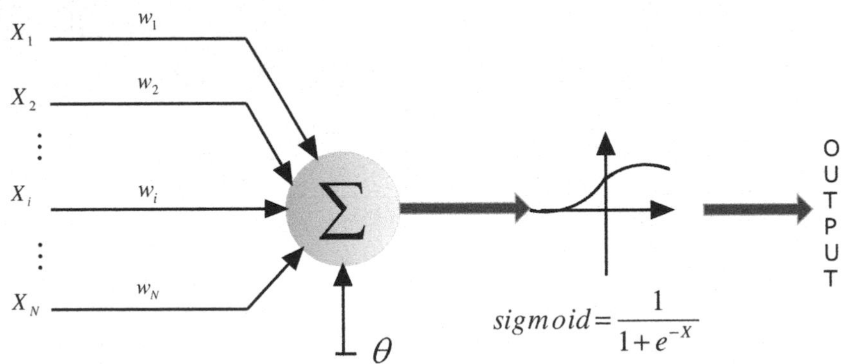

Fig. 12.17 Activation function in the case of MLP.

A multi layer network (with one or more hidden layers) can learn any continuous mapping with an arbitrary accuracy. For some applications more than one hidden layer might be useful, but, in general, one hidden layer is sufficient.

For learning a multilayer neural network, several learning algorithms have been proposed. The backpropagation learning algorithm is one of the most common training methods. We will discuss the backpropagation learning in this chapter.

The central idea behind this solution is that the errors for the units of the hidden layer are determined by back-propagating the errors of the units of the output layer. For this reason the method is often called the back-propagation learning rule. Back-propagation can also be considered as a generalization of the delta rule for non-linear activation functions and multilayer networks [20].

12.5.1 Backpropagation Learning Algorithm

The training of a network by backpropagation involves three stages [11]:

(i) the feedforward of the input training patterns;
(ii) the calculation and backpropagation of the associated error;
(iii) the adjustment of the weights.

In what follows we present the backpropagation learning algorithm for a neural network with one hidden layer.

12.5.1.1 Backpropagation Learning: Network with One Hidden Layer

Consider the network in Figure 12.17 with the following notations:

- $x = (x_1, x_2, \ldots, x_i, \ldots, x_n)$ - input training vector;
- $t=(t_1, t_2, \ldots, t_k, \ldots, t_m)$ – output target vector;

- $X_1, X_2, \ldots, X_i, \ldots, X_n$ - neurons in the input layer;
- $Z_1, Z_2, \ldots, Z_j, \ldots, Z_p$ - neurons in the hidden layer;
- $Y_1, Y_2, \ldots, Y_k, \ldots, Y_m$ - neurons in the output layer;

- θ_{hid_j}, $j=1,\ldots, p$ – threshold of each neuron in the hidden layer;
- θ_{out_k}, $k=1,\ldots, m$ – threshold of each neuron in the output layer;

- v_{ij}, $i=1,\ldots,n$, $j=1,\ldots,p$ – weights between neurons in the input layer and neurons in the hidden layer
- w_{jk}, $j=1,\ldots,p$, $k=1,\ldots,m$ – weights between neurons in the hidden layer and neurons in the output layer

- net input for neurons in the hidden layer:
$$z\_input_j = \theta_{hid\_j} + \sum_i x_i v_{ij}\ ,j=1, 2, \ldots, p$$
- net input for neurons in the output layer:
$$y\_input_k = \theta_{out\_k} + \sum_j z_j w_{jk}\ , k=1, 2,\ldots,m$$

- output signal (activation) of neurons in the hidden layer:
  $z_j = f(z\_input_j), j=1, 2, \ldots, p$

- output signal (activation) of neurons in the output layer:
  $y_k = f(y\_input_k), k=1, 2, \ldots, m$

- δ_k , $k = 1, 2, \ldots, m$ - portion of error correction weight adjustment for weights between neurons in the hidden layer and output layer w_{jk}, due to an error at neurons in the output layer (Y_k for the weight w_{jk}). The error at neuron Y_k is propagated back to the neurons in the hidden layer that feed into neuron Y_k.
- δ_j , $j=1, 2, \ldots, p$ – portion of error correction weight adjustment for weights between neurons in the input layer and hidden layer v_{ij}, due to the backpropagation of error from the output layer to the hidden neuron.

- α - learning rate.

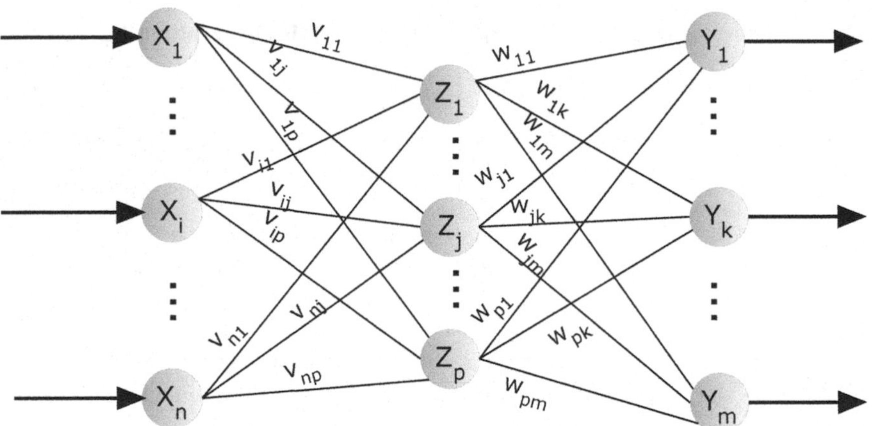

Fig. 12.18 Neural network with one hidden layer for backpropagation.

The activation function for backpropagation network should be continuous, differentiable and monotonically non-decreasing. Its derivative should be easy to compute.

One of the most general used functions is sigmoid function:

$$f(x) = \frac{1}{1+e^{-x}}$$

whose derivative is:

$$f'(x) = f(x)[1 - f(x)].$$

Other two common used activation functions are bipolar sigmoid:

$$f(x) = \frac{2}{1+e^{-x}} - 1$$

and hypertangent:

$$tahn(e) = \frac{e^{x} - e^{-x}}{e^{x} + e^{-x}}.$$

12.5.1.1.1 Backpropagation Learning: Main Steps

The main steps of the backpropagation algorithm are:

1) ***Randomly initialize the weights***
   The weights initialization influences the speed of the network in reaching the goal. It is important to avoid choices of initial weights that will make either activation functions or their derivatives to be zero.
   There are two situations which should be avoided:
   (i) Too large values for the initial weights will make the initial input signals to each neuron in the hidden layer and output layer respectively fall in the region where the derivative of the sigmoid function has a very small value.
   (ii) Too small values for the initial weights will make the net input to a hidden or output neuron be close to zero which causes slow learning.

   A standard way to initialize the weights is to use random values between [-0.5, 0.5]. The weights values can be either positive or negative [11]. Haykin [2][26] suggests initial random values for weights within the range $\left(-\frac{2.4}{F_i}, \frac{2.4}{F_i}\right)$, where F_i is the total number of inputs of neuron i in the network.

2) ***Feedforward***
   During the feedforward, each input unit X_i receives an input signal and broadcasts this signal to each of the neurons Z_1, Z_2, …, Z_p in the hidden layer. Each output neuron Y_k computes its activation to produce the response of the network to the given input pattern.

Each output neuron compares its computed activation y_k with the corresponding target value t_k to determine the associated error for that pattern with that neuron.

Once the error is calculated for each neuron, it backpropagates from layer to layer [11].

3) Backpropagation of the error

The backpropagation starts from the output layer and influence all the neurons in the hidden layers. In our case with just one hidden layer we have the following steps:

(i) For each neuron Y_k in the output layer, the term δ_k is computed based on the associated error. δ_k is used to distribute the error at output neuron Y_k back to all the neurons in the hidden layer which are connected with the neuron Y_k. It is also used to update the weights between the hidden layer and the output layer.

(ii) For each neuron Z_j in the hidden layer, the term δ_j is computed. δ_j is used to update the weights between the neuron in the input and hidden layer. In our case, since we only have one hidden layer, it is not necessary to propagate the error back to the input layer [11].

4) Weights update

Once all the δ terms have been computed, the weights for all layers can be adjusted. This is also done in two steps:

(i) The adjustment of the weights w_{jk}, $j = 1, 2, \ldots, p$, $k = 1, 2, \ldots, m$ between neurons in the hidden layer and output layer are modified based on the term δ_k and the activation z_j, $j = 1, 2, \ldots, m$ of the neurons in the hidden layer.

(ii) The adjustment of the weights v_{ij}, $i = 1, 2, \ldots, n$, $j = 1, 2, \ldots, p$, between neurons in the input layer and neurons in the hidden layer are modified based on the term δ_j and the activation x_j, $i = 1, 2, \ldots, n$ of the neurons in the input layer [11].

12.5.1.1.2 Backpropagation Learning: The Algorithm

The main steps of the backpropagation algorithm are presented in detail below [11]:

Step1. Randomly initialize the weights: v_{ij}, $i = 1, 2, \ldots, n$, $j = 1, 2, \ldots, p$;
w_{jk}, $j = 1,\ldots,p$, $k=1,\ldots,m$;
and thresholds: θ_{hid_j}, $j=1,\ldots, p$;
θ_{out_k}, $k=1,\ldots, m$

Repeat

Step 2. Feedforward

Step 2.1. Each neuron X_i, $i = 1, 2,\ldots,n$ in the input layer receives input signal x_i and broadcasts this signal to all neurons in the hidden layer.

Step 2.2. Neurons in the hidden layer
Each neuron Z_i, $j = 1, 2,\ldots,p$ in the hidden layer:

1. sums its weighted input signals:

$$z\_input_h = \theta_{hid1\_h} + \sum_{i=1}^{n} x_i \cdot u_{ih}$$

2. applies its activation function to compute its output signal:

$z_j = f(z\_input_j)$

3. sends a signal to all the neurons in the layer above (output layer) with whom it is connected.

Step 2.3. Neurons in the output layer
Each neuron Y_k, $k = 1, 2,\ldots,m$ in the output layer:

1. sums its weighted input signals:

$$y\_input_k = \theta_{out\_k} + \sum_{j=1}^{p} z_j wjk$$

2. applies its activation function to compute its output signal:

$y_k = f(y\_input_k)$

Step 3. Backpropagation

Step 3.1. Neurons in the output layer
Each neuron Y_k, $k = 1, 2,\ldots,m$ in the output layer receives a target pattern corresponding to the input training pattern.

1. If the output of the neuron is different for the target, then the error term is computed:

$\delta_k = (t_k - y_k) \cdot f\,'(y\_input_k)$

2. Calculate the weights correction term used for updating the weights:

$$\Delta w_{jk} = \alpha \cdot \delta_k \cdot z_j$$

3. Calculate the threshold correction term used for updating the threshold:

$$\Delta \theta_{out\_k} = \alpha \cdot \delta_k$$

4. Send δ_k to units in the layer below (hidden layer in our case) to which it is connected.

Step 3.2. Neurons in the hidden layer
Each neuron Z_j, $j = 1, 2, \ldots, p$ in the hidden layer:

1. Sums its delta inputs from neurons in the layer above (output layer):

$$\delta\_inputs_j = \sum_{k=1}^{m} \delta_k w_{jk}$$

2. Multiplies this term with the derivative of its activation function to calculate its error information term:

$$\delta_j = \delta\_inputs_j \cdot f'(z\_input_j)$$

3. Calculates its weight correction term used to update v_{ij}:

$$\Delta v_{jk} = \alpha \cdot \delta_j \cdot x_i$$

4. Calculates the threshold correction term used to update θ_{hid_j}:

$$\Delta \theta_{hid\_j} = \alpha \cdot \delta_j$$

Step 4. Weights and thresholds update

Step 4.1. Neurons in the output layer
Each neuron Y_k, $k = 1, 2, \ldots, m$ in the output layer updates its thresholds and weights:

$$w_{jk} = w_{jk} + \Delta w_{jk},\ j = 1, 2, \ldots, p$$
$$\theta_{out\_k} = \theta_{out\_k} + \Delta \theta_{out\_k}$$

Step 4.2. Neurons in the hidden layer

Each neuron Z_j, $j = 1, 2,\ldots,p$ in the hidden layer updates its thresholds and weights:

$v_{ij} = v_{ij} + \Delta v_{ij}$, $i = 1, 2, \ldots, n$
$\theta_{hid\_j} = \theta_{hid\_j} + \Delta\theta_{hid\_j}$

Until stopping condition.

Remarks

(i) One cycle through the entire set of training vectors is known as an *epoch*. Usually, several epochs are required to train a backpropagation neural network. In the algorithm above, the weights are updated after each training pattern is presented. Another common approach is to update the weights cumulated oven an entire epoch.

(ii) A common stopping condition for ending the training algorithm is when the total squared error reaches a minimum. But there are certain situations in which it is not efficient to continue training until the error reaches its minimum. Hecht – Nielsen [24] suggests using two sets of disjoint data during training: a set of training patterns and a set of training-testing patterns. Weights adjustment is based on the training data but at certain intervals during training the error is computed using the training –testing patterns. Training continues as long as the error for the training-testing patterns decreases and it stops when this error starts increasing.

12.5.1.1.3 Application of a Backpropagation Neural Network

After training, a backpropagation neural network is applied by using only the feedforward phase of the training algorithm.

The algorithm is [11]:

Step 1. Initialize the weights and thresholds from the training algorithm

Step 2. Set activation of input unit x_i, $i = 1, 2, \ldots, n$;

Step 3. For all neurons in the hidden layer calculate:

$$z\_input_j = \theta_{hid\_j} + \sum_{i=1}^{n} x_i \cdot v_{ij}$$

$z_j = f(z\_input_j)$

$j = 1, 2, \ldots, p$

Step 4. For all neurons in the output layer calculate:

$$y\_input_k = \theta_{out\_k} + \sum_j z_j w_{jk}$$

$$y_k = f(y\_input_k)$$

$$k = 1, 2, \ldots, m$$

12.5.1.2 Backpropagation Learning: Network with Two Hidden Layers

Consider the multilayer network with two hidden layers given in Figure 12.19. The hidden layers are Z and ZZ. We use the following notations:

- $x = (x_1, x_2, \ldots, x_i, \ldots, x_n)$ - input training vector;
- $t=(t_1, t_2, \ldots, t_k, \ldots, t_m)$ – output target vector;

- $X_1, X_2, \ldots, X_i, \ldots, X_n$ - neurons in the input layer;
- $Z_1, Z_2, \ldots, Z_h, \ldots, Z_q$ - neurons in the first hidden layer;
- $ZZ_1, ZZ_2, \ldots, ZZ_j, \ldots, ZZ_q$ - neurons in the second hidden layer;
- $Y_1, Y_2, \ldots, Y_k, \ldots, Y_m$ - neurons in the output layer;

- θ_{hid1_h}, $h=1,\ldots, q$ – threshold of each neuron in the first hidden layer;

- θ_{hid2_j}, $j=1,\ldots, p$ – threshold of each neuron in the second hidden layer;
- θ_{out_k}, $k=1,\ldots, m$ – threshold of each neuron in the output layer;

- u_{ih}, $i=1,\ldots,n$, $h=1,\ldots,q$ - weights between neurons in the input layer and neurons in the first hidden layer
- v_{hj}, $h=1,\ldots,q$, $j=1,\ldots,p$ – weights between neurons in the first hidden layer and neurons in the second hidden layer
- w_{jk}, $j=1,\ldots,p$, $k=1,\ldots,m$ – weights between neurons in the second hidden layer and neurons in the output layer

- net input for neurons in the first hidden layer:

$$z\_input_h = \theta_{hid1\_h} + \sum_i x_i u_{ih}, \ h=1, 2, \ldots, q$$

- net input for neurons in the second hidden layer:

$$zz\_input_j = \theta_{hid2\_j} + \sum_h z_h v_{hj}, \ j=1, 2, \ldots, p$$

- net input for neurons in the output layer:

$$y\_input_k = \theta_{out\_k} + \sum_j zz_j w_{jk} \text{ , } k=1, 2,\ldots,m$$

- output signal (activation) of neurons in the first hidden layer:
  $z_h = f(z\_input_h)$, $h=1, 2, \ldots, q$

- output signal (activation) of neurons in the second hidden layer:
  $zz_j = f(zz\_input_j)$, $j=1, 2, \ldots, p$

- output signal (activation) of neurons in the hidden layer:
  $y_k = f(y\_input_k)$, $k=1, 2, \ldots, m$

- δ_k, $k = 1, 2, \ldots, m$ - portion of error correction weight adjustment for weights between neurons in the hidden layer and output layer w_{jk}, due to an error at neurons in the output layer (Y_k for the weight w_{jk}). The error at neuron Y_k is propagated back to the neurons in the second hidden layer that feed into neuron Y_k.
- δ_j, $j=1, 2, \ldots, p$ – portion of error correction weight adjustment for weights between neurons in the first hidden layer and second hidden layer v_{hj}, due to the backpropagation of error from the output layer to the hidden neuron.
- δ_h, $h=1, 2, \ldots, q$ – portion of error correction weight adjustment for weights between neurons in the input layer and first hidden layer u_{ih}, due to the backpropagation of error from the output layer to the hidden neuron.

- α - learning rate.

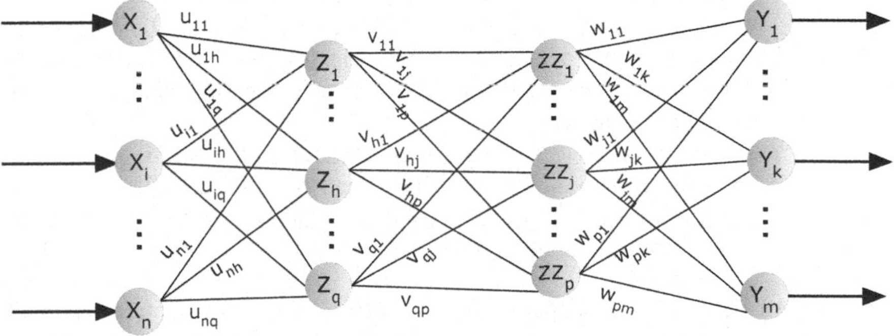

Fig. 12.19 Multilayer network with two hidden layers.

12.5.1.2.2 Backpropagation Learning – Two Hidden Layers: Main Steps

The main steps are same as in the case of a network with a single hidden layer:

1) Weights and thresholds initialization;
2) Feedforward;
3) Backpropagation of error;
4) Weights and thresholds update.

Weights and thresholds are initialized in the same manner as in the case of backpropagation learning for a network with one hidden layer described in the previous sections.

During *feedforward*, the process works as follows [11]:

1) During the feedforward, each input unit receives an input signal and broadcasts it to each of the hidden units (with whom it is connected) in the first hidden layer.
2) Each of the neurons Z_h, h = 1, 2, …, q in the first hidden layer computes their activation and sends its signal to the neurons ZZ_j, j = 1, 2, …, p in the second hidden layer.
3) Each neuron ZZ_j, j = 1, 2, …, p in the second hidden layer computes its activation and sends its signal to the neurons in the output layer.
4) Each neuron Y_k, k= 1, 2, …, m of the output layer computes its activation to get the response of the network to the given input pattern.

During the *backpropagation*, the following steps are performed [11]:

1) Each output neuron compares its computed activation with the target value to determine the error: $e_k = t_k - y_k$. Based on the error, the term δ_k, k=1, 2, …, m is computed. δ_k is used to distribute the error at the output layer back to all the neurons in the previous layer (which is the second hidden layer in our case) and to update the weights w_{jk} between the second hidden layer and the output layer.
2) The term δ_j, j=1, 2, …, p is computed for all the neurons ZZ_j, j = 1, 2, …, p in the second hidden layer. δ_j is used to distribute the error back to all neurons in the next lower layer (which is the first hidden layer in our case) and to update the weights between the first hidden layer and the second hidden layer.
3) The term δ_h, h=1, 2, …, q is computed for all the neurons Z_h, h = 1, 2, …, q in the first hidden layer. Since there is no need to propagate the error back to the input layer, δ_h is used to update the weights between the input layer and the first hidden layer.

The last step of the algorithm, *weights and thresholds update*, starts when all the δ terms have been computed. The weights for all the layers are adjusted simultaneously [11].

1) The adjustment of weight w_{jk}, between neuron ZZ_j in the second hidden layer and the neuron Y_k in the output layer is adjusted based on the factor δ_k and the activation of neuron ZZ_j.

2) The adjustment of weight v_{hj}, between neuron Z_h in the first hidden layer and the neuron ZZ_j in the second hidden layer is adjusted based on the factor δ_j and the activation of neuron Z_h.
3) The adjustment of weight u_{ih}, between neuron X_i in the input layer and the neuron Z_h in the first hidden layer is adjusted based on the factor δ_h and the activation of neuron X_i.

12.5.1.2.2 Backpropagation Learning – Two Hidden Layers: The Algorithm

The main steps of the backpropagation algorithm for network having two hidden layers are presented in detail below [11]:

Step1. Randomly initialize the weights: u_{ih}, $i = 1, 2, \ldots, n$, $h = 1, 2, \ldots, q$;
v_{hj}, $h = 1, 2, \ldots, q$, $j = 1, 2, \ldots, p$;
w_{jk}, $j = 1,\ldots,p$, $k = 1,\ldots,m$;
and thresholds: θ_{hid1_h}, $h = 1,\ldots, q$;
θ_{hid2_j}, $j = 1,\ldots, p$;
θ_{out_k}, $k = 1,\ldots, m$

Repeat

Step 2. Feedforward

Step 2.1. Each neuron X_i, $i = 1, 2,\ldots,n$ in the input layer receives input signal x_i and broadcasts this signal to all neurons in the hidden layer.

Step 2.2. Neurons in the first hidden layer

Each neuron Z_h, $h = 1, 2,\ldots,q$ in the first hidden layer:

1. sums its weighted input signals:

$$z\_input_h = \theta_{hid1\_h} + \sum_{i=1}^{n} x_i \cdot u_{ih}$$

2. applies its activation function to compute its output signal:

$$z_h = f(z\_input_h)$$

3. sends a signal to all the neurons in the layer above (second hidden layer) with whom it is connected.

Step 2.3. Neurons in the second hidden layer

Each neuron ZZ_j, $j = 1, 2,\ldots,p$ in the second hidden layer:

1. sums its weighted input signals:

$$zz\_input_j = \theta_{hid2\_j} + \sum_{h=1}^{q} z_h \cdot v_{hj}$$

2. applies its activation function to compute its output signal:

$zz_j = f(zz\_input_j)$

3. sends a signal to all the neurons in the layer above (output layer) with whom it is connected.

Step 2.4. Neurons in the output layer
Each neuron Y_k, $k = 1, 2,\ldots,m$ in the output layer:

1. sums its weighted input signals:

$$y\_input_k = \theta_{out\_k} + \sum_{j=1}^{p} zz_j w_{jk}$$

2. applies its activation function to compute its output signal:

$y_k = f(y\_input_k)$

Step 3. Backpropagation

Step 3.1. Neurons in the output layer
Each neuron Y_k, $k = 1, 2,\ldots,m$ in the output layer receives a target pattern corresponding to the input training pattern.

1. If the output of the neuron is different for the target, then the error term is computed:

$\delta_k = (t_k - y_k) \cdot f\,'(y\_input_k)$

2. Calculate the weights correction term used for updating the weights:

$\Delta w_{jk} = \alpha \cdot \delta_k \cdot zz_j$

3. Calculate the threshold correction term used for updating the threshold:

$\Delta\theta_{out\_k} = \alpha \cdot \delta_k$

4. Send δ_k to units in the layer below (second hidden layer in our case) to which it is connected.

Step 3.2. Neurons in the second hidden layer
Each neuron ZZ_j, $j = 1, 2,\ldots,p$ in the second hidden layer:

1. Sums its delta inputs from neurons in the layer above (output layer):

$$\delta\_inputs_j = \sum_{k=1}^{m} \delta_k w_{jk}$$

2. Multiplies this term with the derivative of its activation function to calculate its error information term:

$$\delta_j = \delta\_inputs_j \cdot f'(zz\_input_j)$$

3. Calculates its weight correction term used to update v_{hj}:

$$\Delta v_{hj} = \alpha \cdot \delta_j \cdot z_h$$

4. Calculates the threshold correction term used to update θ_{hid2_j}:

$$\Delta\theta_{hid2\_j} = \alpha \cdot \delta_j$$

Step 3.3. Neurons in the first hidden layer
Each neuron Z_j, $j = 1, 2,\ldots,q$ in the first hidden layer:

1. Sums its delta inputs from neurons in the layer above (output layer):

$$\delta\_inputs_h = \sum_{j=1}^{p} \delta_j v_{hj}$$

2. Multiplies this term with the derivative of its activation function to calculate its error information term:

$$\delta_h = \delta\_inputs_h \cdot f'(z\_input_h)$$

3. Calculates its weight correction term used to update u_{ih}:

$$\Delta u_{ih} = \alpha \cdot \delta_h \cdot x_i$$

4. Calculates the threshold correction term used to update θ_{hid1_h}:

$$\Delta\theta_{hid1\_h} = \alpha \cdot \delta_h$$

Step 4. Weights and thresholds update

Step 4.1. Neurons in the output layer
Each neuron Y_k, $k = 1, 2,\ldots,m$ in the output layer updates its thresholds and weights:

$$w_{jk} = w_{jk} + \Delta w_{jk}, j = 1, 2, \ldots, p$$
$$\theta_{out\_k} = \theta_{out\_k} + \Delta\theta_{out\_k}$$

Step 4.2. Neurons in the second hidden layer
Each neuron ZZ_j, $j = 1, 2,\ldots,p$ in the second hidden layer updates its thresholds and weights:

$$v_{hj} = v_{hj} + \Delta\, v_{hj}, h = 1, 2, \ldots, q$$
$$\theta_{hid2\_j} = \theta_{hid2\_j} + \Delta\theta_{hid2\_j}$$

Step 4.3. Neurons in the first hidden layer
Each neuron Z_h, $h = 1, 2,\ldots,q$ in the first hidden layer updates its thresholds and weights:

$$u_{hj} = u_{ih} + \Delta\, u_{ih}, i = 1, 2, \ldots, n$$
$$\theta_{hid1\_j} = \theta_{hid1\_j} + \Delta\theta_{hid1\_j}$$

Until stopping condition.

12.5.2 Relationship between Dataset, Number of Weights and Classification Accuracy

The relationship between the dataset (number of training patterns) and the number of weights to be trained influences the accuracy of the results. This has been proven by Baum and Haussler [25].

Let us use the following notations:

- P – number of patterns;
- W – number of weights to be trained;
- A – accuracy of classification expected.

If there are enough training patterns the network will be able to classify unknown training patterns correctly. The number of training patterns is given by:

$$P = \frac{W}{A}$$

For instance, if we expect accuracy 0.1, a network with 10 weights will require 100 training patterns.

12.5.3 Improving Efficiency of Backpropagation Learning

There are several deficiencies of the backpropagation learning algorithm which make it inefficient for certain classes of applications.

Some of them are related to the weights update procedure.

If the weights are adjusted to very large values, the total input of a hidden neuron or output neuron can reach very high (either positive or negative) values, and because of the sigmoid activation function the neuron will have an activation very close to zero or very close to one [20].

The error surface of a complex network is not al all uniform and it is full of hills and valleys. Because of the gradient descent, the network can get trapped in a local minimum when there is a much deeper minimum nearby. Probabilistic methods can help to avoid this trap, but they tend to be slow. Another suggested possibility is to increase the number of hidden units. Although this will work because of the higher dimensionality of the error space, and the chance to get trapped is smaller, it appears that there is some upper limit of the number of hidden units which, when exceeded, again results in the system being trapped in local minima [20].

Some of the improvements which can help the backpropagation learning work better refer to:

- procedure for updating the weights;
- alternatives to the sigmoid activation function.

One improved variant includes *momentum*. In the backpropagation learning with momentum, the weight change is in a direction that is a combination of the current gradient and previous gradient. The advantage of modifying the gradient descent arises when some training data are very different from the majority of the data. It is advisable to use a small learning rate to avoid a major disruption in the direction of learning when a very unusual pair of training patterns is presented.

In the backpropagation with momentum, the weights from one or more previous training patterns must be saved.

For instance, if the weights at training step $t + 1$ are based on the weights at steps t and $t - 1$, then the formula of backpropagation with momentum for a network with one hidden layer is [11]:

$$w_{jk}(t+1) = w_{jk}(t) + \alpha \cdot \delta_k \cdot z_j + \mu[w_{jk}(t) - w_{jk}(t-1)]$$

and

$$v_{ij}(t+1) = v_{ij}(t) + \alpha \cdot \delta_j \cdot x_i + \mu[v_{ij}(t) - v_{ij}(t-1)]$$

or

$$\Delta w_{jk}(t+1) = \alpha \cdot \delta_k \cdot z_j + \mu \cdot \Delta w_{jk}(t)$$

$$\Delta v_{ij}(t+1) = \alpha \cdot \delta_j \cdot x_i + \mu \cdot \Delta v_{ij}(t)$$

where μ is the momentum parameter and it is a strict positive value between (0, 1).

Momentum allows the network to make large weight adjustments as long as the corrections are in the same general direction for several patterns, while using a smaller learning rate to prevent a large response to the error from any one training pattern. The network with momentum proceeds in the direction of a combination of the current gradient and the previous direction of the weight correction, instead of only proceeding in the direction of the gradient. Momentum forms an exponentially weighted sum with μ as a base and time t as an exponent of the past and present weight changes [11].

Another situation consists on using the *delta-bar-delta update*. The delta-bar-delta algorithm allows each weight to have its own learning rate. The learning rates also vary with time as training progresses.

Two heuristics are used to determine the appropriate change in the learning rate for each weight:

- If the weight change is in the same direction – increase or decrease – for several time steps then the learning rate for that weight should be increased;
- If the direction of the weight change alternates, the learning rate should be decreased.

The weight change will be in the same direction if the partial derivative of the error with respect to that weight has the same sign for several time steps [11].

Summaries

In this chapter we presented artificial neural networks. Most of the chapter concentrates on feedforward neural networks with emphasize on single layer perceptron and multilayer perceptron. The representational power of single layer feedforward networks was discussed and two learning algorithms for finding the optimal weights – perceptron learning rule and delta rule – were presented. The disadvantage of the single layer network is the limited representational power: only linear classifiers can be constructed and in case of function approximation, only linear functions can be represented. The advantage, however, is that because of the linearity of the system, the training algorithm will converge to the optimal solution which is not true in the case multi layer networks.

A multilayer perceptron is a feedforward neural network with one or more hidden layers.

The network consists of an input layer of source neurons, at least one middle or hidden layer of computational neurons, and an output layer of computational neurons.

The input signals are propagated in a forward direction on a layer-by-layer basis.

Learning in a multilayer network proceeds the same way as for a perceptron: a training set of input patterns is presented to the network. The network computes its output pattern, and if there is an error – a difference between actual and desired output patterns – the weights are adjusted to reduce this error.

A hidden layer hides its desired output. Neurons in the hidden layer cannot be observed through the input/output behaviour of the network. There is no obvious way to know what the desired output of the hidden layer should be [2].

The multi layer perceptron and many other neural networks learn using an algorithm called backpropagation. With backpropagation, the input data is repeatedly presented to the neural network. With each presentation the output of the neural network is compared to the desired output and an error is computed. This error is then fed back (backpropagated) to the neural network and used to adjust the neuron's weights such that the error decreases with each iteration and the neural model gets closer and closer to producing the desired output. This process is known as training.

An ANN is designed for a specific application, such as a data classification, through a learning process.

Multi-layered perceptrons have high representational power. They can represent the following [1]:

- *boolean functions*: every boolean function can be represented with a network having two layers of units.
- *continuous functions*: all bounded continuous functions can also be approximated with a network having two layers of units.
- *arbitrary functions*: any arbitrary function can be approximated with a network with three layers of units.

References

1. `http://www.onlinefreeebooks.net/free-ebooks-computer-programming-technology/artificial-intelligence/artificial-intelligence-course-material-pdf.html` (accessed on February 10, 2011)
2. Negnevitsky, M.: Artificial Intelligence: A Guide to Intelligent Systems. Addison-Wesley/Pearson Education, Harlow (2002) ISBN 0201-71159-1
3. Stergiou, C., Siganos, D.: Neural Networks, Imperial College London Report, `http://www-dse.doc.ic.ac.uk/~nd/surprise_96/journal/vol4/cs11/report.html` (accessed on February 10, 2011)
4. Elert, G. (ed.): The Physics Factbook. An Encyclopedia of Scientific Essays (2002)
5. Bear, M.F., Connors, B.W., Pradiso, M.A.: Neuroscience: Exploring the Brain, 2nd edn. Williams and Wilkins, Baltimore (2001)
6. Groves, P.M., Rebec, G.V.: Introduction to Biological Psychology, 4th edn. William C Brown Pub. (1992)
7. Shepherd, G.M.: The Synaptic Organization of the Brain (1998)

8. Koch, C.: Biophysics of Computation. Information Processing in Single Neurons. Oxford University Press, New York (1999)
9. Mitchell, T.: Machine Learning. McGraw-Hill, New York (1997)
10. Rosenblatt, F.: The perceptron: a probabilistic model for information storage and organization in the brain. Psychological Review 65, 386–408 (1958)
11. Fausett, L.V.: Fundamentals of Neural Networks: Architectures, Algorithms And Applications. Prentice-Hall, Englewood Cliffs (1993)
12. Haykin, S.: Neural Networks: A Comprehensive Foundation, 2nd edn. Prentice-Hall, Englewood Cliffs (1998)
13. Widrow, B., Hoff, M., Adaptive, M.: switching circuits. Western Electronic Show and Convention, Institute of Radio Engineers 4, 96–104 (1960)
14. Callan, R.: The Essence of Neural Networks. Prentice-Hall, Englewood Cliffs (1999)
15. Minsky, M., Papert, S.: Perceptrons: an introduction to computational geometry. MIT Press, Cambridge (1969)
16. Kohonen, T.: Associative memory: a system-theoretical approach. Springer, Heidelberg (1977)
17. Kohonen, T.: Self-organized formation of topologically correct feature maps. Biological Cybernetics 43, 59–69 (1982)
18. Kohonen, T.: Self organization and associative memory. Springer, Berlin (1984)
19. Kohonen, T.: Self-Organizing Maps. Springer, Heidelberg (1995)
20. Krose, B., van der Smagt, P.: An introduction to Neural Networks, University of Amsterdam, Eight Edition (1996)
21. Rumelhart, D.E., Hinton, G.E., Williams, R.J.: Learning representations by backpropagating errors. Nature 323, 533–536 (1986)
22. Rumelhart, D.E., McClelland, J.L.: Parallel Distributed Processing: Explorations in the Microstructure of Cognition. MIT Press, Cambridge (1986)
23. Block, H.D.: The perceptron: A model for brain functioning. Reviews of Modern Physics 34, 123–135 (1962)
24. Hecht-Nielsen, R.: Neurocomputing. Addison-Wesley, Reading (1990)
25. Baum, E.B., Haussler, D.: What size net gives valid generalization? Neural Computation 1(1), 151–160 (1989)
26. Haykin, S.: Neural networks: a comprehensive foundation. Macmillan College Publishing Company, New York (1994)

Verification Questions

1. How can you define the perceptron?
2. What the perceptron can be used for?
3. What are the limitations of the perceptron?
4. What is the difference between single layer perceptron and multilayer perceptron?
5. What kind of problems can be solved using a multilayer perceptron?
6. How does a multilayer neural network learn?

Exercises

1. Consider two 2-dimensional problems to be solved. Use 2000 data for training and 500 data for test (generate the data by yourself). Each data consists of 2 inputs and 1 output.

Implement a 2-layer MLNN (MLNN with 1 hidden layer)

Specifications are as follows:

Network structure: 2-N-1

Activation function of the hidden layer: $f(x) = \dfrac{2}{1+\exp(-2x)} - 1$ (tangent sigmoid)

Activation function of the output layer: linear

Termination condition: RMSE ≤ 0.02 (root mean square error)

Initial weights: uniformly random within [-0.5, 0.5]

You should determine the number of hidden neurons (N) for each function. How many hidden neurons are required for good performance?

The functions are:

- F1 (sine function)

$$f_1(x_1, x_2) = \sin(4x_1x_2), \quad x_1, x_2 \in [-1,1]$$

- F3 (Mexican hat)

$$f_2(x_1, x_2) = \sin\left(2\pi\sqrt{x_1^2 + x_2^2}\right), \; x_1, x_2 \in [-1,1]$$

2. Solve the parity 4 problem using minimum number of neurons of layered feed forward network. Use bipolar activation function. Parity problem means that output must be +1 if there is odd number of +1s on inputs and output should be -1 when there is even number of +1s on inputs. In the case of parity 4 you have four inputs, one output and 16 possible input patterns.

[0 0 0 0] ==> -1	[0 0 0 1] ==> +1
[0 0 1 0] ==> +1	[0 0 1 1] ==> -1
[0 1 0 0] ==> +1	[0 1 0 1] ==> -1
[0 1 1 0] ==> -1	[0 1 1 1] ==> +1
[1 0 0 0] ==> +1	[1 0 0 1] ==> -1
[1 0 1 0] ==> -1	[1 0 1 1] ==> +1
[1 1 0 0] ==> -1	[1 1 0 1] ==> +1
[1 1 1 0] ==> +1	[1 1 1 1] ==> -1

Design a neural network for printed digit recognition. Each character is represented by a *m* X *n* matrix having elements from {0; 1}. For instance, for digits 1 and 3 the representation on a 5 X 5 matrix is:

```
0 0 1 0 0     1 1 1 1 1
0 1 1 0 0     0 0 0 0 1
1 0 1 0 0     1 1 1 1 1
0 0 1 0 0     0 0 0 0 1
0 1 1 1 0     1 1 1 1 1
```

If the digits are represented by *m* X *n* matrices and there are *M* classes (digits) to be identified the network architecture will consist of:

- an input level consisting of $N = n * m + 1$ units (including the fictitious one - for activation threshold)
- a level of *M* functional units (which are also output units).

The input and output levels are fully connected. The activation function for the output units can be linear or logistic.

The network will be trained starting from a set of digits for which the class label is known. The training set will contain pairs containing the digit matrix and the class label.

3. Consider the following sequence of equations:

n = 1: $1^2 + 1 = 2$
n = 2: $2^2 + 2 = 6$
n = 3: $3^2 + 3 = 12$
n = 4 $4^2 + 4 = 20$
Implement a neural network to provide a result for n=5 and n=6.

4. For the following training samples:

$x_1 = (0, 0)$
$x_2 = (0, 1)$
$x_3 = (1, 0)$
$x_4 = (1, 1)$

Apply the perceptron learning rule to the above samples one-at-a-time to obtain weights that separate the training samples. Set θ to 0.2. Write the expression for the resulting decision boundary.

5. Consider a two-layer feedforward artificial neural network with two inputs *a* and *b*, one hidden unit *c*, and one output unit *d*. This network has three weights (w_{ac}, w_{bc}, w_{cd}), and two thresholds θ_c and θ_d.

Assume that these weights are initialized to (0.1, 0.1, 0.1) and the thresholds have the value 0.2.

Assume a learning rate of 0.3. Also assume a sigmoid threshold function *f*.

Use the following approximate values for f where necessary below:

x	$f(x)$
$5.0 \leq x < 2.5$	0.001
$2.5 \leq x < 0.05$	0.20
$0.05 \leq x < 0$	0.49
0	0.5
$0 < x \leq 0.05$	0.51
$0.05 < x \leq 2.5$	0.80
$2.5 < x \leq 5.0$	0.999

Consider the following training example for the network described above: $a = 1$, $b = 0$, $d = 1$.

(a) Show the output for each node during the feedforward pass.

(b) Show the error for each node as it would be computed using the Backpropagation algorithm.

Chapter 13
Advanced Artificial Neural Networks

13.1 Introduction

The networks discussed in the previous Chapter – perceptron and multilayer perceptron – are feedforward networks in the sense that the information is processed forward from layer to layer and no cycles are presented in the network. *Recurrent* (or *feedback*) networks contain cycles which can connect neurons in the hidden layers with neurons in the input layers, hidden neurons between them or, in the most general case, all the neurons between them. A network may be fully recurrent, i.e., all units are connected back to each other and to themselves, or some part of the network may be fed back in recurrent links.

One of the earliest recurrent neural networks is the auto-associative neural network proposed by Hopfield in 1982 [12] and known as *Hopfield network*. Standard feedforward networks are appropriate for mapping where the output is dependent only on the current input. They are not able to encode mappings, which are depending not only on the current input but on the previous ones. Feedforward networks have no memory. Jordan [10] [11] and Elman [9] proved that by adding further neurons to encode context with feedback links from either the output neurons or the hidden neurons to these context neurons such mappings can be performed [8].

These context neurons (units) act as a memory for the network's state and allow the output of the network to be sensitive to its state on the previous input cycle [8].

This Chapter presents recurrent networks, which are extensions of the feedforward networks presented in the Chapter12.

13.2 Jordan Network

The Jordan neural network has been proposed by Jordan in 1986 [10] [11]. In this type of network the context neurons receive a copy from the output neurons and from themselves. In the Jordan network, the set of extra input neurons (units) is called the *state units*. There are as many state units as there are output units in the network. The connections between the output and state units have a fixed weight of +1; learning takes place only in the connections between input and hidden units

C. Grosan and A. Abraham: Intelligent Systems, ISRL 17, pp. 325–344.
springerlink.com

as well as hidden and output units. Thus all the learning rules derived for the multi-layer perceptron can be used to train this network [7]. An example of Jordan network is depicted in Figure 13.1.

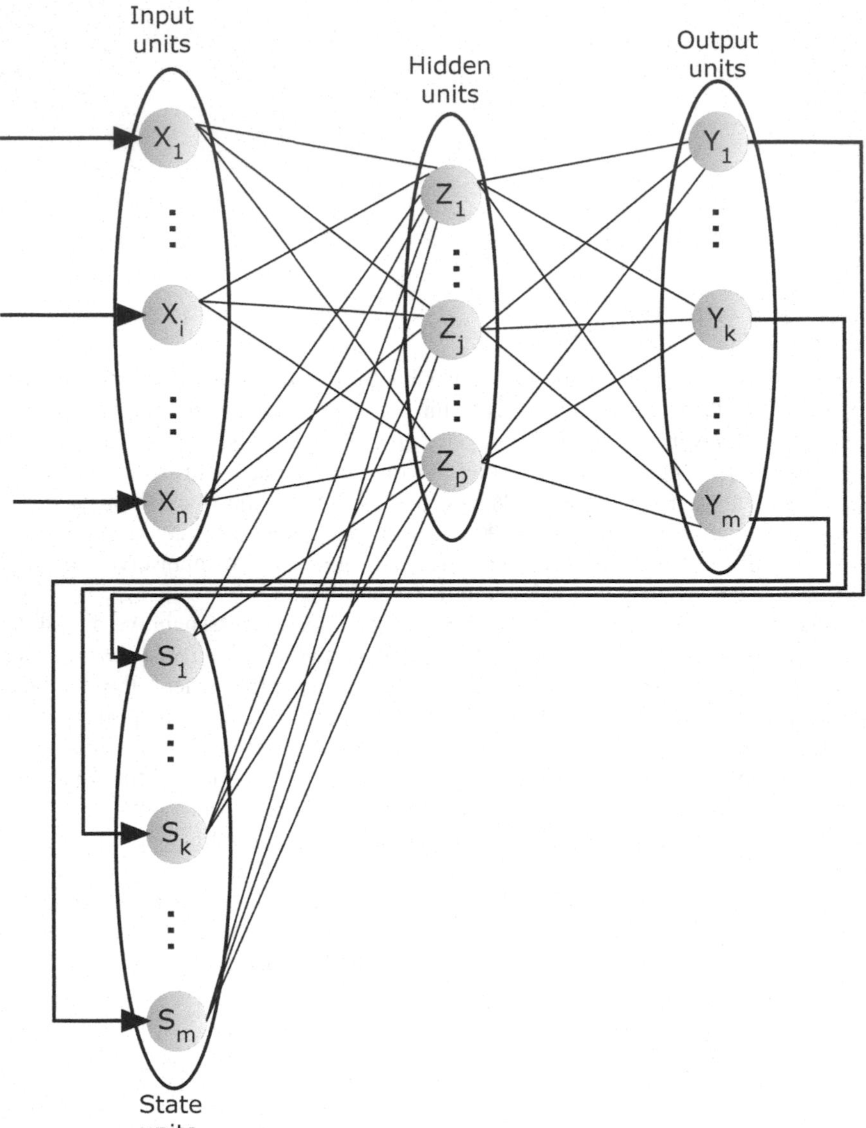

Fig. 13.1 The Jordan network

13.3 Elman Network

The Elman network was introduced by Elman [9]. In this network a set of *context neurons* are introduced and they represent extra input units whose activation values are fed back from the hidden units. The network is very similar to the Jordan network, except that the hidden neurons instead of the output units are fed back.

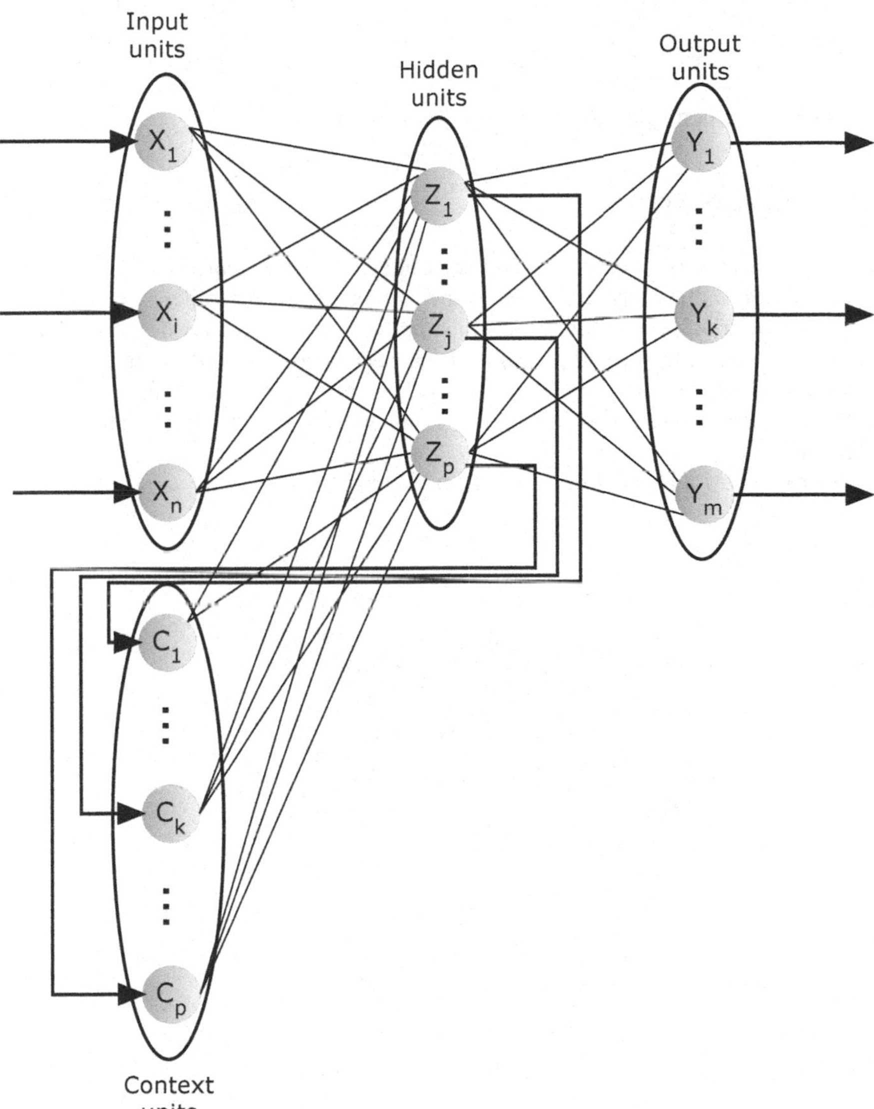

Fig. 13.2 The Elman network

The training mechanism is as follows:

Step 1. The activations of the context neurons are initialized to zero at the initial instant.
Step 2. The pattern $x(t)$ at instant t is presented and the network and the calculations are propagated towards the output of the network, obtaining therefore the prediction at instant t+1.
Step 3. The back propagation algorithm is applied to modify the weights of the network
Step 4. The time variable is increased in one unit and the procedure goes to step 2.

The schematic structure of this network is shown in Figure 13.2.

13.4 Hopfield Network

The Hopfield network consists of a set of N interconnected neurons which update their activation values asynchronously and independently of other neurons. All neurons are both input and output neurons, i.e., a pattern is clamped, the network iterates to a stable state, and the output of the network consists of the new activation values of the neurons. The activation values are binary. Originally, Hopfield chose activation values of 1 and 0, but using values +1 and -1 presents some advantages [7].

The structure of a Hopfield network is given in Figure 13.3.

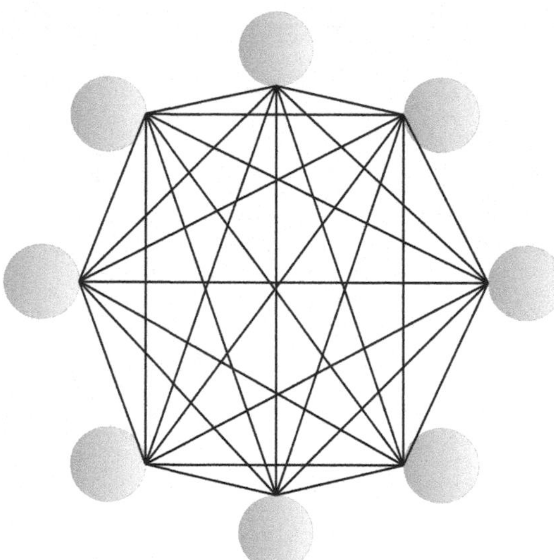

Fig. 13.3 The Hopfield network

The connections between neurons are bidirectional: the connection between neuron i and neuron j (the weight w_{ij}) is same as the connection from neuron j to neuron i.

Let us use the following notations [2]:

- u_i – internal activity of neuron i;
- $v_i = g(u_i)$ – output signal;
- w_{jk} - connection weight between neurons j and k;

The energy function is defined by:

$$E = \frac{1}{2} \cdot \sum_{i=1}^{n} \sum_{j=1}^{n} w_{ij} v_i v_j + \sum_{i=1}^{n} \theta_i v_i$$

The network will converge to a stable configuration (that is the minimum of the energy function) as long as [2]:

$$\frac{d}{dt} u_i = -\frac{\partial E}{\partial v_i} = -\sum_{j=1}^{n} w_{ij} v_j - \theta_i$$

13.5 Self Organizing Networks

In the previous sections we have presented networks which learn with the help of an external teacher (supervised learning). In contrast to supervised learning, unsupervised or self-organised learning does not require an external teacher. During the training session, the neural network receives a number of different input patterns, discovers significant features in these patterns and learns how to classify input data into appropriate categories.

Unsupervised learning algorithms aim to learn rapidly and can be used in real-time [2].

For certain problems, the training data consisting of input and desired output pairs are not available, the only information is provided by a set of input patterns. In these cases the relevant information has to be found within the (redundant) training samples.

In this chapter we describe Hebb networks and self organizing maps.

13.5.1 Hebb Networks

The earliest and simplest learning rule for a neural network has been proposed by Hebb in 1949 and it is known as Hebb's law or Hebb's rule. Hebb's law states that "When the axon of cell A is near enough to excite a cell B and repeatedly or persistently takes part in firing it, some growth process or metabolic change takes place in one or both cells such that A's efficiency, as one of the cells firing B, is increased[13]". This means that learning occurs by modification of weights in such a way that if two interconnected neurons are both *on* at the same time then the weight between those neurons should be increased.

If neuron i is near enough to excite neuron j and repeatedly increses its activation, the weight between these two neurons is strengthen and the neuron j becomes more sensitive to slimuli from neuron i [1]. In its initial form, the rule only talks about neurons firing at the same time and do not state anything about neurons that do not fire at the same time.

McClelland and Rumelhart[14] extended the rule such as the weights are also increased if the neurons are both *off* in the same time.

Hebb's law can be described as [1]:

(i) If two neurons on either side of a connection are activated synchronously (or deactivated synchronously) then the weight of that connection is increased;

(ii) If two neurons on either side of a connection are activated asynchronously then the weight of that connection is decreased.

The hebbian learning in a neural network is presented in Figure 13.22.

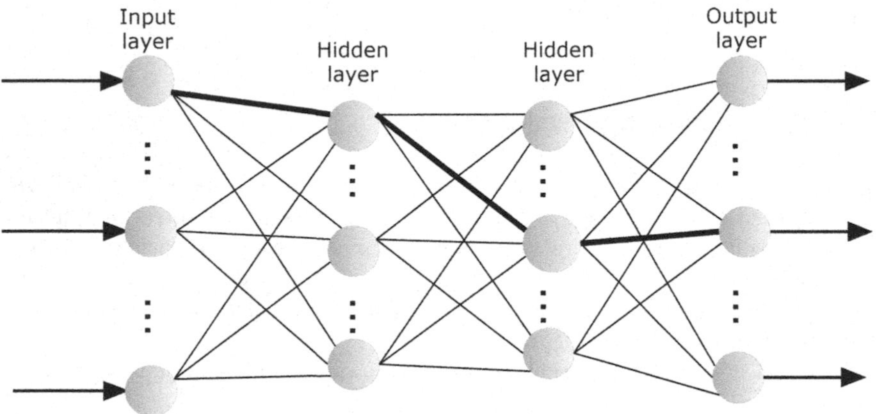

Fig. 13.4 Hebbian learning.

Using Hebb's law we can express the adjustment applied to the weight w_{ij} at iteration t in the following form [1]:

$$\Delta w_{ij}(t) = F[y_j(y), x_i(t)]$$

where $F[y_j(t), x_i(t)]$ is a function of both postsynaptic and presynaptic activities.

As a special case, Hebb's Law can be represented as:

$$\Delta w_{ij}(p) = \alpha \; y_j(p) \; x_i(p)$$

where α is the *learning rate* parameter.

This equation is referred to as the *activity product rule* [1].

Hebbian learning implies that weights can only increase. To resolve this problem, a limit on the growth of synaptic weights can be imposed. This can be done by introducing a non-linear *forgetting factor* into Hebb's Law [5]:

$$\Delta w_{ij}(t) = \alpha \ y_j(t) \ x_i(t) - \varphi \ y_j(t) \ w_{ij}(t)$$

where φ is the forgetting factor.

Forgetting factor usually falls in the interval between 0 and 1, typically between 0.01 and 0.1, to allow only a little "forgetting" while limiting the weight growth. If the forgetting factor is 0, the network is capable only of strengthening its synaptic weights (which grow towards infinity). If the forgetting factor is 1 or close to 1, the network remembers very little of what it learns. The above equation may also be written in the *generalized activity product rule* form:

$$\Delta w_{ij}(t) = \phi \ y_j(t) \ \lfloor \lambda x_i(t) - w_{ij}(t) \rfloor$$

where $\lambda = \dfrac{\alpha}{\phi}$.

The steps of the generalized Hebbian learning algorithm are [1]:

Step 1: Initialisation.

Set initial synaptic weights and thresholds to small random values, in the interval [0, 1].

Step 2: Activation.

Compute the neuron output at iteration t

$$y_j(t) = \sum_{i=1}^{n} x_i(t) \, w_{ij}(t) - \theta_j$$

where n is the number of neuron inputs, and θ_j is the threshold value of neuron j.

Step 3: Learning.

Update the weights in the network:

$$w_{ij}(t+1)=w_{ij}(t)+\Delta w_{ij}(t)$$

where $\Delta w_{ij}(t)$ is the weight correction at iteration t.

The weight correction is determined by the generalised activity product rule:

$$\Delta w_{ij}(t) = \phi \ y_j(t) \ \lfloor \lambda x_i(t) - w_{ij}(t) \rfloor$$

Step 4: Iteration.

Increase iteration t by one and go to Step 2.

13.5.2 Self Organizing Maps

Self Organizing Maps (SOM) fall into category of competitive learning. In competitive learning, neurons compete among themselves to be activated. The basic idea of competitive learning was introduced in the early 1970s. In the late 1980s, Teuvo Kohonen introduced a special class of artificial neural networks called *self-organising feature maps* [3][4][5][6]. These maps are based on competitive learning [1].

The map consists of a regular grid of neurons. A simple example of a Kohonen network is depicted in Figure 13.5. The network is a 2D lattice of nodes each of which is fully connected to the input layer. Figure 13.5 shows a very small Kohonen network of 5×6 nodes connected to the input layer represented by a two dimensional vector.

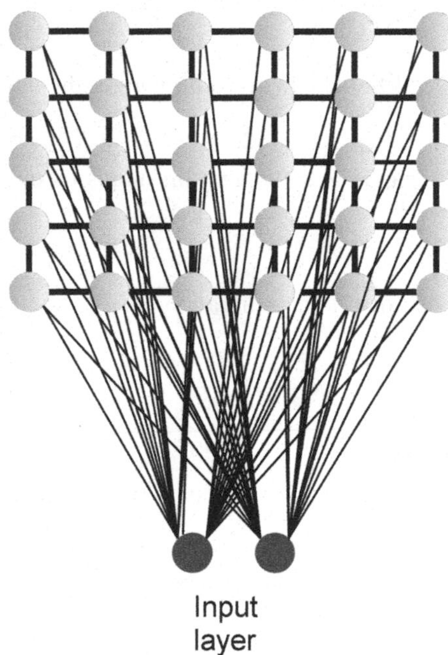

Fig. 13.5 Kohonen network.

A model of some multidimensional observation, eventually a vector consisting of features, is associated with each unit. The map attempts to represent all the available observations with optimal accuracy using a restricted set of models. At the same time the models become ordered on the grid so that similar models are close to each other and dissimilar models far from each other.

The objective of a Kohonen network is to map input vectors (patterns) of arbitrary dimension N onto a discrete map with 1 or 2 dimensions. Patterns close to one another in the input space should be close to one another in the map: they should be topologically ordered. A Kohonen network is composed of a grid of output units and N input units. The input pattern is fed to each output unit. The input lines to each output unit are weighted. These weights are initialized to small random numbers.

In a Kohonen network, the neurons are presented with the inputs, which calculate their *net* (weighted sum) and neuron with the closest output magnitude is chosen to receive additional training. Training, though, does not just affect the one neuron but also its neighbors. Each node has a set of neighbors. Examples of neighborhoods with radii 0, 1, 2, and 3 for a two-dimensional grid and 0, 1 and 2 for a one-dimensional grid are depicted in Figure 13.6.

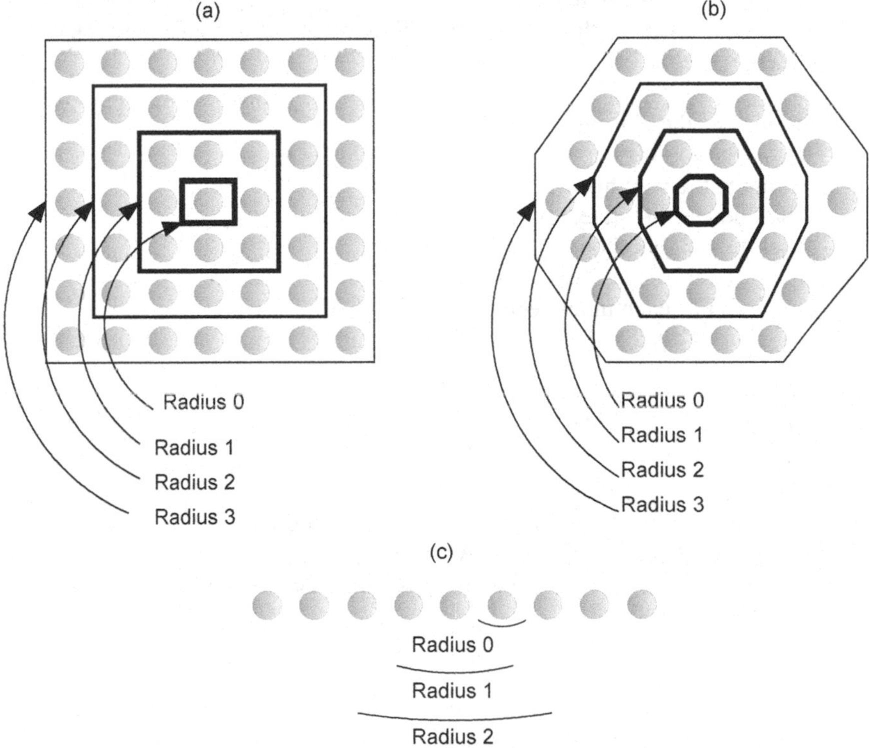

Fig. 13.6 Example of neighborhoods with radii 0, 1, 2, and 3 for a two-dimensional grid (a) and (b) and 0, 1 and 2 for a one dimensional grid (c).

The learning process has the following main steps:

1) initialise the weights for each output unit
2) repeat

- for each input pattern:
  - present the input pattern
  - find the winning output unit
  - find all units in the neighborhood of the winner
  - update the weight vectors for all those units
- reduce the size of neighborhood if required

until weight changes are negligible

The winning output neuron is the unit with the weight vector that has the smallest Euclidean distance to the input pattern.

13.5.2.1 Kohonen Self Organizing Maps: The Algorithm

Let us use the following notations:

- x = $(x_1, x_2, \ldots, x_n)$ – input vector;
- w_{ij} – weight between i-th input and the j-th node (neuron) in the network;
- α - learning rate

The main steps of the algorithm are [2]:

Step 1. Initialize the weights w_{ij}.
Set the learning rate.
Set the neighborhood.

Repeat

Step 2. For each input vector do

Step 2.1. For each node j calculate:

$$D(j) = \sum_{i=1}^{n} \left(w_{ij} - x_i\right)^2$$

Step 2.2. Find J for which $D(J)$ is minimum.

Step 2.3. Update the weights:

For all units j within the neighborhood and for all i:

$$w_{ij} = w_{ij} + \alpha(x_i - w_{ij})$$

Step 3. Update learning rate

Step 4. Reduce the neighborhood.

Until stopping condition.

When a node wins a competition not only are its weights adjusted, but those of the neighbors are also changed. The weight of the neighbors are not changed that much and the changes are based on the distance to the winning neuron: the further the neighbor is from the winner, the smaller its weight change. As training goes on, the neighborhood gradually shrinks. At the end of training, the neighborhoods have shrunk to zero size. The learning rate α is also a function slowly decreasing with time. Kohonen [2] [4] indicates that a linearly decreasing function is satisfactory for practical computations. A geometric decrease will produce similar results.

13.6 Neocognitron

The neocognitron [15] [16] is a hierarchical network, which has many layers with sparse and localized pattern of connectivity between layers.

Hierarchical feature extraction is the basic principle of the neocognitron. Hierarchical feature extraction consists in distribution of extracted features to several stages. The simplest features (usually only rotated lines) are extracted in the first stage and in each of the following stages the more complex features are extracted. In this process it is important fact that only information obtained in the previous stage are used for feature extraction in the certain stage. An example is given in Figure 13.7.

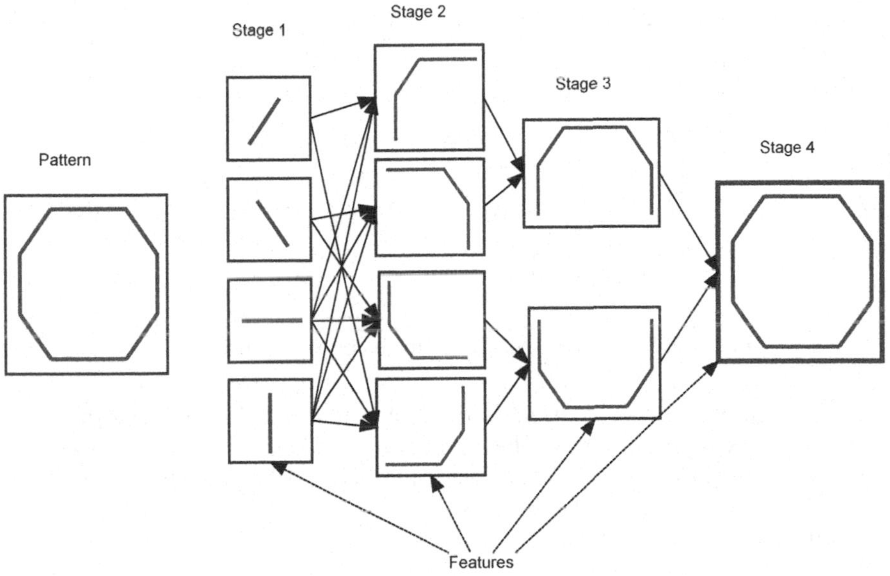

Fig. 13.7 Hierarchical feature extraction

A good description of the neocognitron including applets can be found in [17]. In what follows we will also present the architecture of the neocognitron for the above example. Structure of the neocognitron arises from a hierarchy of extracted features. One appropriate stage of the neocognitron is created for each stage of the hierarchy of extracted features. The network contains one additional stage, stage 0, which is not used, in contrast to higher stages, for feature extraction.

All the stages of the neocognitron and a part of features extracted by them, corresponding to the hierarchy in Figure 13.7 are shown in Figure 13.8. Total number of stages of the neocognitron depends on the complexity of recognized patterns. The more complex recognized patterns are, the more stages of hierarchy of extracted features we need and the higher number of stages of the neocognitron is.

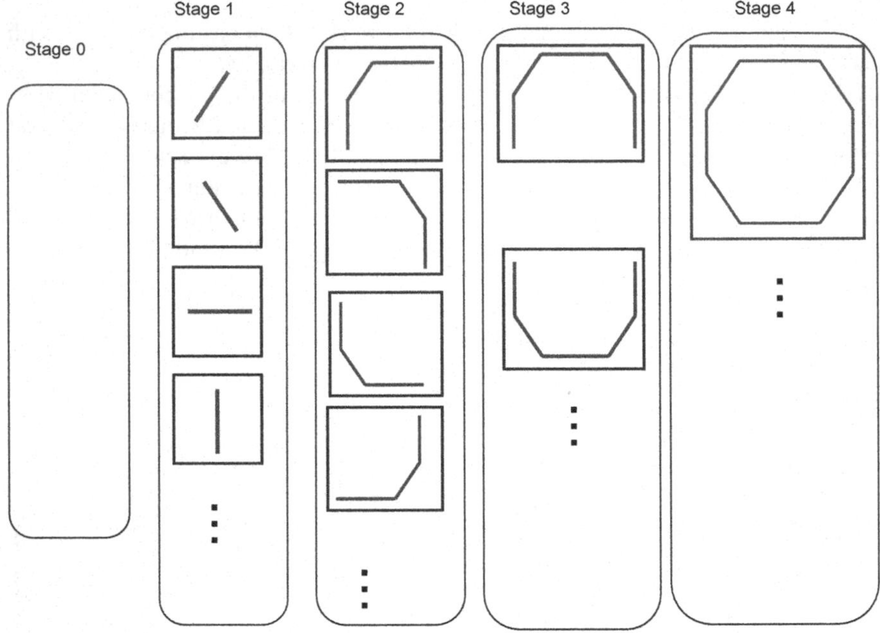

Fig. 13.8 Stages corresponding to the hierarchy in Figure 13.7.

Each stage of the neocognitron consists of certain number of layers of given type. For the example considered above, four types of layers exist in the neocognitron. Stage 0 always consists of only one input layer. All higher stages consist of one S-layer, one V-layer and one C-layer.

Let us use the following notations:

- S_i - S-layer in the *i*-th stage of the network;
- V_i - V-layer in the *i*-th stage of the network;
- C_i - C-layer in the *i*-th stage of the network.

With these notations, the layers for the example are depicted in Figure 13.9.

Each layer in the neocognitron consists of certain number of cell *planes* of the same type except from the input layer. Number of cell planes in each S-layer and C-layer depends on the number of features extracted in corresponding stage of the network. Each V-layer always consists of only one cell plane. Structure of the network in our example after drawing of cell planes is presented in

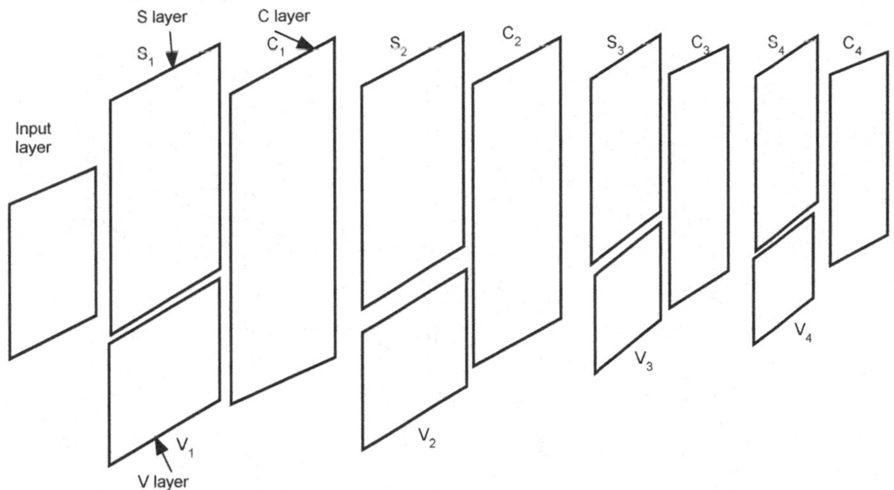

Fig. 13.9 Layers corresponding to the hierarchy in Figure 13.7

Figure 13.9. The basic component of the neocognitron is the *cell*. The neocognitron is made of large amount of cells of several distinct types. There are 4 types of cells:

- Receptor cell;
- S cell;
- V cell;
- C cell.

The cells are organized in cell planes, layers and stages. All the cells, regardless of their type, process and generate analog values. Figure 13.10 presents the cells in the network structure for our example.

Size of cell arrays is the same for all cell planes in one layer and it decreases with increasing of the network stage. Each C-plane in the highest stage of the network contains only one cell. Its output value indicates a measure of belonging of presented pattern into the category represented by this cell. Size of cell array in each V-plane is the same as size of cell arrays in S-planes in the same stage of the network.

Each V-cell in the neocognitron evaluates outputs of C-cells (or receptor cells) from the certain connection areas from previous C-layer (or input layer). Size of connection areas is the same for all V-cells and S-cells in one stage of the network and it is determined at construction of the network.

V-cell output value represents average activity of cells from connection areas and it is used for inhibition of corresponding S-cell activity. Exact specification of V-cell function is described in mathematical description of its behavior. Each S-cell in the neocognitron evaluates outputs of C-cells (or receptor cells)

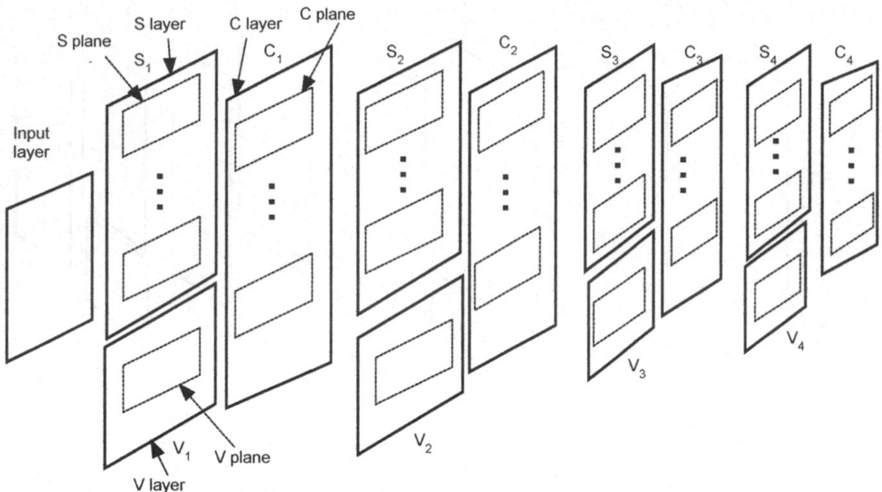

Fig. 13.10 Cell planes corresponding to the hierarchy in Figure 13.7.

from the certain connection areas from previous C-layer (or input layer). Size of connection areas is the same for all S-cells in one S-layer and it is determined at construction of the network. Function of each S-cell is to extract the certain feature at the certain position in the input layer. For extraction of this feature S-cell uses only information obtained from its connection areas and information about average activity in these areas obtained from corresponding V-cell. All S-cells in one S-plane always extract the same feature.

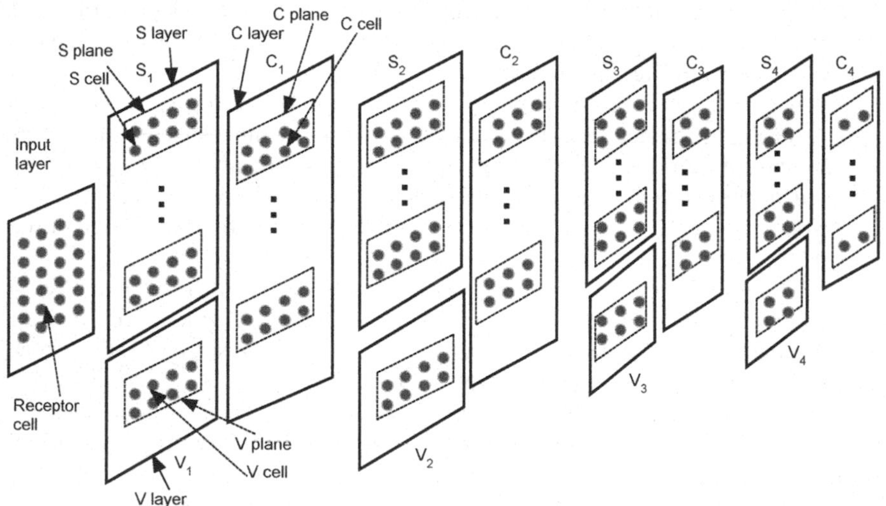

Fig. 13.11 Cells corresponding to the hierarchy in Figure 13.7

The feature extracted by S-cell is determined by weights for this cell. The meaning of weights is best for cells from layer S_1. Each S-cell in this layer has only one connection area and this area is S-cell's receptive field at the same time. Weights contain directly representation of the certain feature. In higher S-layers correspondence between extracted feature and its representation by the weights is already not so obvious. S-cell is activated only if this feature is present in S-cell's receptive field (it is identical with connection area here). When incorrect feature is presented the cell becomes inactive.

S-cell output value is determined exactly by the equation described in mathematical description:

$$output_S = \varphi * \left[\frac{E(a, output_C)}{I(r, b, output_V)} \right]$$

where:

$output_S$: S cell output value;
φ*: nonlinear function;
E: excitatory part;
a, b: weights;
$output_C$: output values of C cells from connection areas;
$output_S$: V cell output value;
I: inhibitory part;
r: selectivity.

The S-cells ability to extract not only learned features but also deformed representations of these features is influenced by the choice of parameter denoted as selectivity to a great extent.

The process of feature extraction is influenced by selectivity to a great extent. For each S-layer in the neocognitron we can set different amount of selectivity at construction of the network. By the change of selectivity we change the effect of inhibitory part on the S-cell output value. Decreasing of selectivity involves decreasing of effect of inhibition part. Decreased S-cell ability to distinguish learned feature exactly is the result of it. In other words it means that S-cell considers also more deformed features to be correct.

Each C-cell in the neocognitron evaluates outputs of S-cells from the certain connection area from one of S-planes from previous S-layer. Number of S-planes, however, can be greater in some cases. Size of connection areas is the same for all C-cells in one C-layer and it is determined at construction of the network

C-cell output value depends on activity of S-cells from connection area. The greater number of active S-cells is or the greater their activities are the greater C-cell output value is. C-cell function is exactly described in mathematical description. For C-cell to be active it is sufficient that at least one active S-cell is present in its connection area. With regard to overlapping of neighboring C-cell connection areas activity of one S-cell affects activity of greater number of C-cells.

Ability of C-cell to compress content of connection area in the certain way is the next consequence of C-cell function. Hence we can decrease the density of cells in C-layer to the half of density of cells in previous S-layer in some cases. The neocognitron is characteristic not only by large number of cells but also by large number of connections. These connections serve for transfer of information between cells in adjoining layers. Particular cell obtains by means of connections information from all cells, which are located in its connection areas.

For each connection there is a weight by means of it we can affect amount of transferred information. If we imagine a connection as a pipeline with a valve we can compare weight assigned to the connection to a degree of opening of this valve. Four types of weights (a-weights, b-weights, c-weights and d-weights) exist in the neocognitron. Each of these types of weights is used for connections between two layers of different types (see Figure 13.12).

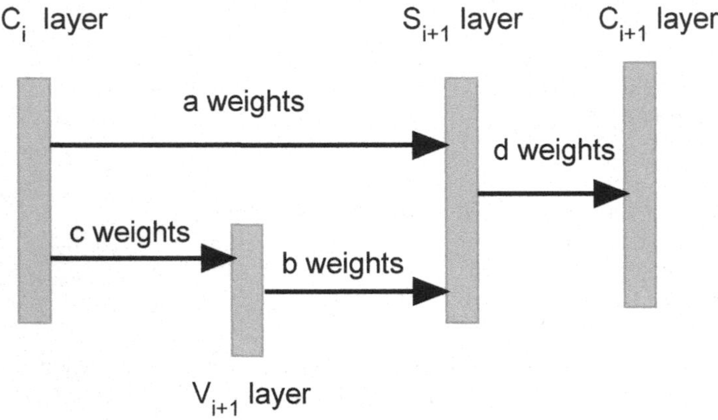

Fig. 13.12 The four type of weights.

Some of the weights will be adjusted by learning and some of the weights will remain fixed. a and b weight modify by learning while c and d are fixed. The reader is advised to consult [17] which is a very good source of examples and detailed explanations of the neocognitron. The main advantage of neocognitron is its ability to recognize correctly not only learned patterns but also patterns, which are produced from them by using of partial shift, rotation or another type of distortion.

13.7 Application of Neural Networks

Neural networks have been applied in almost all-important domains. A list of domains and possible applications is listed below [19]:

- Data mining
    - Prediction
    - Classification

- Knowledge discovery
- Time series analysis

- Medical domain
  - Diagnosis
  - Detection and evaluation of medical phenomena
- Industry
  - Process control
  - Quality control
  - Electric Power
  - VLSI
- Finance
  - Stock Market Prediction
  - Credit Worthiness
  - Credit Rating
  - Bankruptcy Prediction
  - Property Appraisal
  - Fraud Detection
  - Price Forecasts
  - Economic Indicator Forecasts
- Science
  - Pattern Recognition
  - Chemical Compound Identification
  - Physical System Modeling
  - Ecosystem Evaluation
  - Polymer Identification
  - Recognizing Genes
  - Signal Processing: Neural Filtering
  - Biological Systems Analysis
- Operational analysis
  - Scheduling optimization
  - Managerial decision making
- Criminology
- Games
  - Backgammon
  - Bridge
  - Checkers
  - Go
  - Go-Moku
  - Othello
  - Tic-Tac-Toe
- Sports
- Gambling

Hopfield neural networks have been used for a large number of optimization problems from object recognition to graph planarization. However, the fact that the Hopfield energy function is of quadratic order limits the problems to which it can be applied. Some objective functions that cannot be reduced to Hopfield's quadratic energy function can still be reasonable approximated by a quadratic energy function. For other problems the objective function must be modeled by a higher order energy function [38].

Summaries

In this chapter we presented recurrent networks – Jordan network, Elman network and Hopfield network – and self organizing networks – Hebb network, Kohonen self organizing maps and neocognitron.

A feedforward network does not maintain a short-term memory. Any memory effects are due to the way past inputs are re-presented to the network. A simple recurrent network has activation feedback, which embodies short-term memory. A state layer is updated not only with the external input of the network but also with activation from the previous forward propagation. The feedback is modified by a set of weights as to enable automatic adaptation through learning.

Two fundamental ways can be used to add feedback into feedforward multilayer neural network: Elman introduced feedback from the hidden layer to the context portion of the input layer while Jordan uses feedback from the output layer to the context nodes of the input layer. The Hopfield network represents an autoassociative type of memory. The network has two phases: storage and retrieval. In the first phase the network is required to store a set of states determined by the current outputs of the neurons. In the second phase, an unknown, incomplete or corrupted version of the fundamental memory is presented to the network. The network output is calculated and fed back to adjust the input and the process is repeated until the output becomes constant [1].

Hebb network is a feedforward neural network trained using the Hebb rule. The network can learn (without a teacher) to associate stimuli commonly presented together.

Kohonen self organizing map consists of a single layer of computation neurons but has two types of connections: forward connections from the neurons in the input layer to the neurons in the output layer and lateral connections between neurons in the output layer. The lateral connections are used to create a competition between the neurons. In a Kohonen self organizing map, a neuron learns by shifting its weights from inactive connections to active ones. Only the winning neuron and a defined neighborhood of it are allowed to learn.

The neocognitron was designed to recognize handwritten characters – specifically the Arabic numerals 0, 1, …, 9. The purpose of the network is to make its response insensitive to variations in the position and style in which the digit is written [2].

The advantage of neural networks lies in their ability to represent both linear and non-linear relationships and in their ability to learn these relationships directly from the data being modeled. Traditional linear models are simply inadequate when it comes to modeling data that contains non-linear characteristics.

References

1. Negnevitsky, M.: Artificial Intelligence: A Guide to Intelligent Systems. Addison-Wesley/Pearson Education, Harlow (2002) ISBN 0201-71159-1
2. Fausett, L.V.: Fundamentals of Neural Networks: Architectures, Algorithms And Applications. Prentice-Hall, Englewood Cliffs (1993)
3. Kohonen, T.: Associative memory: a system-theoretical approach. Springer, Heidelberg (1977)
4. Kohonen, T.: Self-organized formation of topologically correct feature maps. Biological Cybernetics 43, 59–69 (1982)
5. Kohonen, T.: Self organization and associative memory. Springer, Berlin (1984)
6. Kohonen, T.: Self-Organizing Maps. Springer, Heidelberg (1995)
7. Krose, B., van der Smagt, P.: An introduction to Neural Networks, University of Amsterdam, Eight Edition (1996)
8. Green, D.W., et al.: Cognitive Science: An Introduction. Blackwell Publishers Ltd., Oxford (1998)
9. Elman, J.L.: Finding structure in time. Cognitive Science 14, 179–211 (1990)
10. Jordan, M.I.: Attractor dynamics and parallelism in a connectionist sequential machine. In: Proceedings of the Eighth Annual Conference of the Cognitive Science Society, pp. 531–546. Erlbaum, NJ (1986)
11. Jordan, M.I.: Serial order: A parallel distributed processing approach, Technical report, Institute for Cognitive Science, University of California (1986)
12. Hopfield, J.J.: Neural networks and physical systems with emergent collective computational abilities. Proceedings of the National Academy of Sciences 79, 2554–2558 (1982)
13. Hebb, D.O.: The organization of behavior: a neuropsychological theory. John Wiley, New York (1949)
14. McClelland, J.L., Rumelhart, D.E.: Explorations in parallel distributed processing. MIT Press, Cambridge (1988)
15. Fukushima, K.: Neocognitron: A self-organizing neural network model for a mechanism of pattern recognition unaffected by shift in position. Biological Cybernetics 36(4), 193–202 (1980)
16. Fukushima, K., Miyake, S., Ito, T.: Neocognitron: a neural network model for a mechanism of visual pattern recognition. IEEE Transactions on Systems, Man, and Cybernetics 13(3), 826–834 (1983)
17. `http://neuron.felk.cvut.cz/courseware/data/chapter/velinsky2000`
18. Medsker, L.R., Jain, L.C. (eds.): Recurrent neural networks: design and applications. CRC Press, Boca Raton (1999)
19. `http://www.faqs.org/faqs/ai-faq/neural-nets/part7/section-2.html`

Verification Questions

1. What are recurrent neural networks?
2. What is the difference between different types of recurrent neural networks?

3. What does Hebb law represents?
4. How a Hebb network works?
5. How does a Kohonen self organizing map learn?
6. For what task has been the neocognitron proposed?
7. What is the main advantage of the neocognitron?
8. What is the basic principle of the neocognitron?
9. Mention all the types of cells used in the neocognitron.

Chapter 14
Evolutionary Algorithms

14.1 Introduction

In nature, evolution is mostly determined by natural selection of different individuals competing for resources in the environment. Those individuals that are better are more likely to survive and propagate their genetic material. The encoding for genetic information (genome) is done in a way that admits ascxual reproduction, which results in offspring that are genetically identical to the parent. Sexual reproduction allows some exchange and re-ordering of chromosomes, producing offspring that contain a combination of information from each parent. This is the recombination operation, which is often referred to as crossover because of the way strands of chromosomes cross over during the exchange. The diversity in the population is achieved by mutation operation.

Usually grouped under the term *evolutionary computation* or *evolutionary algorithms* [1][4], we find the domains of genetic algorithms [9], evolution strategies [26][27], evolutionary programming [28] and genetic programming [29]. They all share a common conceptual base of simulating the evolution of individual structures via processes of selection, recombination and mutation reproduction and thereby producing better solutions. The processes depend on the perceived performance of the individual structures as defined by the problem. The procedure is then iterated as illustrated in Figure 14.1.

Darwinian evolutionary theory principles of reproduction and natural selection (survival of the fittest) are the base of the evolutionary theory:

- individuals who survive are the ones best adapted to exist in their environment due to the possession of variations;
- individuals that survive will reproduce and transmit these variations to their offspring;
- as time and generations continue, many adaptations are perpetuated in individuals until new species evolve in forms different from the common ancestor;
- traits which are beneficial to the survival of an organism in a particular environment tend to be retained and passed on, increasing in frequency within the population;

C. Grosan and A. Abraham: Intelligent Systems, ISRL 17, pp. 345–386.
springerlink.com

- trait which have low survival tend to decrease in frequency;
- when environmental conditions change, traits that were formally associated with low survival may have greater survival.

The correspondence between natural evolution and problem solving inspired by it is given in Figure 14.2.

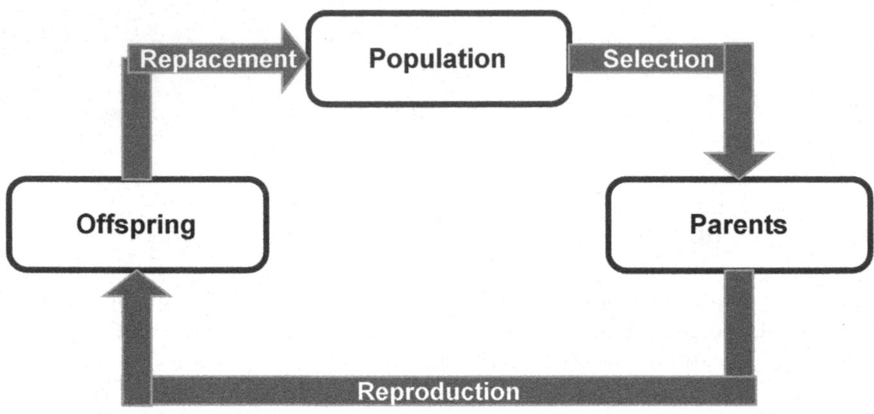

Fig. 14.1 Evolutionary scheme.

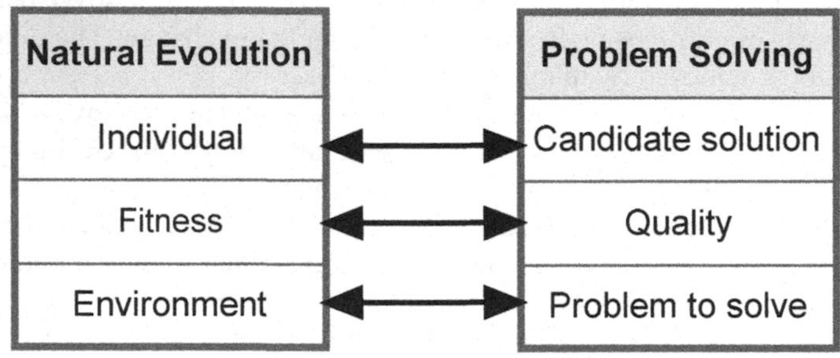

Fig. 14.2 Correspondence between natural evolution principles and problem solving.

The basic evolutionary algorithm is described below.

Evolutionary Algorithm
Step 1. Set $t= 0$;

Step 2. Randomly initialize the population $P(t)$.
Repeat
  Step 3.1. Evaluate individuals from $P(t)$;
  Step 3.2. Selection on $P(t)$
    Let P'(t) be the set of selected individuals.

Step 3.3. Crossover on *P'*(*t*);
Step 3.4. Mutation on *P'*(*t*).
Step 3.5. Survival on P'(t).
Step 3.6. *t*=*t*+1.
Until *t*=*number of iterations*

14.2 How to Build an Evolutionary Algorithm?

In order to build an evolutionary algorithm there are a number of steps that we have to perform:

- design a representation;
- find a way to initialize the population;
- design a way of mapping a genotype to a phenotype;
- design a way of evaluating an individual;
- decide how to select individuals to be parents (design a selection mechanism);
- design suitable mutation operators;
- design suitable recombination operators;
- decide how to select individuals to be replaced;
- decide when to stop the algorithm.

There are many different variants of evolutionary algorithms, but they all share the same evolutionary structure: given a population of individuals (candidate solutions), the natural selection will influence the survival of the fittest. Based on the fitness (or an evaluation function), the best individuals (candidate solutions) will be selected to seed the population of the next generation. Recombination and/or mutation operators are applied to them. The new obtained candidate solutions (offspring) will compete with the old ones (based on their fitness) to be part of the next generation.

The process will be iterated until a candidate solution with sufficient quality is found or until the available resources are finished.

The variants of evolutionary algorithms differ only in technical details [2][4][5]:

- in the case of Genetic Algorithms (GAs) the solutions are represented as strings over a finite alphabet;
- solutions are represented as real-valued vectors in Evolutionary (Evolution) Strategies (ES);
- in the case of Evolutionary Programming (EP), solutions are finite state machines;
- Genetic Programming (GP) uses trees for solution representation.

Genetic Algorithms, Evolution Strategies, Evolutionary programming and Genetic Programming are the four main classes of Evolutionary Algorithms and will be independently presented in the following sections.

14.2.1 Designing a Representation

Representation here refers to a way of representing an individual as a genotype. An individual is also known as *chromosome* and it is composed of a set of *genes.*

Objects forming the original problem context are referred to as *phenotypes* and their encoding (which are the individuals in the evolutionary algorithm) are called *genotypes.*

Let us consider the problem of minimizing the function $(x\text{-}2)^2$, $x \in \{-5, -4, -3, -2, -1, 0, 1, 2, 3, 4, 5\}$.

The set of integers represents the set of phenotypes. If we use a binary code to represent them, then 1 can be seen as a phenotype and 000001 as the genotype representing it (in this case we used a string of size 6 to represent it). The phenotype space can be different from the genotype space but the whole search process takes place in the genotype space.

There are many ways to design a representation and the way we choose must be relevant to the problem that we are solving. When choosing a representation, we have to bear in mind how the genotypes will be evaluated and what the genetic operators might be. For one problem, there can be multiple representations which can fit. We should also be careful to select the most efficient one (in terms of resources consuming).

There are a few standard representations:

- binary representation;
- real representation;
- order based representation (permutations);
- tree representation.

14.2.2 Initializing the Population

A *population* is a set of genotypes (some of the population's members can have multiple copies). Initializing the population means specifying the number of individuals in it (which is given by population size).

Each member of the population represents a possible solution for the problem to solve. The *diversity* of a population is measured as the number of different solutions of that population. During the search process, the algorithm should be able to preserve diversity in the population. If the problem has multiple solutions (for example, the n-queens problem) then the final population should contain as many different solutions as possible.

If the problem has just a single solution, then the final population's individuals should be as similar among them as possible.

The population should be initialized (if possible) uniformly on the search space.

Each individual of the population is (unless otherwise mentioned) randomly initialized (defined) over the given domain (or search space).

14.2.3 Evaluating an Individual

In order to evaluate an individual, a *fitness function* – also called *evaluation function* or *quality function* – is employed.

The role of the evaluation function represents the requirements to adapt to [2]. Evaluation function is required for the selection process.

From the problem solving perspective it represents the task to solve in evolutionary context. This function actually assigns a quality measure to the genotypes.

For our example – minimization of $(x\text{-}2)^2$ – the fitness of the genotype 000001 is $(1\text{-}2)^2 = 1$.

Example

Let us consider the 8-Queens problem. The problem requires arranging 8 queens on an 8×8 chess board such as there will be no attacks among the queens.

The representation used is order-based representation, which is a vector of size 8 containing a permutation of the numbers 1 to 8 (whose meaning is: the value of position i in the vector represents the column of queen in line i).

If our representation is (graphically depicted in Figure 14.3):

(2 4 1 6 7 8 3 5)

then the fitness function's value can be calculated as:0+3+0+3+3+3+1+0=13 because first queen does not attack any other queen, second queen attacks three other queens on the diagonal, third queen attacks no queens and so on.

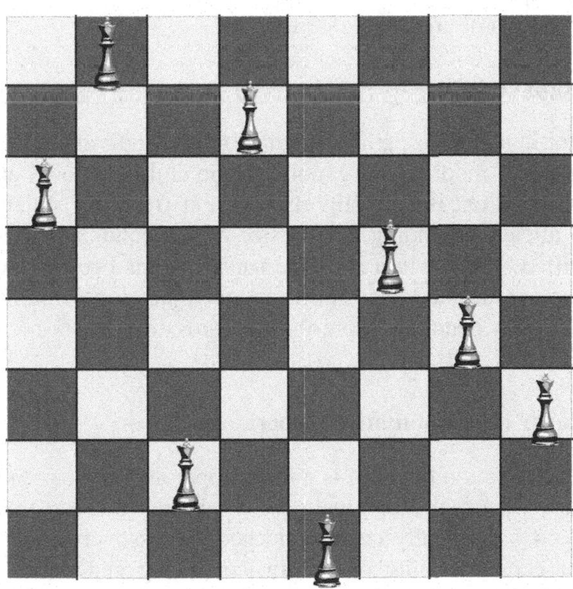

Fig. 14.3 Graphical representation of the solution (2 4 1 6 7 8 3 5) for the 8-queens problem.

14.2.4 Selection Mechanism

The role of selection mechanism is to distinguish among individuals based on their fitness in order to allow better individuals to become parents of the next generation. In evolutionary computation, parent selection is probabilistic: high quality individuals get higher chance to become parents. Low quality individual are also given a low change to become parents in order to avoid getting stuck in a local optima. There are several selection mechanisms, which will be discussed in detail in the following sections.

Remark
If, in the beginning of the evolutionary algorithms, the highest chances were given to best individuals, nowadays approaches also give chances to individuals, which do not have a high quality.

14.2.5 Designing Suitable Variation Operators

The role of variation operators is to create new individuals from the existing ones. The new created individuals should represent candidate solutions in the corresponding phenotype space. There are several known variation operators, which can be applied to individuals of an evolutionary algorithm population, but two of them are the most important and widely used:

- mutation operator and
- recombination or crossover operator.

14.2.5.1 Mutation Operator

Mutation operator is a unary operator (applied to one individual only) and it usually affects (or slightly modify) one genotype. The child obtain by applying mutation operator to his parent is only slightly different from it. Mutation can affect one or multiple alleles depending on the size of the candidate solution. The affected alleles will be chosen in a random manner.It has been proved that, given sufficient time, an evolutionary algorithm can reach the global optimum relying on the propriety that each genotype representing a possible solution can be reached by variation operators [2][3].

14.2.5.2 Crossover (Recombination) Operator

Crossover or recombination operator is a binary operator (there are some rare situations where more than two individuals are combined; this is sound mathematically but it has no biologically correspondence). Crossover merges information from two parent genotypes into one or two offspring genotypes. The operator chooses what parts of each parent are combined and the way in which these are combined in a random manner. Recombination operator is never used in evolutionary programming algorithms. The reason behind recombination is that by mating two individuals with different features the offspring will combine both of these features (which is very successful in nature, especially for plants and livestock).

Variation operators are representation dependant: for various representations different variation operators have to be designed [2].

14.2.6 Designing a Replacement Scheme

Survivor selection mechanism or replacement has the role to distinguish among individuals, which will be kept for the next generation based on their quality. Since the size of the population is usually kept constant during the evolution process, there should exist a way to select among the existing and new obtained (by applying variation operators) candidate solutions.

The survivor selection process takes place after having created the offspring of the selected parents. This selection is based on the candidate solutions' quality (or fitness).

14.2.7 Designing a Way to Stop the Algorithm

There are two main cases of a suitable termination condition or stopping criterion for an evolutionary algorithm [2]:

(i) If the problem has a known optimal fitness level (for example, the optimum of an optimization problem is known, or the value of the fitness function for the expected solution is known) then reaching this level (with a given sufficiently small positive precision ε>0) should be used as termination condition.

(ii) If the problem to solve doest not has a known optimum then the previous stopping condition cannot be used. In this case the termination condition may be one of the following:

a. a given number of generations is reached;
b. a given number of fitness evaluations is reached;
c. the available resources are overloaded (the maximally allowed CPU time elapses);
d. there is no improvement for the fitness function for a given number of consecutive generations or fitness evaluations;
e. the population diversity drops under a given threshold.

In these situations the termination condition may consists of two criteria: either the optimum was reach or one of the conditions above was satisfied.

14.3 Genetic Algorithms

The most widely known type of evolutionary algorithms and probable the most used are genetic algorithms. Simple to implement and use, genetic algorithms are one of the most suitable techniques for any kind of optimization problem. The chapter presents the most important ways to design a representation, selection mechanisms, variation operators – recombination and mutation, survival schemes and population models.

14.3.1 Representing the Individuals

Representation of individuals is one of the most important steps in designing a genetic algorithm. Representation should be adequate for the problem to solve and should consume as less resources as possible. For one problem there can be multiple possible representations; thus, we have to be able to decide which one is more adequate to satisfy our needs and requirements.

There are some standard types of representing an individual, which have the same form for different problems but might have a different interpretation. Some of the most important ones are presented in what follows:

- binary representation;
- integer representation;
- real valued or floating- point representation;
- order based or permutation representation.

14.3.1.1 Binary Representation

Probable the most common type of representations used by the evolutionary algorithms is the binary representation. An example of a binary individual (chromosome) of size 8 is depicted in Figure 14.4.

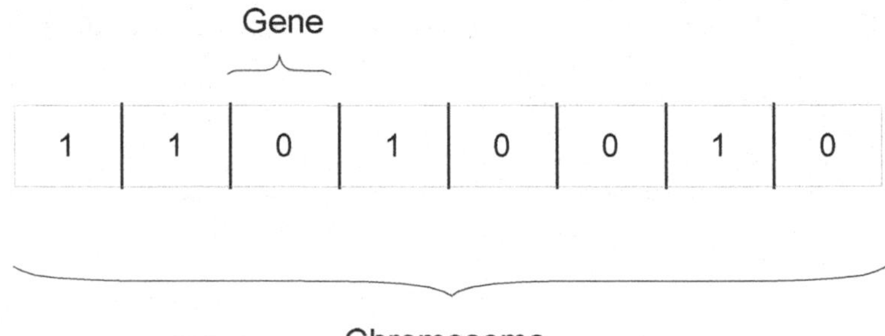

Fig. 14.4 Binary representation.

A binary individual can have different meanings depending on the problem it is used for.

Some of the possible meanings for certain problems are as follows (but the list is not limited to this):

- A real (or integer) number represented over the alphabet 2 (or using the base 2). In our example we have the number:

$$1*2^7+1*2^6+0*2^5+1*2^4+0*2^3+0*2^2+1*2^1+0 = 128+64+0+16+0+0+2+0=210$$

This integer number can be scaled to any interval or can be transformed into a real number in any given interval.

For instance, if we wish to have the real representation within [-1, 1] of the binary number in Figure 14.3, this is given by (we take into account that our binary number is a real number between 0 and $2^8 = 256$):

$$-1+\frac{210}{256}(1-(-1))=0.64$$

Note. We use here the formula of transforming a number x from the interval [0, *Max*] to the interval [min, max] which is (*Max* and max are different):

$$\min+\frac{x}{Max}(\max-\min)$$

- A solution for the 0-1 knapsack problem: the values of 1 represent the selected items and the values of 0 the items which are not selected. For instance, the chromosome in Figure 14.2 will select the items 1, 2, 4 and 7.
- A solution for the graph partitioning problem (partition of the nodes into two sets with various proprieties): the values of zero represent the nodes belonging to the first set and the values of one represent the nodes belonging to the second set. In our example in Figure 14.3, nodes 3, 5, 6 and 8 belong to the first set while the nodes 1, 2, 4 and 7 belong to the second set.

14.3.1.2 Real Representation

Real representation is mostly used for real function optimization. The size of the chromosome will be equal to the number of dimensions (variables).

An example of real representation for n variables is given in Figure 14.5.

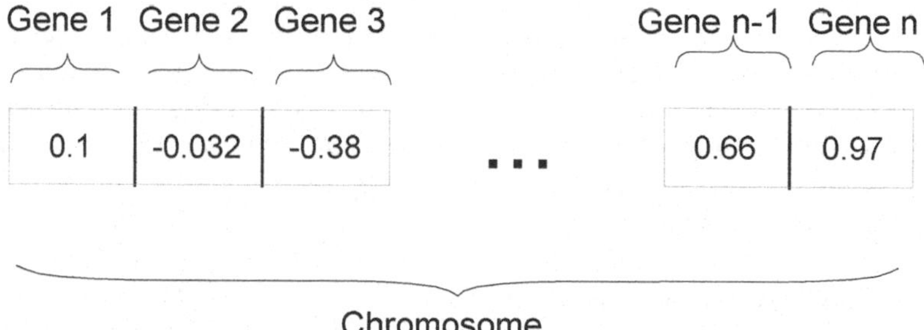

Fig. 14.5 Real representation.

14.3.1.3 Integer Representation

Sometimes real or binary representation is not suitable for a certain problem. For instance, in the case of map coloring problem[1], none of these representations might be used. For this problem the representation is a string of integers, having the size equal to the number of countries and taking values form $\{1, 2, \ldots, k\}$.

An example of integer representation for the map-coloring problem with 10 countries and 6 available colors (denoted $\{1, 2, 3, 4, 5, 6\}$) is given in Figure 14.6 (it can be observed that only 5 of the 6 available colors are used for encoding the individual). In the case of integer representation, the individual is a string of integers whose values can be restricted to a given domain or set of possible values or unrestricted.

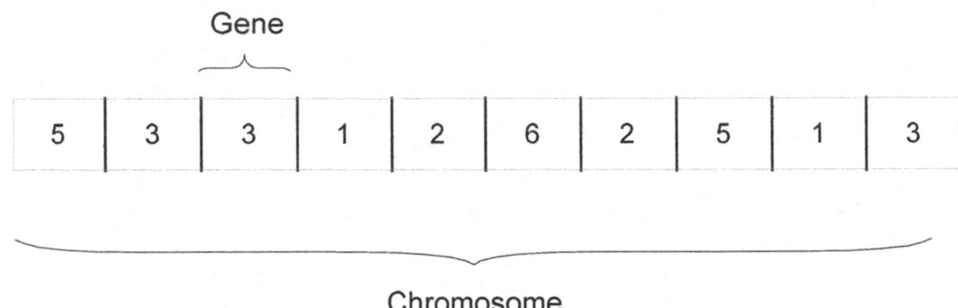

Fig. 14.6 Integer representation.

14.3.1.4 Order-Based Representation

In this case, the individuals are represented as permutations. This kind of representation is mostly used for ordering/sequencing problems. An example of chromosome using order-based representation is given in Figure 14.7.

Some famous example of problems which use permutation representation are:

- Traveling Salesman Problem (TSP): in this problem, every city gets assigned a unique number from 1 to n. A solution is a permutation of the numbers 1, 2, …, n representing the order in which the salesman visits the cities.
- n-Queens problem: a solution is a permutation of size n. The indices represent the queen on each row and the values represent the column each queen belongs to (we know that we cannot have two queens on the same row or column and by using this representation this is avoided; the only remaining attacks should be checked on the diagonals)
- Quadratic Assignment Problem;

[1] The map coloring problem we refer to states as follows: given a set of n countries and a set of k available colors, color each country with one of the k colors such as no neighboring countries will have the same color and the number of colors used is minimal (we consider the general case with k colors but it has been proven that 4 colors are enough to color any map).

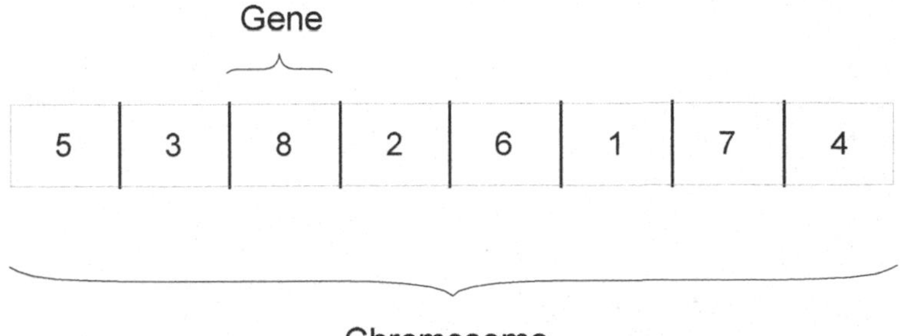

Fig. 14.7 Order-based representation.

14.3.2 Initializing the Population

Population initialization is the next step in developing a genetic algorithm once the representation has been decided. Initialization means (randomly) seeding a set with a given number of individuals having the chosen encoding/representation. While seeding the population, there are chances that some of the individuals might have multiple copies (the population is a multiset):

- For binary representation: an individual is a string of {0, 1}. The genes may be initialized with the values 0 or 1 with the probability 0.5.
- For real representation: an individual is a vector of real numbers. The genes can be initialized with random real values within the given domain (if the domain is finite), or they can be initialized using a distribution: Gaussian, Cauchy, etc.
- For order based representation: the individual is a permutation. Suppose the permutation has the size n. Each gene i will be initialized with a value between $\{1, \ldots, n\}$ which does not occur on the previous i-1 already initialized positions.
- For tree based representation: the individual is a tree. We have a set of terminals and a set of functions. Each node of the tree is randomly initialized:
    - o if the node is a terminal then a random value from the terminals set will be taken (certain terminals can be used multiple times while others might not be used at all);
    - o if the node is not a terminal, then a random function from the set of functions will be selected.

Examples

Binary representation: vector of size 10

0 1 1 0 0 0 1 1 0 1

Real representation: vector of size 10 on the domain [0, 1]

{0.004, 0.92, 0.34, 0.11, 0.26. 0.63, 0.0001, 0.019, 0.82, 0.0056}

Order based representation: vector of size 10

4 9 3 1 6 2 8 10 5 7

14.3.3 Selection Mechanisms

Selection process in a genetic algorithm occurs when parents who will be further used for crossover are to be selected. In a standard way, high priority and chances are given to fittest individuals.

Some of the most used selection mechanisms are:

- tournament selection;
- fitness proportional selection;
- roulette wheel selection;
- rank based selection.

14.3.3.1 Tournament Selection

Tournament selection is one of the simplest selection schemes. It is suitable when the population size is very large and it is not practical to compare or rank all the individuals at a time. Thus, this selection does not require any global knowledge of the population.

It relies on an ordering relation that can rank any two individuals.

There are two known versions of the tournament selection:

(i) binary tournament
(ii) *k*-tournir (or *k*-tournament).

In the case of *binary tournament*, two individuals are randomly selected from the population and the best among them (in terms of fitness value) is kept in a separate set. The procedure is repeated until the number of selected individuals equals the required number of individuals, which are to be selected.

The *k - tournament* is a generalization of the binary tournament in the sense that k individuals are randomly selected from the population and the best individual among all of them is kept in a separate set. The process is then again repeated until the required number of parents is selected from the whole population.

The probability that an individual will be selected as a result of a tournament depends on four factors [2][6][7][8]:

(i) its *rank in the population*: this is estimated without the need for sorting the whole population;
(ii) *the tournament size k*: the larger the tournament the more chance that it will contain members whose fitness is above average and the less that it will consist entirely of low-fitness members;

(iii) the *probability* that the most fit member of the tournament is selected; usually this probability is 1 (deterministic tournament) but there are stochastic versions which use a probability less than 1;

(iv) whether the individuals are chosen *with or without replacement.* In the second case, with deterministic tournaments, the k-1 least fit members of the population can never be selected. If the tournament candidates are picked with replacement, it is always possible for even the least-fit member of the population to be selected.

14.3.3.2 Fitness Proportional Selection

This selection mechanism was introduced in [9] (see also [10]). Let us denote by f_i the fitness of the i-th individual in the population and by N the population size (number of individuals in the population). Thus, the probability p_i that individual i is selected for mating (for recombination) is given by:

$$p_i = \frac{f_i}{\sum_{i=1}^{N} f_i}$$

This means that the selection probability depends on the absolute fitness value of the individual compared to the absolute fitness values of the rest of the population.

There are some problems with this selection mechanism [2]:

- individuals that are much better than the rest take over the entire population very quickly and this leads to *premature convergence*;
- when fitness values are very close there is almost no selection pressure;
- the mechanism behaves differently on transposed versions of the same fitness function.

14.3.3.3 Roulette Wheel Selection

The roulette-wheel selection is also called *stochastic sampling with replacement* and has been introduced in [11]. This is a stochastic algorithm and involves the following technique: the individuals are mapped to contiguous segments of a line, such that each individual's segment is equal in size to its fitness. A random number is generated and the individual whose segment spans the random number is selected. The process is repeated until the desired number of individuals is obtained. This technique is analogous to a roulette wheel with each slice proportional in size to the fitness. Figure 14.8 shows an example containing 9 individuals for which the fitness value and the selection probability is displayed. Individual 1 is the fittest individual and occupies the largest interval, whereas individual 9 as the least fit individual has the smallest interval on the line.

Number of individual	1	2	3	4	5	6	7	8	9
Fitness value	5.0	4.5	4.0	3.5	3.0	2.5	2.0	1.5	1.0
Selection probability	0.19	0.17	0.15	0.13	0.11	0.09	0.07	0.05	0.04

Fig. 14.8 Example of individuals, their fitness values and the corresponding selection probability.

The line segment and the roulette wheel corresponding to this example are shown in Figures 14.9 and 14.10.

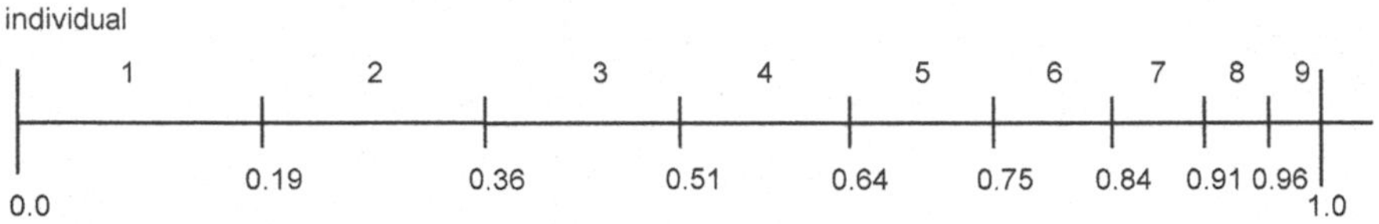

Fig. 14.9 Line segment corresponding to the example.

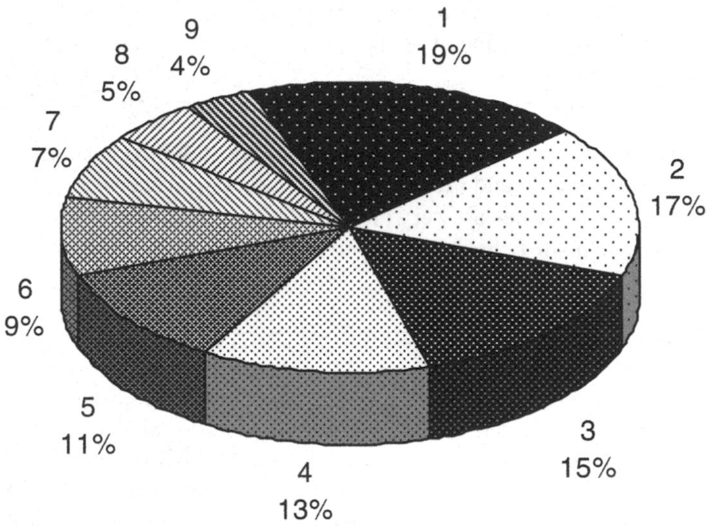

Fig. 14.10 Roulette wheel corresponding to the example.

For selecting the mating population the appropriate number of uniformly distributed random numbers (uniform distributed between 0.0 and 1.0) is independently generated.

For example, if the following 5 random numbers are generated:

```
0.61, 0.33, 0.95, 0.11, 0.45
```

the resulting selected individuals at each trial can be observed in Figure 14.11 which shows the selection process of the individuals for the example in table together with the above sample trials.

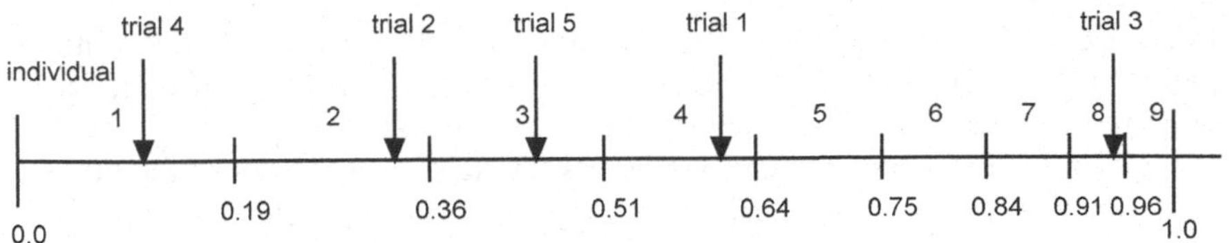

Fig. 14.11 Selection process of individuals for the example.

After selection the chosen individuals are:

```
4, 2, 8, 1, 3.
```

The roulette-wheel selection algorithm provides a zero bias but does not guarantee minimum spread.

14.3.3.4 Stochastic Universal Sampling

Stochastic universal sampling [11] provides zero bias and minimum spread. The individuals are mapped to contiguous segments of a line, such that each individual's segment is equal in size to its fitness exactly as in roulette-wheel selection.

Equally spaced pointers are placed over the line as many as there are individuals to be selected. Consider *NP* the number of individuals to be selected, then the distance between the pointers is 1/*NP* and the position of the first pointer is given by a randomly generated number in the range [0, 1/*NP*].

For 5 individuals to be selected, the distance between the pointers is 1/5=0.2. Figure 14.12 shows the selection for the above example with the random starting point 0.1.

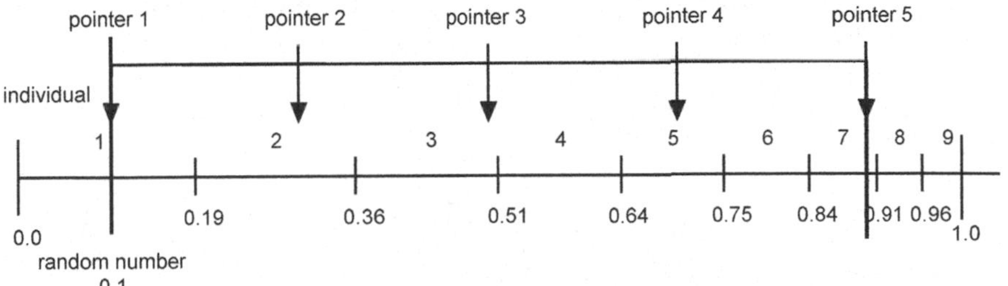

Fig. 14.12 Stochastic universal sampling.

After selection the mating population consists of the individuals:

```
1, 2, 3, 5, 7.
```

Stochastic universal sampling ensures a selection of offspring which is closer to what is deserved then roulette wheel selection.

14.3.3.5 Rank Based Selection

In rank-based fitness assignment, the population is sorted according to the fitness (quality) values. The fitness assigned to each individual depends only on its position in the individual's rank and not on the actual fitness value.

Rank-based fitness assignment overcomes the scaling problems of the proportional fitness assignment. (Stagnation in the case where the selective pressure is too small or premature convergence where selection has caused the search to narrow down too quickly.) The reproductive range is limited, so that no individuals generate an excessive number of offspring. Ranking introduces a uniform scaling across the population and provides a simple and effective way of controlling selective pressure [12]. Rank-based fitness assignment behaves in a more robust manner than proportional fitness assignment and, thus, is the method of choice [12][13][14].

Linear Ranking

Consider N the number of individuals in the population, i the position of an individual in this population (least fit individual has i =1, the fittest individual i =N) and SP the selective pressure. The fitness value for an individual is calculated as:

$$fitness(i) = 2 - SP + 2 \cdot (SP - 1) \cdot \frac{(i-1)}{(N-1)}$$

Linear ranking allows values of selective pressure in [1.0, 2.0].

Non-linear ranking

A new method for ranking using a non-linear distribution was introduced in [15]. The use of non-linear ranking permits higher selective pressures than the linear ranking method.

$$fitness(i) = \frac{N \cdot X^{i-1}}{\sum_{i=1}^{N} X^{i-1}}$$

X is computed as the root of the polynomial:

$$0 = (SP - N) \cdot X^{N-1} + SP \cdot X^{N-2} + \ldots + SP \cdot X + SP .$$

Non-linear ranking allows values of selective pressure in the interval [1, N - 2].

14.3.3.6 Local Selection

In local selection every individual resides inside a constrained environment called the local neighborhood. In the other selection methods the whole population or subpopulation is the selection pool or neighborhood. Individuals interact only with individuals inside this region. The neighborhood is defined by the structure in which the population is distributed. The neighborhood can be seen as the group of potential mating partners. The first step is the selection of the first half of the mating population uniform at random (or using one of the other mentioned selection algorithms, for example, stochastic universal sampling or truncation selection). Now a local neighborhood is defined for every selected individual. Inside this neighborhood the mating partner is selected (best, fitness proportional, or uniform at random) [12].

The structure of the neighborhood can be:

- linear:
  - full ring
  - half ring (see Figure 14.13)
- two-dimensional
  - full cross
  - half cross (see Figure 14.14, top)
  - full star
  - half star (see Figure 14.14, bottom)
- three-dimensional and more complex with any combination of the above structures [12].

The distance between possible neighbors together with the structure determines the size of the neighborhood.

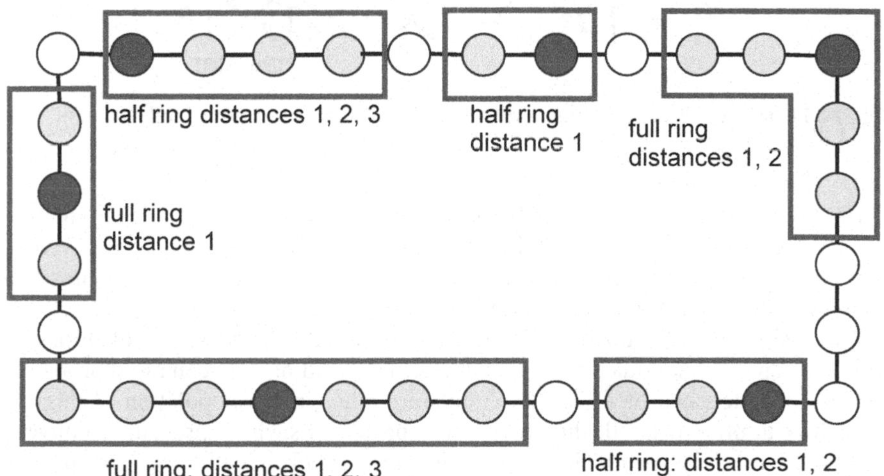

Fig. 14.13 Linear neighborhood: full and half rings.

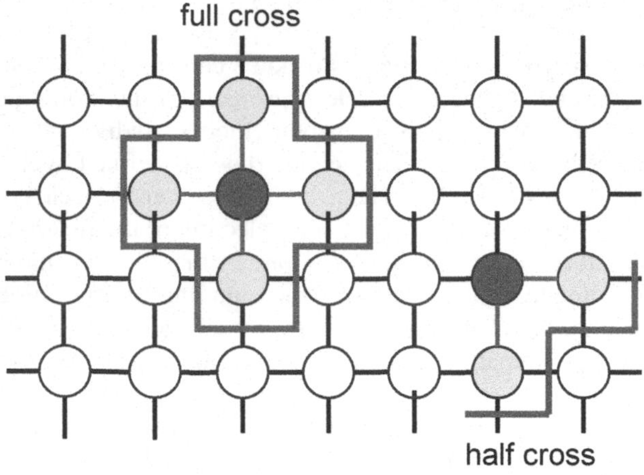

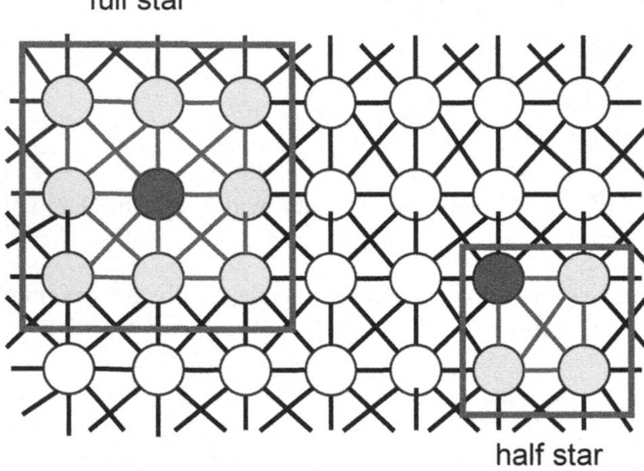

Fig. 14.14 Two-dimensional neighborhood: full and half cross (top); full and half star (bottom).

Between individuals of a population isolation by distance exists. The smaller the neighborhood, the bigger the isolation distances. However, because of overlapping neighborhoods, propagation of new variants takes place. This assures the exchange of information between all individuals.

The size of the neighborhood determines the speed of propagation of information between the individuals of a population, thus deciding between rapid propagation and maintenance of a high diversity/variability in the population. A higher variability is often desired, thus preventing problems such as premature convergence to a local minimum. Local selection in a small neighborhood performed better than local selection in a bigger neighborhood. Nevertheless, the interconnection of the whole population must still be provided. Two-dimensional neighborhood with structure half star using a distance of 1 is recommended for local selection. However, if the population is bigger (>100 individuals) a greater distance and/or another two-dimensional neighborhood should be used [12].

14.3.4 Variation Operators

The main variation operators are recombination or crossover and mutation. Each of them has specific forms for different individual representations and will be presented in what follows.

14.3.4.1 Crossover or Recombination

The role of recombination operator is to produces new individuals by combining the information contained in two or more parents. This is done by combining the variable values of the parents. Depending on the representation of the variables different methods must be used.

14.3.4.1.1 Recombination for Binary Representation

This section describes recombination methods for individuals with binary variables. During the recombination of binary variables only parts of the individuals are exchanged between the individuals. Depending on the number of parts, the individuals are divided before the exchange of variables (the number of cross points).

Single point crossover

Let us denote by *nrvar* the length of the binary string used to encode an individual (the number of variables). In single-point crossover [9] one crossover position (cutting point) $k \in [1, 2, ..., nrvar\text{-}1]$, is selected uniformly at random. Two new offspring are produced by exchanging variables between the individuals about this point.

Consider the following two individuals of size (length) 10:

```
parent 1     1  1  1  0  0  0  1  1  0  1
parent 2     1  0  0  0  1  1  0  1  1  0
```

and the chosen crossover position 4.

After crossover the two new individuals created are (see Figure 14.15):

```
offspring 1     1  1  1  0  1  1  0  1  1  0
offspring 2     1  0  0  0  0  0  1  1  0  1
```

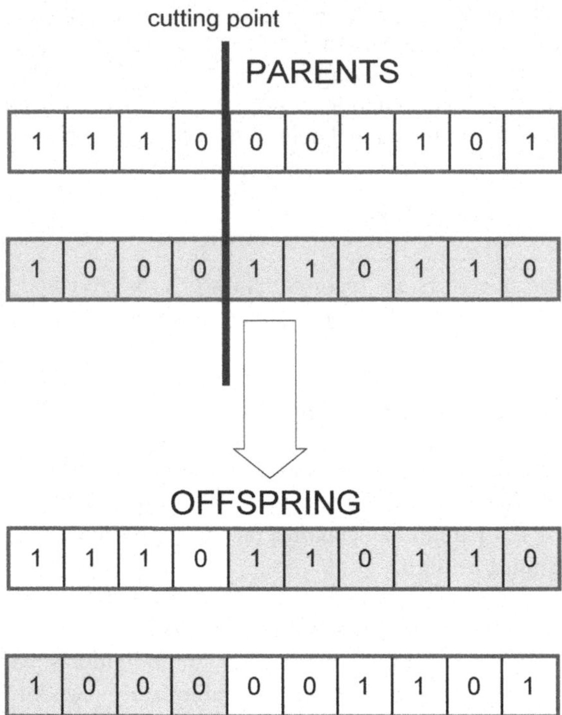

Fig. 14.15 Single point crossover

Double point / multi point crossover

In the case of double-point crossover two crossover positions are selected uniformly at random and the variables are exchanged between the individuals between these points. Two new offspring are produced. An example of double point crossover is shown in Figure 14.16.

Single-point and double-point crossover are special cases of the general method multi-point crossover.

For multi-point crossover, m crossover positions $k_i \in [1, 2, \ldots, nrvar-1]$, $i=1,\ldots,m$, are chosen at random with no duplicates and sorted into ascending order. Then, the variables between successive crossover points are exchanged between the two parents to produce two new offspring.

Consider the following two individuals with 15 binary variables each:

```
parent 1     1  1  1  1  1  1  1  1  1  1  1  1  1  1  1
parent 2     0  0  0  0  0  0  0  0  0  0  0  0  0  0  0
```

We choose 5 crossover points; the chosen crossover positions are 2, 5, 8, 11, 13.

After crossover the two new individuals created are (see Figure 14.17):

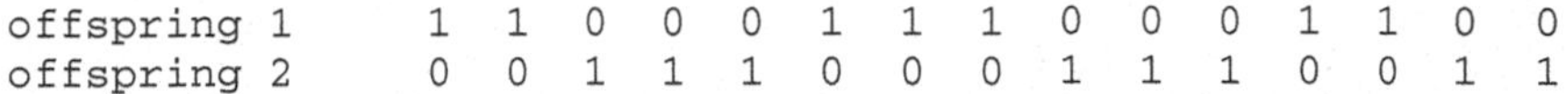

```
offspring 1      1  1  0  0  0  1  1  1  0  0  0  1  1  0  0
offspring 2      0  0  1  1  1  0  0  0  1  1  1  0  0  1  1
```

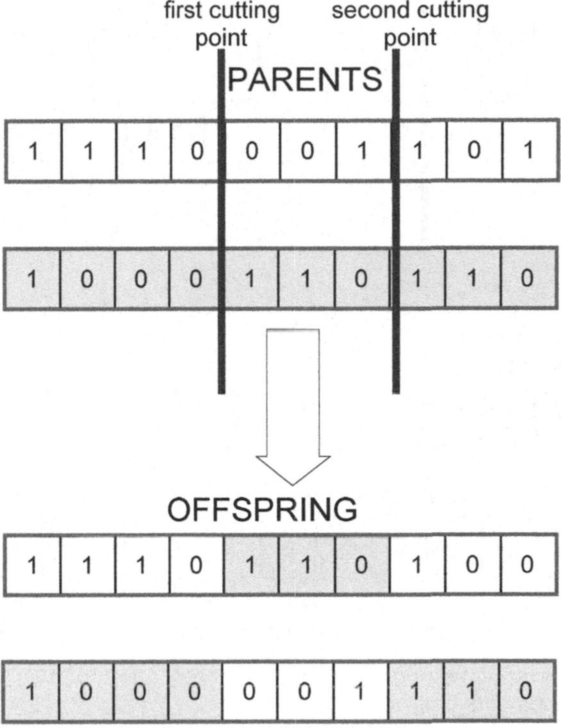

Fig. 14.16 Double point crossover.

Uniform crossover

Uniform crossover [16] generalizes the multi point crossover to make every locus a potential crossover point. For each variable the parent who contributes its variable to the offspring is chosen randomly with equal probability. Uniform crossover works by treating each gene independently and making a random choice as to which parent it should be inherited.

This is implemented by generating a string of random variables from a uniform distribution over [0, 1] whose size is equal to the size of an individual from the population. In each position, if the value is below a parameter p (usually 0.5), the gene is inherited from the first parent; otherwise from the second. The second offspring is created using the inverse mapping.

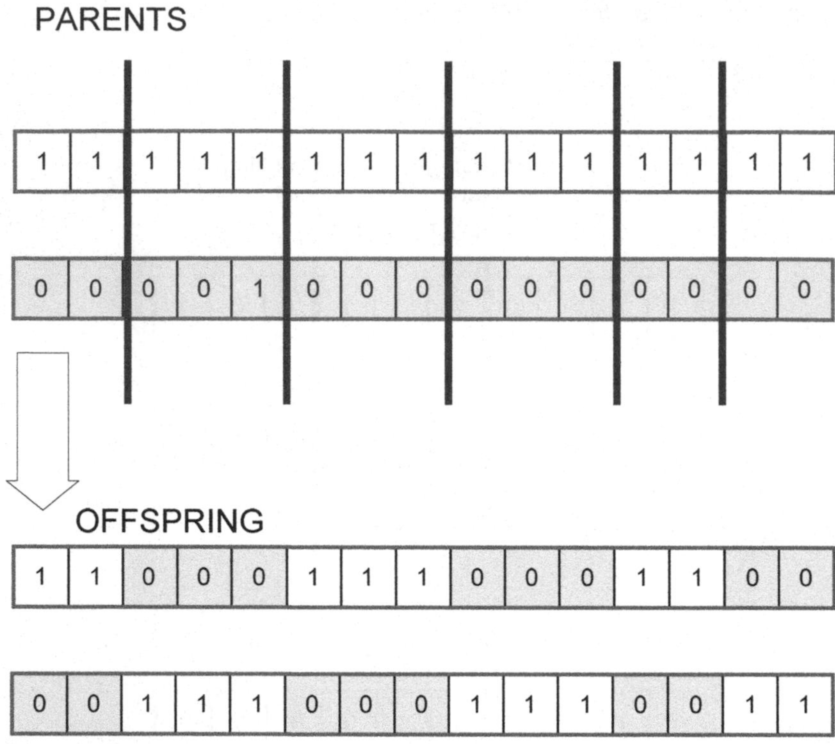

Fig. 14.17 Multi point crossover (5 cutting points on the positions 2, 5, 8, 11 and 13).

If we have the parents of size 10:

```
parent 1      1  1  1  1  1  1  1  1  1  1  1
parent 2      0  0  0  0  0  0  0  0  0  0  0
```

and the string of random variables:

```
[0.17, 0.80, 0.33, 0.45, 0.51, 0.97, 0.12, 0.66, 0.73, 0.23]
```

then the offspring obtained after crossover are:

```
offspring 1      1  0  1  1  0  0  1  0  0  1
offspring 2      0  1  0  0  1  1  0  1  1  0
```

Uniform crossover, like multi-point crossover, has been claimed to reduce the bias associated with the length of the binary representation used and the particular coding for a given parameter set. This helps to overcome the bias in single-point crossover towards short substrings without requiring precise understanding of the significance of the individual bits in the individual's representation. In [17] it is demonstrated how uniform crossover may be parameterized by applying a probability to the swapping of bits. This extra parameter can be used to control the amount of disruption during recombination without introducing a bias towards the length of the representation used [12].

14.3.4.1.2 Recombination for Real Representation

There are two ways to perform recombination for real values representation [2]:

(i) Using an operator similar to the one used for binary representation. This has the disadvantage that only mutation can insert new values in the population since the recombination only gives new combinations of the existing floats. Recombination operators of this type are known as discrete recombination.

(ii) Using an operator that, in each gene position, creates a new value in the offspring that lies between those of the parents. If we have the parents x and y and the offspring z, then we have $z_i = \alpha x_i + (1-\alpha)y_i$, $\alpha \in [0, 1]$. Operators for this type are known as arithmetic or intermediate recombination.

Arithmetic recombination

There are three types of arithmetic recombination [2][18]: simple, single arithmetic and whole arithmetic. The choice of parameter α is made at random between [0, 1] but in practice it is common to use the value 0.5 for it (in this case we have *uniform arithmetic recombination*).

Simple recombination

In this case a crossover position k is randomly selected between {1, 2, …, *nrvar*-1}. For the first child, the first k floats of the first parent are taken. The rest is the arithmetic average of parent 1 and 2. Child 2 is analogue with the parents reversed.

parent 1: $x_1, x_2, \ldots, x_k, x_{k+1} \ldots, x_{nrvar}$
parent 2: $y_1, y_2, \ldots, y_k, y_{k+1} \ldots, y_{nrvar}$

offspring 1: $x_1, x_2, \ldots, x_k, \alpha \cdot x_{k+1} + (1-\alpha) \cdot y_{k+1} \ldots, \alpha \cdot x_{nrvar} + (1-\alpha) \cdot y_{nrvar}$
offspring 2: $y_1, y_2, \ldots, y_k, \alpha \cdot y_{k+1} + (1-\alpha) \cdot x_{k+1} \ldots, \alpha \cdot y_{nrvar} + (1-\alpha) \cdot x_{nrvar}$

An example of simple recombination is shown in Figure 14.18. The value of k is 5 and the value of α is 0.5. The size of the individual (chromosomes) is 8.

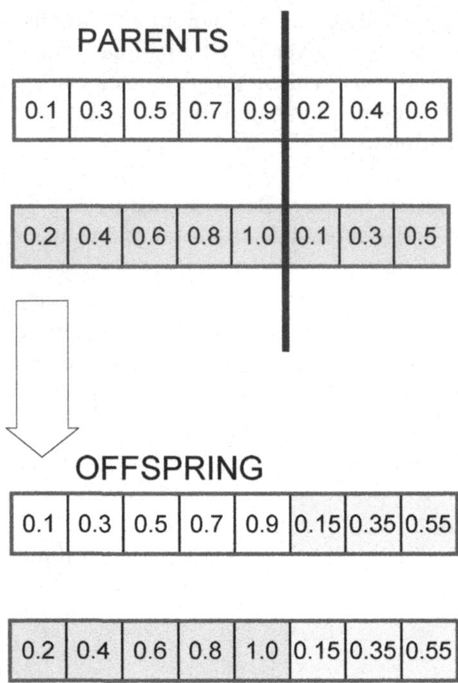

Fig. 14.18 Simple arithmetic recombination.

Single arithmetic recombination

In this case one gene k of the chromosome is picked at random. At that position, the arithmetic average of the two parents is taken. Rest of the chromosome remains the same. The second child is created in the same way with the parents reversed.

$$\text{parent 1: } x_1,\ x_2,\ \ldots,\ x_k,\ \ldots,\ x_{nrvar}$$
$$\text{parent 2: } y_1,\ y_2,\ \ldots,\ y_k,\ \ldots,\ y_{nrvar}$$

$$\text{offspring 1: } x_1,\ x_2,\ \ldots,\ \alpha\cdot x_k+(1-\alpha)\cdot y_k\ \ldots,\ x_{nrvar}$$
$$\text{offspring 2: } y_1,\ y_2,\ \ldots,\ \alpha\cdot y_k+(1-\alpha)\cdot x_k\ \ldots,\ y_{nrvar}$$

An example for $k = 5$ and $\alpha = 0.5$ is shown in Figure 14.19.

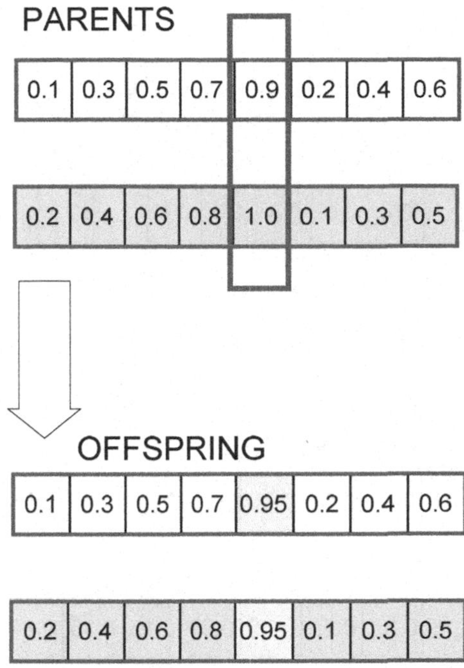

Fig. 14.19 Single arithmetic crossover.

Whole arithmetic recombination

This is the most common used recombination operator for real representation and works by taking the weighted average sum of the two parents for each gene:

parent 1: $x_1, x_2, \ldots, x_k, \ldots, x_{nrvar}$
parent 2: $y_1, y_2, \ldots, y_k, \ldots, y_{nrvar}$

offspring 1: $\alpha \cdot x_1+(1-\alpha)\cdot y_1, \alpha \cdot x_2+(1-\alpha)\cdot y_2, \ldots, \alpha \cdot x_{nrvar}+(1-\alpha)\cdot y_{nrvar}$ offspring 2: $\alpha \cdot y_1+(1-\alpha)\cdot x_1, \alpha \cdot y_2+(1-\alpha)\cdot x_2, \ldots, \alpha \cdot y_{nrvar}+(1-\alpha)\cdot x_{nrvar}$

An example for $\alpha = 0.5$ is shown in Figure 14.20.

14.3.4.1.3 Recombination for Order-Based Representation

For order based representation it is difficult to apply any of the operators discussed above suitable for binary and real value encodings due to the fact that the new individuals obtained will not remain a permutation. There are some specific recombination operators for permutations, which will be discussed in what follows.

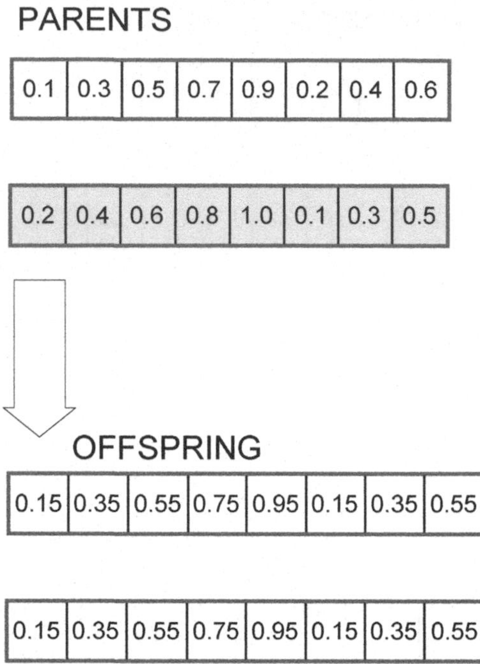

Fig. 14.20 Whole arithmetic recombination.

Partially mapped crossover

Partially mapped crossover (PMX) has been proposed in [19] as a recombination operator for the TSP. There exist now several variants of it. The steps of PMX are as follows [2][20]:

1) Choose two crossover points at random and copy the segment between them from the first parent into the first offspring.
2) Starting from the first crossover point, look for elements in that segment of the second parents that have not been copied.
3) For each i of these, look in the offspring to see what element (j) has been copied in its place from the first parent.
4) Place i into the position occupied by j in the second parent.
5) If the place occupied by j in the second parent has already been filled in the offspring by another element k, put i in the position occupied by k in the second parent.
6) Once the elements from the crossover segment have been dealt with, the rest of the offspring can be filled from the second parent.
7) The second offspring is created in a similar manner with the parents reversed.

A graphical illustration of the PMX operator is presented in Figure 14.21.

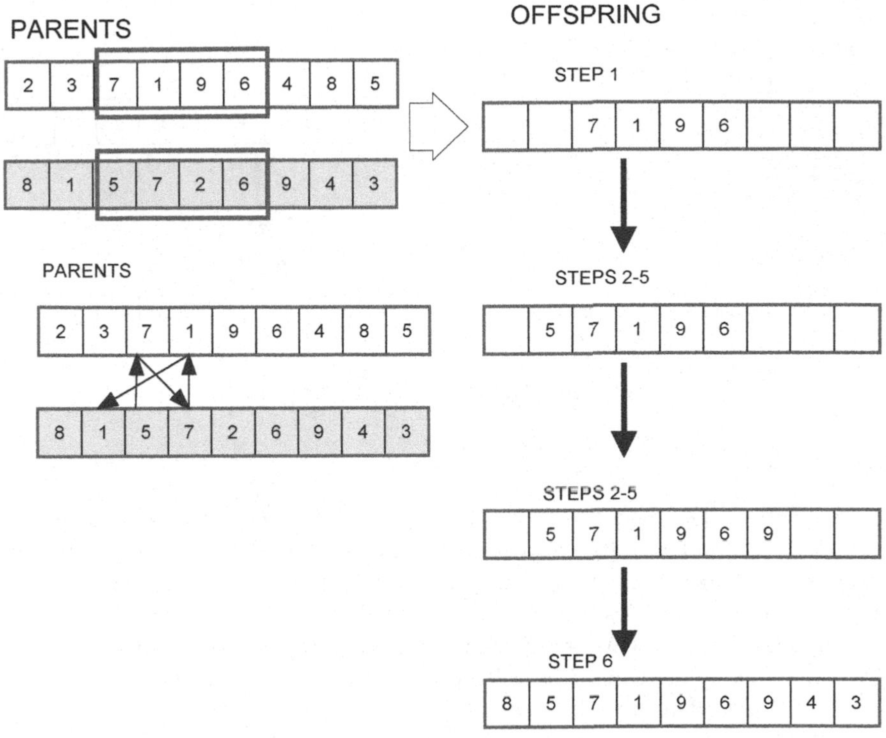

Fig. 14.21 PMX operator.

Order crossover

The order crossover operator [21] is in a way similar to PMX and has the following steps [2]:

1) Choose two crossover points at random and copy the segment between them from the first parent into the first offspring.
2) Starting from the second crossover point in the second parent, copy the remaining unused elements into the first offspring in the order that they appear in the second parent, wrapping around at the end of the list (treating string as toroidal).
3) Create the second offspring in an analogous manner, reversing the parents.

An example of the order crossover operator is shown in Figure 14.22.

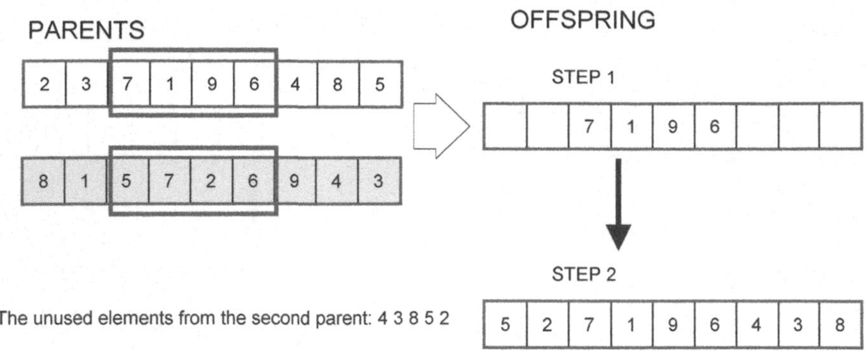

Fig. 14.22 Order crossover.

Cycle crossover

Cycle crossover [22] is concerned with preserving as much information as possible about the absolute position in which elements occur. The operator works by dividing the elements into cycles. A cycle is a subset of elements that has the propriety that each element always occurs paired with another element of the same cycle when the two parents are aligned. Once the permutations are divided in cycles, the offspring are created by selecting alternate cycles from each parent. The steps of the procedure are [2]:

1) Start with the first unused position of the first parent.
2) Look at the allele in the same position in the second parent.
3) Go to the position with the same allele in the first parent.
4) Add this allele to the cycle.
5) Repeat the steps 2-4 until you arrive at the first allele of the first parent.

An example of Cycle crossover is presented in:

- Figure 14.23 – identifying the cycles;
- Figure 14.24 – building the offspring.

```
Cycle 1: 2, 9, 8, 4 in the first parent (8, 2, 9, 4
respectively in the second parent)

Cycle 2: 3, 7, 1, 5 in the first parent (1, 5, 7, 3
respectively in the second parent).

Cycle 3: 6.
```

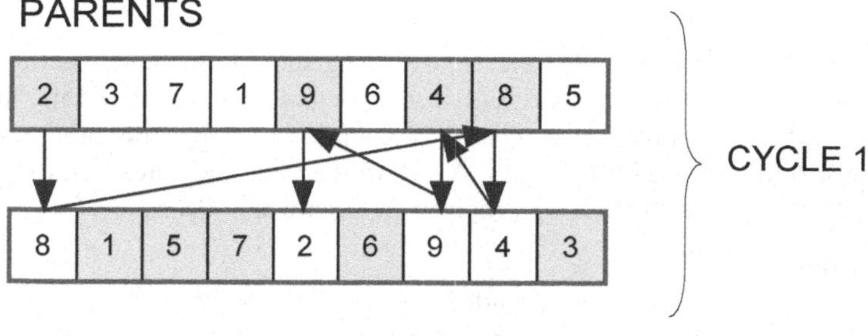

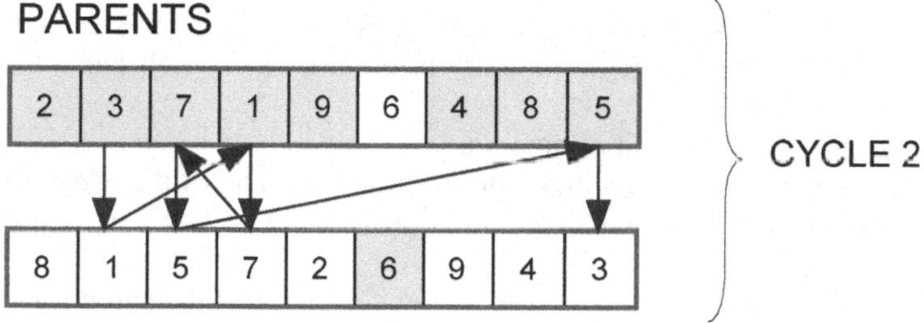

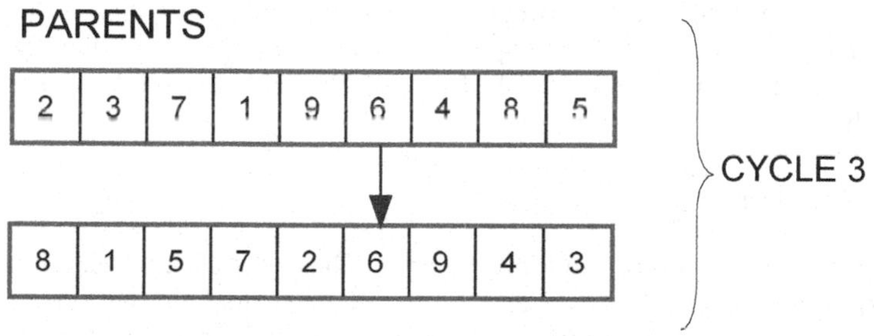

Fig. 14.23 Cycle crossover: identifying the cycles.

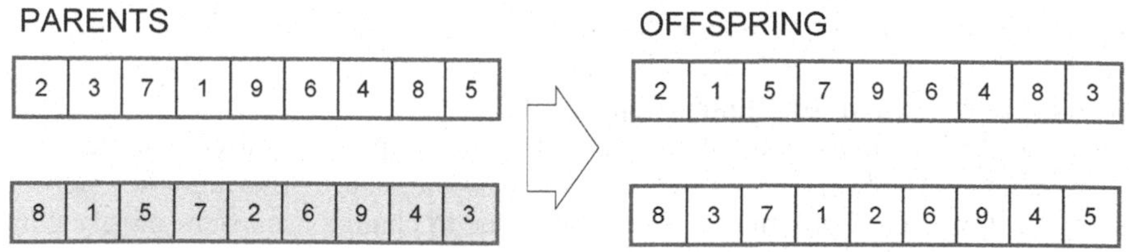

Fig. 14.24 Cycle crossover: building the offspring.

Edge crossover

Edge recombination uses the idea that an offspring should be created as far as possible using only edges that are present in one or both parents. For this, an edge table (adjacent list) is constructed which, for each element, lists the other elements that are linked to it in the two parents. A "+" in the table indicates that the edge is present in both parents. The steps of the procedure are as follows [2]:

1) Construct edge table.
2) Pick an initial element at random and put it in the offspring.
3) Set the variable current_elem = entry.
4) Remove all references to current_elem from the table.
5) Examine the list of current_elem:
   a. If there is a common edge, pick that to be the next element.
   b. Otherwise pick the entry in the list which itself has the shortest list.
   c. Ties are split at random.
6) In the case of reaching an empty list, the other end of the offspring is examined for extension. Otherwise a new element is chosen at random.

Figure 14.25 illustrates an example taken from [2] for which the parents are:

```
parent 1: 1 2 3 4 5 6 7 8 9
parent 2: 9 3 7 8 2 6 5 1 4.
```

14.3.4.1.4 Recombination for Integer Representation

For integer representation, one can apply the same operators as in the case of binary representation (the operators used for real representation might yield to non-integer values).

14.3.4.2 Mutation

By mutation individuals are randomly altered. These variations (mutation steps) are mostly small. Mutation is only applied to one individual and will produce one offspring. They will be applied to the variables of the individuals with a low probability (mutation probability or mutation rate). Normally, offspring are mutated after being created by recombination.

As in the case of recombination, mutation operator takes various forms depending on the individual's representation used.

14.3.4.2.1 Mutation for Binary Representation

For binary valued individuals mutation means the flipping of variable values, because every variable has only two states. Thus, the size of the mutation step is always 1. For every individual the variable value to change is chosen (mostly uniform at random). Figure 14.26 shows an example of a binary mutation for an individual with 10 variables, where variable 4 is mutated.

EDGE TABLE

Element	Edges
1	2, 5, 4, 9
2	1, 3, 6, 8
3	2, 4, 7, 9
4	1, 3, 5, 9
5	1, 4, 6+
6	2, 5+, 7
7	3, 6, 8+
8	2, 7+, 9
9	1, 3, 4, 8

PERMUTATION CONSTRUCTION

Choices	Element selected	Reason	Partial result
All	1	Random	1
2, 5, 4, 9	5	Shortest list	1 5
4, 6	6	Common edge	1 5 6
2, 7	2	Random	1 5 6 2
3, 8	8	Shortest list	1 5 6 2 8
7, 9	7	Common edge	1 5 6 2 8 7
3	3	Only element in list	1 5 6 2 8 7 3
4, 9	9	Random	1 5 6 2 8 7 3 9
4	4	Last element	1 5 6 2 8 7 3 9 4

Fig. 14.25 Edge crossover.

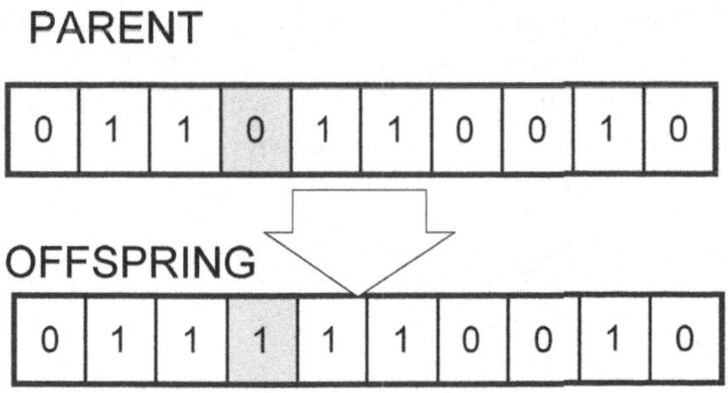

PARENT

0	1	1	0	1	1	0	0	1	0

OFFSPRING

0	1	1	1	1	1	0	0	1	0

Fig. 23.26 One bit mutation example.

The most common mutation operator considers each gene separately and allows each bit to flip with a small probability. The actual number of values changed is thus not fixed but depends on the sequence of random numbers drawn. Let us consider the probability 0.5 for a bit to flip and the string of probabilities for the example above as being:

0.2, 0.35, 0.17, 0.76, 0.52, 0.27, 0.13, 0.88, 0.95, 0.12

Thus, the offspring obtained after mutation is depicted in Figure 14.27 (the genes 1, 2, 3, 6, 7 and 10 are flipped).

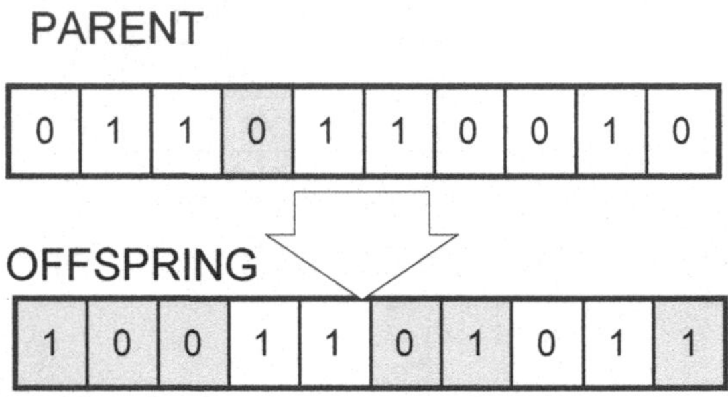

Fig. 14.27 Mutation for binary representation.

14.3.4.2.2 Mutation for Real Representation

Mutation of real variables means that randomly created values are added to the variables with a low probability. Thus, the probability of mutating a variable (mutation rate) and the size of the changes for each mutated variable (mutation step) must be defined.

The probability of mutating a variable is inversely proportional to the number of variables (dimensions). The more dimensions one individual has, the smaller is the mutation probability. Different papers reported results for the optimal mutation rate. In [23] it is mentioned that a mutation rate of 1/*nrvar* (*nrvar* represents the number of variables of an individual) produced good results for a wide variety of test functions. That means that only one variable per individual is mutated. Thus, the mutation rate is independent of the size of the population.

Two types of mutation can be distinguished according to the probability distribution from which the new gene values are drawn [2]:

- uniform mutation and
- non-uniform mutation.

Uniform Mutation

In this case the values of the genes in the offspring are drawn uniformly randomly from the definition domain. This option is analogue to bit-flipping for binary representation and the random resetting for integer representation. It is normally used with a position-wise mutation probability.

Nonuniform Mutation with a Fixed Distribution

This form of mutation is designed so that the amount of change introduced is small. This is achieved by adding to the current gene value an amount drawn randomly from a Gaussian (or normal) distribution with mean zero and user specified standard deviation. The obtained value is then scaled to the definition domain if necessarily. The Gaussian distribution has the propriety that approximately two thirds of the samples drawn lie within one standard deviation. This means that most of the changes made will be small but there is nonzero probability of generating very large changes since the tail of the distribution never reaches zero. This operator is usually applied with probability one per gene. An alternative to Gaussian distribution is to use the Cauchy distribution. The probability of generating larger values is slightly bigger than for Gaussian distribution with the same standard deviation [2][24].

14.3.4.2.3 Mutation for Order-Based Representation

For permutations it is not possible to use any of the forms of the mutation operators presented above. There are four common forms of the mutation operator for order-based representation [2][25]:

- *(i)* swap mutation;
- *(ii)* insert mutation;
- *(iii)* scramble mutation;
- *(iv)* inversion mutation.

Swap mutation

This form of mutation works by randomly picking two genes in the string and swapping their allele values. Figure 14.28 shows an example of this mutation.

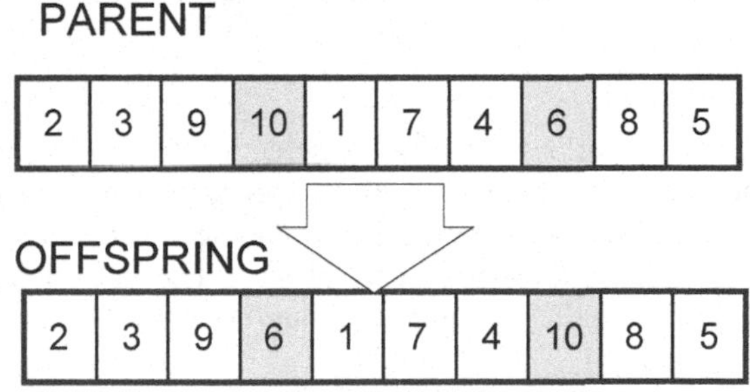

Fig. 14.28 Swap mutation.

Insert Mutation

This operator works by picking two alleles at random and moving one so that it is next to the other, shuffling along the others to make room.

An example of insert mutation is illustrated in Figure 14.29.

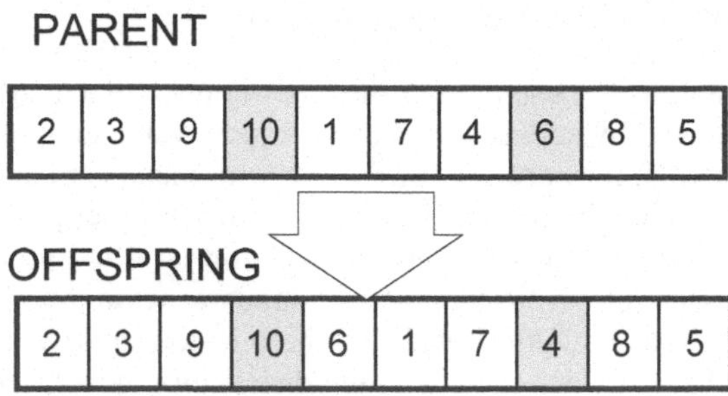

Fig. 14.29 Insert mutation.

Scramble Mutation

In this case the entire string or a subset of it (randomly chosen) has their values scrambled. An example is shown in Figure 14.30 where the selected subset is between positions 2 and 5.

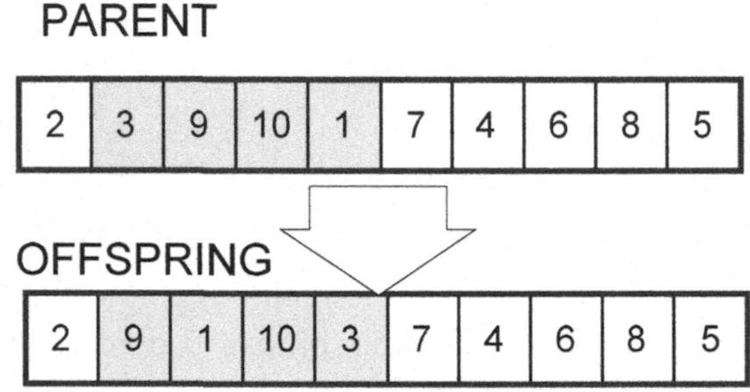

Fig. 14.30 Scramble mutation.

Inversion Mutation

This mutation operator works by randomly selecting two positions in the string and reversing the order in which the values appear between these positions. It breaks the string into three parts with all links inside a part being preserved and only the two links between the parts being broken. The inversion of a randomly chosen substring is the smallest change that can be made to an adjacency based problem.

An example is shown in Figure 14.31 with the selected positions 2 and 7.

14.3.4.2.4 Mutation for Integer Representation

There are two main forms of mutation used for integer representation; both mutate each gene independently with a defined probability:

- random resetting and
- creep mutation [2].

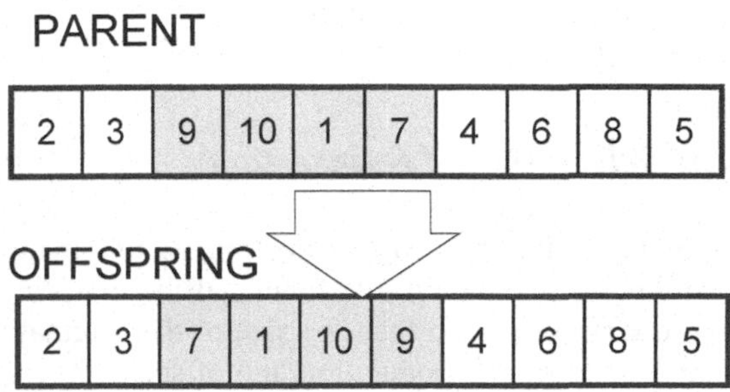

Fig. 14.31 Inversion mutation.

Random Resetting

The bit-flipping mutation of binary representation is extended to random resetting so that with a probability a new value is chosen at random from the set of permissible values in each position.

Creep Mutation

This type of mutation works by adding – with a given probability – a small (positive or negative) value to each gene. Usually, these values are sampled randomly for each gene from a distribution that is symmetric about zero and it is more likely to generate small changes than big ones.

14.3.5 Population Models

There are two main population models used by genetic algorithms:

1) Generational model and
2) Steady state model.

Generational Model

The generational model works as follows: in each generation the algorithm starts with a population of size *N*. A mating pool of size *N* is selected from this (some individuals will have multiple copies while other will not be selected at all). *N* offspring are further created by applying the variation operators. After each generation the whole population is replaced by the offspring population which will be the population of the next generation.

Steady-state model

In the steady-state model the population is not changed at once. In this case *M* ($M<N$) old individuals are replaced by *M* new individuals (from the offspring). The percentage of the population that is replaced is called generational gap and it

is equal to M/N. The steady state algorithm has been widely applied especially with M=1 and the corresponding generational gap $1/N$ [2].

14.3.6 Survivor Selection and Reinsertion

Once the offspring have been produced by selection, recombination and mutation of individuals from the old population, the fitness of the offspring may be determined. If less offspring are produced than the size of the original population then to maintain the size of the original population, the offspring have to be reinserted into the old population. Similarly, if not all offspring are to be used at each generation or if more offspring are generated than the size of the old population then a reinsertion scheme must be used to determine which individuals are to exist in the new population [12].

There are two main reinsertion strategies:

1) local reinsertion and
2) global reinsertion.

14.3.6.1 Local Reinsertion

In local selection, individuals are selected in a bounded neighborhood. The reinsertion of offspring takes place in exactly the same neighborhood. Thus, the locality of the information is preserved. The parent of an individual is the first selected parent in this neighborhood.

For the selection of parents to be replaced and for selection of offspring to reinsert the following schemes are possible [12]:

- insert every offspring and replace individuals in neighborhood uniformly at random;
- insert every offspring and replace weakest individuals in neighborhood;
- insert offspring fitter than weakest individual in neighborhood and replace weakest individuals in neighborhood;
- insert offspring fitter than weakest individual in neighborhood and replace parent;
- insert offspring fitter than weakest individual in neighborhood and replace individuals in neighborhood uniformly at random;
- insert offspring fitter than parent and replace parent.

14.3.6.2 Global Reinsertion

Different schemes of global reinsertion exist:

- produce as many offspring as parents and replace all parents by the offspring (pure reinsertion);
- produce less offspring than parents and replace parents uniformly at random (uniform reinsertion);

- produce less offspring than parents and replace the worst parents (elitist reinsertion);
- produce more offspring than needed for reinsertion and reinsert only the best offspring (fitness-based reinsertion).

Pure reinsertion is the simplest reinsertion scheme. Every individual lives one generation only. This scheme is used in the simple genetic algorithm. However, it is very likely, that very good individuals are replaced without producing better offspring and thus, good information is lost.

The elitist combined with fitness-based reinsertion prevents losing of information and is the recommended method. At each generation, a given number of the least fit parents are replaced by the same number of the most fit offspring. The fitness-based reinsertion scheme implements a truncation selection between offspring before inserting them into the population (i.e. before they can participate in the reproduction process). On the other hand, the best individuals can live for many generations. However, with every generation some new individuals are inserted. It is not checked whether the parents are replaced by better or worse offspring.

Because parents may be replaced by offspring with a lower fitness, the average fitness of the population can decrease. However, if the inserted offspring are extremely bad, they will be replaced with new offspring in the next generation [12].

14.3.7 The Basic Genetic Algorithm

The basic form of a general genetic algorithm is:

Step 1 Generate random population of N chromosomes.

Step 2 Evaluate each chromosome in the population using the fitness function

Step 3 Create a new population by repeating following steps until the new population is complete

Step 3.1 Selection

Select two parent chromosomes from a population according to their fitness (the better fitness, the higher the chance to be selected)

Step 3.2 Crossover

With a crossover probability cross over the parents to form new offspring (children). If no crossover was performed, offspring is the exact copy of parents.

Step 3.3 Mutation

With a mutation probability mutate new offspring at each locus (position in chromosome).

Step 3.4 replacement

Place new offspring in the new population

Step 4 Use new generated population for a further generation (iteration) of the algorithm

Step 5 If the termination condition is satisfied, stop, and return the best solution in current population

Step 6 Go to Step 2

Summaries

This chapter presented an evolutionary computation method with a focus on genetic algorithms. Genetic algorithms represent the most used techniques in practice among all of them. They can work with any representation and can be applied for a large variety of problems. There are any ways to speed up and improve a GA-based application as knowledge about problem domain is gained.

Some of the GAs advantages are (and these are also valid for the other EAs):

- concept is easy to understand and implement;
- modular, separate from application;
- supports multi-objective optimization;
- can be easily adapted for parallel machines;
- good for noisy and dynamic environments;
- easy to exploit previous or alternate solutions;
- flexible building blocks for hybrid applications.

When to use GAs:

- alternate solutions are too slow or overly complicated;
- need an exploratory tool to examine new approaches;
- problem is similar to one that has already been successfully solved by using a GA;
- want to hybridize with an existing solution;
- benefits of the GA technology meet key problem requirements.

References

1. Bäck, T., Fogel, D., Michalewicz, Z.: Handbook of evolutionary computation. IOP Publishing and Oxford University Press, New York (1997)
2. Eiben, A.E., Smith, J.E.: Introduction to Evolutionary Computing. Springer, Heidelberg (2003)
3. Eiben, A.E., Aarts, E.H.L., van Hee, K.M.: Global convergence of genetic algorithms: a markov chain analysis. In: Schwefel, H.-P., Männer, R. (eds.) PPSN 1990. LNCS, vol. 496, pp. 4–12. Springer, Heidelberg (1991)
4. Bäck, T.: Evolutionary algorithms in theory and practice. Oxford University Press, New York (1996)
5. Fogel, D.B.: Evolutionary Computation. IEEE Press, Los Alamitos (1995)

6. Bäck, T.: Generalized convergence models for tournament and (μ, λ) selection. In: Proceedings of the 6th International Conference on Genetic Algorithms, pp. 2–8 (1995)
7. Blickle, T., Thiele, L.: A comparison of selection schemes used in genetic algorithms. Evolutionary Computation 4(4), 361–394 (1996)
8. Goldberg, D.E., Deb, K.: A comparative analysis of selection schemes used in genetic algorithms. Foundations of Genetic Algorithms 1, 69–93 (1991)
9. Holland, J.H.: Adaptation in natural and artificial systems. MIT Press, Cambridge (1992)
10. Goldberg, D.E.: Generic algorithms in search, optimization and machine learning. Addison-Wesley, Reading (1989)
11. Baker, J.E.: Reducing Bias and Inefficiency in the Selection Algorithm. In: Proceedings of the Second International Conference on Genetic Algorithms and their Application, pp. 14–21. Lawrence Erlbaum Associates, Hillsdale (1987)
12. Evolutionary algorithms tutorial, `http://www.geatbx.com/`
13. Bäck, T., Hoffmeister, F.: Extended Selection Mechanisms in Genetic Algorithms. In: Bäck, T., Hoffmeister, F. (eds.) Proceedings of the Fourth International Conference on Genetic Algorithms, San Mateo, California, USA, pp. 92–99 (1991)
14. Whitley, D.: The GENITOR Algorithm and Selection Pressure: Why Rank-Based Allocation of Reproductive Trials is Best. In: Proceedings of the Third International Conference on Genetic Algorithms, San Mateo, California, USA, pp. 116–121 (1989)
15. Pohlheim, H.: Ein genetischer Algorithmus mit Mehrfachpopulationen zur Numerischen Optimierung. at-Automatisierungstechnik 3, 127–135 (1995)
16. Syswerda, G.: Uniform crossover in genetic algorithms. In: Proceedings of the Third International Conference on Genetic Algorithms, San Mateo, California, USA, pp. 2–9 (1989)
17. Spears, W.M., De Jong, K.A.: On the Virtues of Parameterised Uniform Crossover. In: Proceedings of the Fourth International Conference on Genetic Algorithms, San Mateo, California, USA, pp. 230–236 (1991)
18. Michalewicz, Z.: Genetic Algorithms + Data Structures = Evolution Programs. Springer, Berlin (1996)
19. Goldberg, D.E., Lingle, R.: Alleles, loci and the traveling salesman problem. In: Proceedings of the First International Conference on Genetic Algorithms and Their Applications, pp. 154–159. Lawrence Erlbaum, Hillsdale (1985)
20. Whitley, D.: Permutations, In Evolutionary Computation 1: Basic Algorithms and Operators. In: Bäck, T., Fogel, D.B. (eds.), pp. 274–284. Institute of Physics Publishing, Bristol (2000)
21. Davis, L. (ed.): Handbook of Genetic Algorithms. Van Nostrand Reinhold, New York (1991)
22. Olivier, L.M., Smith, D.J., Holland, J.: A study of permutation crossover operators on the traveling salesman problem. In: Proceedings of the Second International Conference on Genetic Algorithms and Their Applications, pp. 224–230. Lawrence Erlbaum, Hillsdale (1987)
23. Mühlenbein, H., Schlierkamp-Voosen, D.: Predictive Models for the Breeder Genetic Algorithm: I. Continuous Parameter Optimization. Evolutionary Computation 1(1), 25–49 (1993)
24. Yao, X., Liu, Y.: Fast evolutionary programming, In. In: Proceedings of the Fifth Annual Conference on Evolutionary Programming, pp. 451–460. The MIT Press, Cambridge (1996)

25. Mühlenbein, H., Pass, G.: From recombination of genes to the estimation of distributions I. Binary parameters. In: Proceedings of the 4th Conference on Parallel Problems Solving from Nature, pp. 188–197 (1996)
26. Rechenberg, I.: Evolutionsstrategie: Optimierung technischer Systeme nach Prinzipien der biologischen Evolution. Fromman-Holzboog, Stuttgart (1973)
27. Schwefel, H.P.: Numerische Optimierung von Computermodellen mittels der Evolutionsstrategie. Birkhaeuser, Basel (1977)
28. Fogel, L.J., Owens, A.J., Walsh, M.J.: Artificial intelligence through simulated evolution. Wiley, Chichester (1966)
29. Koza, J.R.: Genetic Programming: On the Programming of Computers by Means of Natural Selection. MIT Press, Cambridge (1992)

Verification Questions

1. What are the main steps in building an evolutionary algorithm?
2. Enumerate a few standard representations.
3. Enumerate and explain the selection mechanisms
4. Define the main variants of crossover operators for different representations.
5. Explain the crossover operators for permutations.
6. Define the main variants of the mutation operator.
7. Explain the mutation operator for permutations.
8. Explain the main population models which can be used by a generic algorithm.
9. Explain survival selection and reinsertion mechanisms.
10. Explain local and global reinsertion.
11. Present and explain the main structure of a genetic algorithm.

Exercises

We propose a list of problems, which can be easily approached with GA.

1. **Vertex Coloring**
Given a graph $G(V, E)$, with n vertex and the connections between them and a set of k colors, color the vertices of a graph such that no two adjacent vertices share the same color.

2. **Edge Coloring**
Given a graph $G(V, E)$, with n vertex and the connections between them and a set of k colors, color the edges of a graph such that no two adjacent edges share the same color.

3. **Monochromatic triangle**
Given a graph $G(V, E)$, with n vertex and the connections between them partition it into two disjoint sets $E1$ and $E2$, such that neither of the two graphs $G1(V,E1)$ and $G2(V,E2)$ contain a triangle. That is: for all nodes in $E1$ or $E2$ there does not exist a set $\{u, v, w\}$ such that $\{u, v\}$, $\{u, w\}$, $\{v, w\}$ are all edges.

4. **Graph partitioning problem**
Given a graph $G(V, E)$ and an integer $k > 1$, partition V into k parts (subsets) V_1, V_2, ... V_k such that the parts are disjoint and have equal size, and the number of edges with endpoints in different parts is minimized.

5. **Traveling salesman problem (TSP)**
Given a list of cities and their pairwise distances, find a shortest possible tour that visits each city exactly once.

6. **Quadratic Assignment Problem (QAP)**
There are a set of n facilities and a set of n locations. For each pair of locations, a *distance* is specified and for each pair of facilities a *weight* or *flow* is specified (e.g., the amount of supplies transported between the two facilities). The problem is to assign all facilities to different locations with the goal of minimizing the sum of the distances multiplied by the corresponding flows.

Given two sets, P (facilities) and L (locations), of equal size, together with a weight function $w : P \times P \to \boldsymbol{R}$ and a distance function $d : L \times L \to \boldsymbol{R}$. Find the bijection $f : P \to L$ ("assignment") such that the cost function:

$$\sum_{a,b \in P} w(a,b) \cdot d\big(f(a), f(b)\big)$$

is minimized.

7. **Subset sum problem**
Given a set of n integers, find a subset whose sum equals S (for S given).

8. **Knapsack problem (one of the variants)**
Given a set of items, each with a weight and a value, determine the number of each item to include in a collection so that the total weight is less than a given limit and the total value is as large as possible.

9. **Partition problem**
Given a multiset S of integers, find a way to partition S into two subsets S_1 and S_2 such that the sum of the numbers in S_1 equals the sum of the numbers in S_2. The subsets S_1 and S_2 must form a partition in the sense that they are disjoint and they cover S.

10. Shortest common supersequence
Given two sequences $X = < x_1,...,x_m >$ and $Y = < y_1,...,y_n >$, a sequence $U = < u_1,...,u_k >$ is a common supersequence of X and Y if U is a supersequence of both X and Y.

The shortest common supersequence is a common supersequence of minimal length. For X and Y given find the shortest common supersequence.

11. Evolutionary algorithm for **sudoku** game.

12. Evolutionary algorithm for **magic squares**.

13. Crossword puzzle
Given a crossword square and an alphabet which can be used (whose size is much higher that the number of words to be filled in the puzzle), find a valid solution for the crossword.

14. *n*-Queens problem
Given an $n \times n$ chess board, place n queens on it so that none of them can hit any other.

15. Coin Problem
Let there be $n \geq 0$ integers $0< a_1 < \ldots <a_n$. The values a_i represent the denominations of n different coins, where these denominations have greatest common divisor of 1. Find a way to pay the sum S by using the smallest number of coins.

Chapter 15
Evolutionary Metaheuristics

15.1 Introduction

Evolution Strategies (ES) were developed in [3][4]. ES tend to be used for empirical experiments that are difficult to model mathematically. The system to be optimized is actually constructed and ES are used to find the optimal parameter settings. Evolution strategies merely concentrate on translating the fundamental mechanisms of biological evolution for technical optimization problems. The parameters to be optimized are often represented by a vector of real numbers. Another vector of real numbers defines the strategy parameters, which controls the mutation of the objective parameters. Both object and strategic parameters form the data-structure for a single individual.

The classical ES works as follows: a single parent produces a single child by mutation.

The child is compared with parent and the better survives. This is a local search procedure, which is essentially hill-climbing. However, the mutation rate is part of the chromosome and the update strategy for the standard deviation of the mutation distribution is updated too. Recombination was also introduces in the ES and several variants were produced.

Given a current solution x^t in the form of a vector of length n, a new candidate solution x^{t+1} is created by adding a random number to each of the n components. A Gaussian distribution is used with mean zero and standard deviation σ. σ is a parameter of the algorithms that determines the extend to which, given values x_i are perturbed by the mutation operator. σ is called *mutation step size*. Theoretical results motivated the adjustment of σ by using the 1/5 success rule, which states as follows:

- Determine percentage p_s of successful mutations in past k iterations.
- Update σ after every k iterations by:

$$\begin{array}{ll} \sigma = \sigma / c & \text{if } p_s > 1/5 \\ \sigma = \sigma \cdot c & \text{if } p_s < 1/5 \\ \sigma = \sigma & \text{if } p_s = 1/5 \end{array}$$

C. Grosan and A. Abraham: Intelligent Systems, ISRL 17, pp. 387–407.
springerlink.com

15.2 Representation

Since ES are typically used for continuous parameter optimization, standard representation of the object variables x_1, x_2, …, x_n is straightforward, each x_i representing a floating point variable. The vector x is only a part of the ES genotype. Individuals also contain some strategy parameters, which are parameters of the mutation operator. Strategy parameters can be divided into two sets:

- σ parameters: represent the mutation step size and their number n_σ is either 1 or n. For any reasonable self-adaptation mechanism at least one σ should be considered.
- α values: represent interaction between step sizes used for different variables and are not always used [1].

The general form of an individual in ES is:

$$\left(\underbrace{x_1, x_2, \ldots x_n}_{x}, \underbrace{\sigma_1, \sigma_2, \ldots \sigma_{n_\sigma}}_{\sigma}, \underbrace{\alpha_1, \alpha_2, \ldots \alpha_k}_{\alpha}\right)$$

where:

$$n_\alpha = \left(n - \frac{n_\sigma}{2}\right)(n_\sigma - 1)$$

15.3 Mutation

The mutation operator in ES is based on Gaussian distribution requiring two parameters: the mean ξ and the standard deviation σ. Mutations are then realized by adding Δx_i to each x_i, where the Δx_i values are randomly drawn using the given Gaussian $N(\xi, \sigma)$ with the corresponding probability density function (p.d.f):

$$p(\Delta x_i) = \frac{1}{\alpha\sqrt{2\pi}} \cdot e^{-\frac{(\Delta x_i - \xi)^2}{2\sigma^2}}$$

In practice, the mean ξ is always set to zero and the vector x is mutated by replacing x_i values by:

$$x_i' = x_i + N(0, \sigma)$$

where $N(0, \sigma)$ denotes a random number drawn from Gaussian distribution with zero mean and standard deviation σ.

15.3.1 Uncorrelated Mutation with One σ

In this case the individual has the form:

$$x_1, x_2, \ldots, x_n, \sigma$$

Mutation occurs in the following steps:

- $\sigma' = \sigma \cdot \exp(\tau \cdot N(0,1))$
- $x_i' = x_i + \sigma' \cdot N(0,1)$

where τ is the learning rate and $\tau \cong 1/\ n^{½}$

There is also a boundary rule to force the step size to be no smaller than a threshold:

$$\sigma' < \varepsilon_0 \Rightarrow \sigma' = \varepsilon_0$$

15.3.2 Uncorrelated Mutation with n σ's

The structure of the chromosome in this case is:

$$x_1, x_2, \ldots, x_n, \sigma_1, \sigma_2 \ldots, \sigma_n$$

The mutation works as follows:

- $\sigma_i' = \sigma_i \cdot \exp(\tau' \cdot N(0,1) + \tau \cdot N_i\ (0,1))$
- $x_i' = x_i + \sigma_i' \cdot N_i\ (0,1)$

where we have two learning rates:

- τ' overall learning rate
- τ coordinate-wise learning rate

with:

- $\tau' \cong 1/(2\ n)^{½}$ and
- $\tau \cong 1/(2\ n^{½})\ ^{½}$

and the boundary rule:

$$\sigma_i' < \varepsilon_0 \Rightarrow \sigma_i' = \varepsilon_0,\ i=1, 2, \ldots, n$$

15.3.3 Correlated Mutation

The chromosome has the most general form in this case:

$$x_1, x_2\ldots, x_n, \sigma_1, \sigma_2\ldots, \sigma_n, \alpha_1, \alpha_2\ldots, \alpha_k$$

The covariance matrix C is defined as:

- $c_{ii} = \sigma_i$
- $c_{ij} = 0$ if i and j are not correlated
- $c_{ij} = \frac{1}{2}\left(\sigma_i^2 - \sigma_j^2\right) \cdot \tan(2 \cdot \alpha_{ij})$ if i and j are correlated.

The mutation mechanism is then:

- $\sigma_i' = \sigma_i \exp(\tau' \cdot N(0,1) + \tau \cdot N_i(0,1))$
- $\alpha_j' = \alpha_j + \beta \cdot N(0,1)$
- $\mathbf{x'} = \mathbf{x} + \mathbf{N(0,C')}$

where:

- x stands for the vector $x_1, x_2\ldots,x_n$
- C' is the covariance matrix C after mutation of the α values
- $\tau' \cong 1/(2n)^{1/2}$ and $\tau \cong 1/(2n^{1/2})^{1/2}$
- $\beta \approx 5°$
- $\sigma_i' < \varepsilon_0 \Rightarrow \sigma_i' = \varepsilon_0$
- $|\alpha_j'| > \pi \Rightarrow \alpha_j' = \alpha_j' - 2\pi \cdot \text{sign}(\alpha_j')$ (boundary rule for α_j values).

15.4 Recombination

The basic recombination scheme in ES involves two parents that create one child. To obtain N offspring the recombination operator is applied N times. There are two recombination variants:

(i) discrete recombination: one of the parent alleles is randomly chosen with equal chance for either parents;

(ii) intermediate recombination: the values of the parent alleles are averaged.

Given two parents x and y, the child z is created where:

$$z_i = \begin{cases} \dfrac{x_i + y_i}{2}, & \text{intermediate recombination} \\ x_i \text{ or } y_i \text{ chosen randomly}, & \text{discrete recombination} \end{cases}, \forall i \in \{1, 2, \ldots n\}$$

An extension of this scheme allows the use of more than two recombinants because the two parents x and y are drawn randomly for each position i in the offspring. This multiparent variant is called *global recombination*. The original variant is called *local recombination*.

15.5 Controlling the Evolution: Survival Selection

Let P be the number of parents in generation i and let C be the number of children in generation i. There are basically four different types of evolution strategies:

1) P, C;
2) P+C;
3) P/R, C;
4) P/R+C.

They mainly differ in how the parents for the next generation are selected and the usage of crossover operators.

15.5.1 P, C Strategy

The P parents produce C children using mutation. Fitness values are calculated for each of the C children and the best P children become next generation parents. The best individuals of C children are sorted by their fitness value and the first P individuals are selected to be next generation parents ($C \geq P$).

15.5.2 P + C Strategy

The P parents produce C children using mutation. Fitness values are calculated for each of the C children and the best P individuals of both parents and children become next generation parents. Children and parents are sorted by their fitness value and the first P individuals are selected to be next generation parents.

15.5.3 P/R, C Strategy

The P parents produce C children using mutation and crossover. Fitness values are calculated for each of the C children and the best P children become next generation parents. The best individuals of C children are sorted by their fitness value and the first P individuals are selected to be next generation parents (C ≥ P). Except the usage of recombination (crossover) operator this is exactly the same as *P*, *C* strategy.

15.5.4 *P/R + C Strategy*

The P parents produce C children using mutation and recombination. Fitness values are calculated for each of the C children and the best P individuals of both parents and children become next generation parents. Children and parents are sorted by their fitness value and the first P individuals are selected to be next generation parents. Except the usage of crossover operator, this is exactly the same as $P + C$ strategy.

15.6 Evolutionary Programming

Evolutionary programming (EP) [2][5] was originally developed to stimulate evolution as a learning process with the aim of generating artificial intelligence. Traditional EP is typically applied to machine learning tasks by finite state machines while contemporary EP can be also applied for (numerical) optimization. Classical EP use real value representation, Gaussian mutation and no recombination. Modern EP have predefined representation in general, thus no predefined mutation (must match representation). It often applies self-adaptation of mutation parameters.

15.6.1 *Representation*

In a similar manner to ES, the EP's chromosomes consist of two parts:

- Object variables: $x_1, x_2 \ldots, x_n$
- Mutation step sizes: $\sigma_1, \sigma_2 \ldots, \sigma_n$

Thus, the general form of an EP individual is given by:

$$\left(\underbrace{x_1, x_2, \ldots x_n}_{x}, \; \underbrace{\sigma_1, \sigma_2, \ldots \sigma_n}_{\sigma} \right)$$

15.6.2 *Mutation*

Having the chromosome:

$$\langle x_1, x_2 \ldots, x_n, \sigma_1, \sigma_2 \ldots, \sigma_n \rangle$$

we have to obtain, by mutation, the chromosome:

$$\langle x_1', x_2' \ldots, x_n', \sigma_1', \sigma_2' \ldots, \sigma_n' \rangle$$

This is achieved in the following manner:

- $\sigma_i' = \sigma_i \cdot (1 + \alpha \cdot N(0,1))$
- $x_i' = x_i + \sigma_i' \cdot N_i(0,1)$

with:

- $\alpha \approx 0.2$
- and the boundary rule: $\sigma' < \varepsilon_0 \Rightarrow \sigma' = \varepsilon_0$

There exist several other variants of EPs whose difference consists in one of the following:

- using lognormal scheme as in ES;
- using variance instead of standard deviation;
- mutate σ-last;
- using other distributions, e.g, Cauchy instead of Gaussian.

15.6.3 Survival Selection

Each individual produces one offspring by mutation.

If $P(t)$ is the parents population containing N parents and $P'(t)$ is offspring population of size N, the survival is done by pairwise competitions in round-robin format:

- Each solution x from $P(t) \cup P'(t)$ is evaluated against q other randomly chosen solutions.
- For each comparison, a "win" is assigned if x is better than its opponent.
- The N solutions with the greatest number of wins are retained to be parents of the next generation.
- Parameter q allows tuning selection pressure.
- Typically $q = 10$.

15.7 Genetic Programming

Genetic programming (GP) [6][7] is an evolutionary technique used for breeding a population of computer programs. If GA wants to evolve only solutions for particular problems GP evolves complex computer programs. GP individuals are represented and manipulated as nonlinear entities, usually trees in the standard approach. Nowadays at least 10 different representations have been developed so far.

15.7.1 Representation

GP individuals (chromosomes) are represented by trees. Any expression can be drawn as a tree of functions and terminals. Depending on the problem to solve, a GP chromosome can be the syntax of an arithmetic expression, formulas in first order predicate logic or code written in a programming language. Some examples of such types of expressions:

- an arithmetic formula:

$$\pi^2 + (2*y - x)$$

- a logical formula:

$$(x \vee \text{true}) \rightarrow (y \wedge \text{true}) \vee (z \wedge \text{true})$$

- a program (in C++ programming language):

```
i=1;
while (i<n)
i=i+1;
```

The trees corresponding to these expressions are presented in Figures 15.32-34.

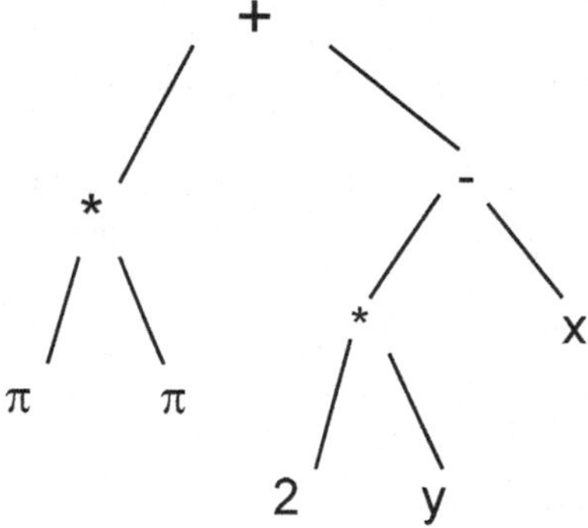

Fig. 15.32 Tree representation of the expression $\pi^2 + 2*y-x$.

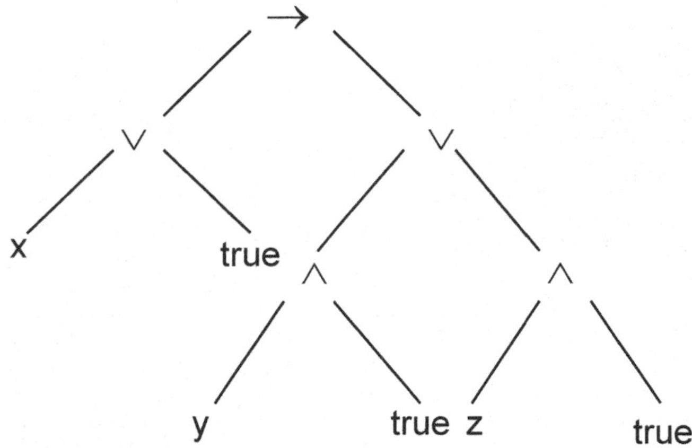

Fig. 15.33 Tree representation of the expression $(x \vee \text{true}) \rightarrow (y \wedge \text{true}) \vee (z \wedge \text{true})$.

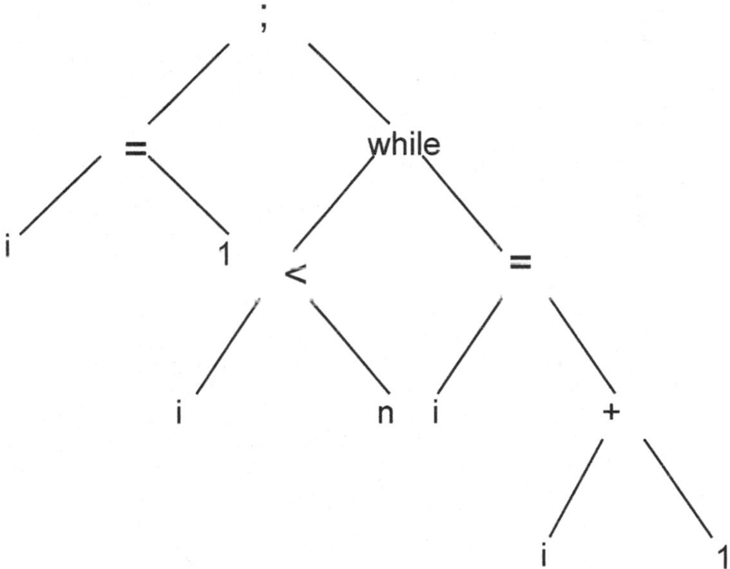

Fig. 15.34 Tree representation of the C++ source code: `i=1; while (i<n) i=i+1.`

The representation of a GP individual uses a defined set of function and a set of terminals. Elements of the terminal set are allowed as leaves while symbols from the functions set are internal nodes. These functions and terminals can be anything. Example:

- functions: {+, -, *, /, sine, cosine, ln, log, tan, If-Then-Else, Turn...}
- terminals: {x, y, 1, 2, 3 (constants), true, false, …}

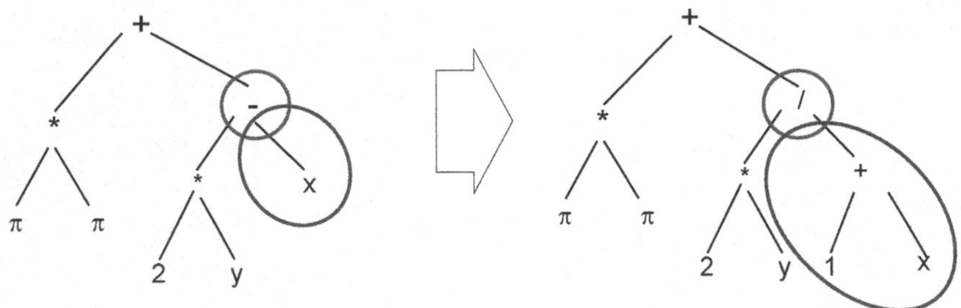

Fig. 15.35 GP mutation.

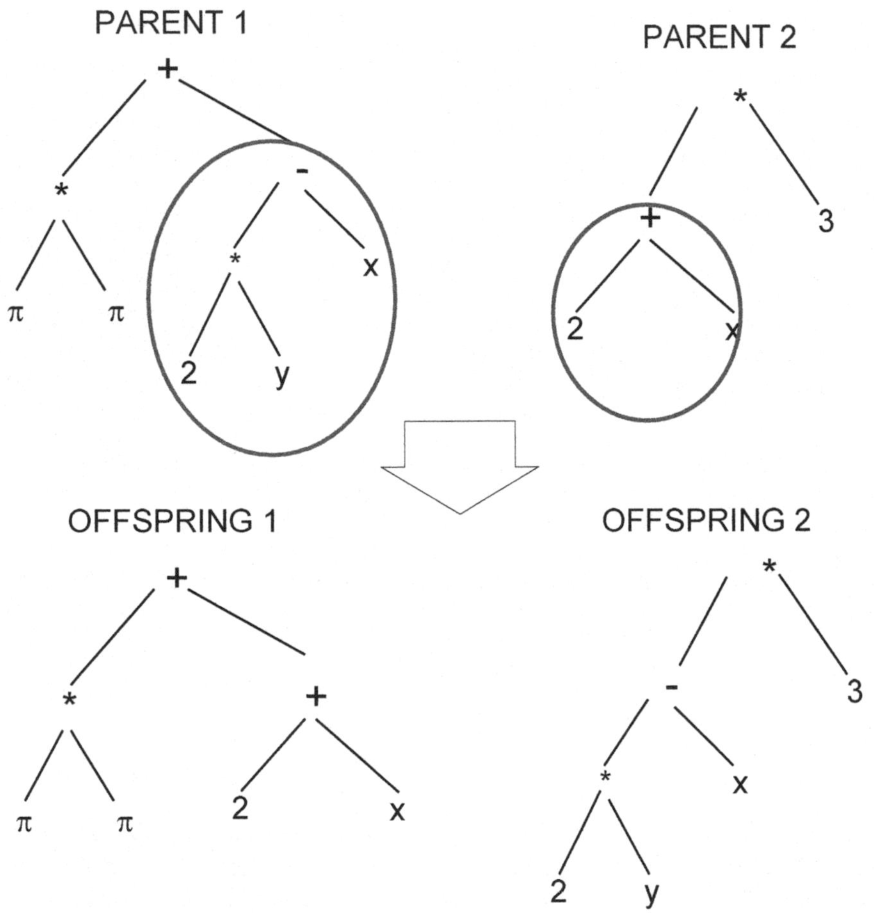

PARENT 1

PARENT 2

OFFSPRING 1

OFFSPRING 2

Fig. 15.37 GP Crossover.

15.7.2 *Variation Operators*

As in the case of GAs, the variation operators used by GP are mutation and recombination. Another operator, which is less used is branch duplication.

15.7.2.1 Mutation

The most common form of the mutation operator works by replacing a subtree of the tree (of the chromosome) starting at a randomly selected node by a randomly generated subtree. The new created subtree is usually generated in the same way as the tree in the population. The size of the offspring can exceed the size of the parent tree.

An example of the mutation operator is presented in Figure 15.35. The expression $\pi^2 + (2*y - x)$ is modified by mutation and the new expression: $\pi^2 + (2*y / (1+x))$ is obtained.

15.7.2.2 Recombination

Recombination in GP works by swapping subtrees among the parents. The operation is done by interchanging the subtrees starting at two randomly selected nodes in the given parents. The size of the offspring can exceed the size of the parents. An illustration of the crossover operator in GP is shown in Figure 15.36.

15.7.2.3 Branch Duplication

This operator works by simply duplicating a subtree of the GP chromosome as shown in Figure 15.38. This operator is not used very often.

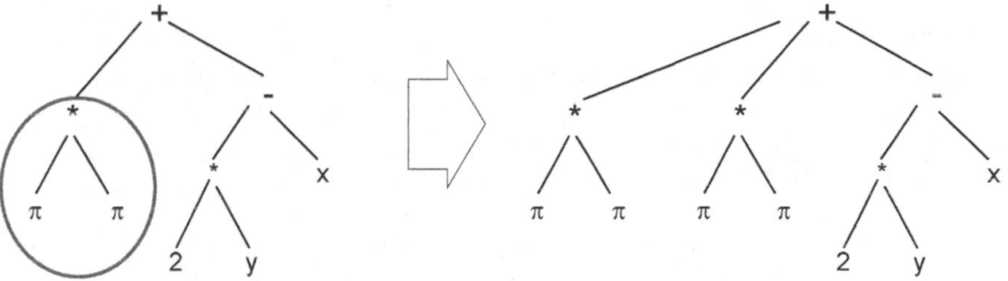

Fig. 15.38 GP branch duplication.

15.7.3 *Fitness Function*

Fitness of a GP chromosome is calculated in a slightly different way while compared to the other evolutionary techniques presented above. The general procedure for calculating the fitness is:

First, a training set is required.
Apply GP chromosome to each training data.
Compute the difference between what you want to obtain and what you have actually obtained.
Fitness should be minimized.

Let us consider as an example a regression problem.

We have the following data:

```
x   f(x)
0   0
1   1
2   4
3   9
4   16
5   25
```

And we wish to find the expression of $f(x)$.

Let us consider the GP chromosome for which we calculate the fitness is x+1.

Thus, for the given data we have:

```
x = 0 f(x) = 1
x = 1 f(x) = 2
x = 2 f(x) = 3
x = 3 f(x) = 4
x = 4 f(x) = 5
x = 5 f(x) = 6
```

And the fitness is: $|0 - 1| + |1 - 2| + |4 - 3| + |9 - 4| + |16 - 5| + |25 - 6| = 38$. The aim is to minimize this value (make it as close to zero as possible).

15.7.4 Parent Selection

Selection of parents for recombination can be performed using any of the selection mechanisms described for the selection of parents in genetic algorithms.

15.7.5 Survival Selection

Both generational and steady-state GP (same as in the case of genetic algorithms) can be implemented and used in practice.

15.7.6 GP Variants

There are several variants of the GP which differ among them mainly from the way they represent the individuals and calculate the fitness function. We present four of them in the following subsections.

15.7.6.1 Linear Genetic Programming

Linear Genetic Programming (LGP) [9][10] uses a specific linear representation of computer programs. Instead of the tree-based GP expressions of a functional programming language (like LISP) programs of an imperative language (like C) are evolved. A LGP individual is represented by a variable-length sequence of simple C language instructions. Instructions operate on one or two indexed variables (registers) r, or on constants c from predefined sets. The result is assigned to a destination register, for example, $r_i = r_j * c$ [8].

An example LGP program is:

```
void LGP(double v [8 ])
{
v [0 ] = v [5 ] + 73;
v [7 ] = v [3 ] - 59;
if (v [1 ] > 0)
if (v [5 ] > 21)
v [4 ] = v [2 ] * v [1 ];
v [2 ] = v [5 ] + v [4 ];
v [6 ] = v [7 ] * 25;
v [6 ] = v [4 ] - 4;
v [1 ] = sin(v [6 ]);
if (v [0 ] >v[1 ])
v [3 ] = v [5 ] * v [5 ];
v [7 ] = v [6 ] * 2;
v [5 ] = v [7 ] + 115;
if (v [1 ] < = v [6 ])
v [1 ] = sin(v [7 ]);
}
```

A LGP individual can be turned into a functional representation by successive replacements of variables starting with the last effective instruction. The maximum number of symbols in a LGP chromosome is four times the number of instructions. LGP uses two-point string crossover. A segment of random position and random length is selected in both parents and exchanged between them. If one of the resulting children would exceed the maximum length, crossover is abandoned and restarted by exchanging equally sized segments. An operand or an operator of an instruction is changed by mutation into another symbol over the same set. LGP also employs a special kind of mutation (called macro mutation), which deletes or inserts an entire instruction.

The fitness of a LGP individual is calculated using the formula:

$$\sum_{j=1}^{Nr} \left| O_j - E_j \right|$$

where Nr is the number of data, O_j is the value returned by a chromosome for the fitness case j ,and E_j is the expected value for the fitness case j.

15.7.6.2 Multi-expression Programming

Multi-Expression Programming MEP [11] uses a representation similar to the way in which C and Pascal compilers translate mathematical expressions into machine code. MEP genes are substrings of a variable length. The chromosome length is constant and equal to the number of genes in that chromosome. Each gene encodes a terminal or a function symbol. A gene encoding a function includes pointers towards the function arguments. Function parameters always have indices of lower values than the position of that function in the chromosome. According to this representation scheme, the first symbol of the chromosome must be a terminal symbol.

An example of a chromosome is given below. Numbers to the left stand for gene labels, or memory addresses. Labels do not belong to the chromosome.

```
1:    a
2:    b
3:    +1, 2
4:    c
5:    d
6:    +4, 5
```

When MEP individuals are translated into computer programs (expressions) they are read top-down starting with the first position. A terminal symbol specifies a simple expression. A function symbol specifies a complex expression (made up by linking the operands specified by the argument positions with the current function symbol).

For instance, genes 1, 2, 4, and 5 in the previous example encode simple expressions composed of a single terminal symbol. The expressions associated with genes 1, 2, 4, and 5 are [31]:

E_1 = a
E_2 = b
E_4 = c
E_5 = d.

Gene 3 indicates the operation + on the operands located in positions 1 and 2 of the chromosome. Therefore gene 3 encodes the expression:

E_3 = a +b.

Gene 6 indicates the operation + on the operands located in positions 4 and 5. Therefore gene 6 encodes the expression:

$E_6 = c + d.$

The expression associated with each position is obtained by reading the chromosome bottom-up from the current position and following the links provided by the function pointers.

The maximum number of symbols in a MEP chromosome is given by the formula:

Number of Symbols = (N +1)Number of Genes- N,

where N is the number of arguments of the function symbol with the greatest number of arguments.

Recombination and mutation are the two variation operators used by MEP. By recombination, two parents exchange genetic materials in order to obtain two offspring. Several variants of recombination have been considered and tested within MEP implementation: one-point recombination, two-point recombination, and uniform recombination.

Every MEP gene may be subject to mutation. The first gene of a chromosome must encode a terminal symbol in order to preserve the consistency of the chromosome. There is no restriction in symbols changing for other genes. If the current gene encodes a terminal symbol it may be changed into another terminal symbol or into a function symbol. In the last case, the positions indicating the function arguments are also generated by mutation. If the current gene encodes a function, the former may be mutated into a terminal symbol or into another function (function symbol and pointers towards arguments).

MEP uses a special kind of fitness assignment. The value of each expression encoded in a chromosome is computed during the individual evaluation (a MEP individual encodes a number of expressions equal to the number of its genes). This evaluation is performed by reading the chromosome only once and storing partial results by using dynamic programming. The best expression is chosen to represent the chromosome. Thus, the fitness of a MEP individual is computed using the formula:

$$\min_{k=1,L}\left\{\sum_{j=1}^{Nr}\left|E_j - O_j^k\right|\right\}$$

where Nr is the number of fitness cases, O_j^k is the value returned (for j-th data) by the k-th expression encoded in the chromosome, L is the number of chromosome genes and E_j is the expected value for that data.

15.7.6.3 Gene Expression Programming

Gene Expression Programming (GEP) [13] uses linear chromosomes that store expressions in breadth first form. A GEP gene is a string of terminal and function symbols. GEP genes are composed of a head and a tail. The head contains both function and terminal symbols. The tail may contain terminal symbols only.

For each problem the head length (denoted h) is chosen by the user. The tail length (denoted by t) is evaluated by:

$$t = (n - 1)h + 1,$$

where n is the number of arguments of the function with more arguments. Let us consider a gene made up of symbols in the set S:

$$S = \{*, /, +, - , a, b\}.$$

In this case $n = 2$. If we choose $h = 10$, then we get $t = 11$, and the length of the gene is 10+11 = 21. Such a gene is given below:

```
+* ab - +aab +ababbbababb.
```

The expression encoded by the gene is:

```
E = a +b * ((a +b )- a ).
```

Chromosomes are modified by mutation, transposition, root transposition, gene transposition, gene recombination, one-point recombination, and two-point recombination.

The fitness of a GEP individual is calculated using the formula:

$$\sum_{j=1}^{Nr} \left(M - \left|O_j - E_j\right|\right)$$

where M is the selection range, Nr is the number of data, O_j is the value returned by a chromosome for the fitness case j ,and E_j is the expected value for the fitness case j.

15.7.6.4 Grammatical Evolution

Grammatical Evolution GE [12] uses the Backus–Naur form (BNF) to express computer programs. BNF is a notation that allows a computer program to be expressed as a grammar. A BNF grammar consists of terminal and non-terminal symbols. Grammar symbols may be rewritten in other terminal and non-terminal symbols.

Each GE individual is a variable-length binary string that contains the necessary information for selecting a production rule from a BNF grammar in its codons (groups of eight bits). An example from a BNF grammar is given by the following production rules:

```
S ::= expr| (0)
if-stmt| (1)
loop. (2)
```

These production rules state that the start symbol *S* can be replaced (rewritten) either by one of the non-terminals (expr or if-stmt), or by loop. The grammar is used in a generative process to construct a program by applying production rules, selected by the genome, beginning with the start symbol of the grammar. In order to select a GE production rule, the next codon value on the genome is generated and placed in the following formula:

```
Rule = Codon Value MOD Num Rules.
```

If the next Codon integer value is four, knowing that we have three rules to select from, as in the example above, we get 4 **MOD** 3=1. Therefore, *S* will be replaced with the non-terminal if-stmt, corresponding to the second production rule. Beginning from the left side of the genome codon, integer values are generated and used for selecting rules from the BNF grammar, until one of the following situations arises.

1) A complete program is generated. This occurs when all the non terminals in the expression being mapped are turned into elements from the terminal set of the BNF grammar.
2) The end of the genome is reached, in which case the wrapping operator is invoked. This results in the return of the genome reading frame to the left side of the genome once again. The reading of the codons will then continue unless a higher threshold representing the maximum number of wrapping events has occurred during this individual mapping process.

In the case that a threshold on the number of wrapping events is exceeded and the individual is still incompletely mapped, the mapping process is halted, and the individual is assigned the lowest possible fitness value.

Example

Consider the grammar:

G = {N, T , S, P },

where the terminal set is:

T = { +, - , * ,/,sin, exp, (,)} ,
and the nonterminal symbols are:

N = {expr, op, pre op} .

The start symbol is: S = <expr>.

The production rules P are:

```
< expr> :: <expr>< op><expr> | (0)
(< expr><op><expr>) | (1)
< pre op>(< expr> ) | (2)
< var>. (3)
< op> ::= +| (0)
-| (1)
| (2)
/ (3)
< pre op> ::= sin| (0)
exp. (1)
```

Here is an example of a GE chromosome:

000000000000001000000001000000110000001000000011.

Translated into GE codons, the chromosome is:

0, 2, 1, 3, 2, 3.

This chromosome is translated into the expression:

E =exp(x)* x.

Standard binary genetic operators are used with GE. GE also makes use of a duplication operator that duplicates a random number of codons and inserts them into the penultimate codon position on the genome.

The fitness of a GE chromosome is calculated using the formula:

$$\sum_{j=1}^{Nr} \left| O_j - E_j \right|$$

where Nr is the number of data, O_j is the value returned by a chromosome for the fitness case j ,and E_j is the expected value for the fitness case j.

15.7.7 GP Applications

GP can be applied for any kind of problem where we have some inputs and we want some outputs. There must be a relationship between inputs and outputs.

Some of the main domains of application are:

- regression problems;
- classification problems;
- prediction and forecast;
- computing primitives for a given function;
- evolving digital circuits, etc.

Summaries

This chapter presented three evolutionary metaheuristics namely evolution strategies, evolutionary programming and genetic programming. The techniques presented all share the same structure and evolutionary scheme. Still, there are some noticeable differences among them, most important being the representation used, the variation operators employed and the way of calculating the fitness function.

ES have their origins in numerical optimisation problems. Some of the features are:

- typically work with real valued vectors;
- mutation taken from a Gaussian (normal) distribution;
- evolution of evolutionary parameters (e.g. mutation rate);
- a wide variety of evolutionary strategies are available, in some cases a population size of 1 is adopted.

Some essential characteristics of the ES are [2]:

- ES are typically used for continuous parameter optimization;
- there is a strong emphasis on mutation for creating offspring;
- mutation is implemented by adding some random noise drawn from a Gaussian distribution;
- mutation parameters are changed during a run of the algorithm.

Evolutionary programming is restricted to certain applications such as machine learning tasks by finite state machines while contemporary EP can be also applied for (numerical) optimization. Classical EP use real value representation, Gaussian mutation and no recombination. Modern EP have predefined representation in general, thus no predefined mutation (must match representation).

Genetic programming is an evolutionary technique used for breeding a population of computer programs. GP individuals are represented and manipulated as nonlinear entities, usually trees. GP approaches and variants are suitable for several applications such as classification, regression, predictions which make them useful in several domains from engineering to medicine. A particular GP subdomain consists of evolving mathematical expressions. In that case the evolved program is a mathematical expression, program execution means evaluating that expression, and the output of the program is usually the value of the expression.

References

1. Eiben, A.E., Smith, J.E.: Introduction to Evolutionary Computing. Springer, Heidelberg (2003)
2. Fogel, D.B.: Evolutionary Computation. IEEE Press, Los Alamitos (1995)
3. Rechenberg, I.: Evolutionsstrategie: Optimierung technischer Systeme nach Prinzipien der biologischen Evolution. Fromman-Holzboog, Stuttgart (1973)
4. Schwefel, H.P.: Numerische Optimierung von Computermodellen mittels der Evolutionsstrategie. Birkhaeuser, Basel (1977)
5. Fogel, L.J., Owens, A.J., Walsh, M.J.: Artificial intelligence through simulated evolution. Wiley, Chichester (1966)
6. Koza, J.R.: Genetic Programming: On the Programming of Computers by Means o f Natural Selection. MIT Press, Cambridge (1992)
7. Koza, J.R.: Genetic Programming I I: Automatic Discovery of Reusable Subprograms. MIT Press, Cambridge (1994)
8. Oltean, M., Grosan, C.: A Comparison of Several Linear Genetic Programming Techniques. Complex-Systems 14(4), 285–313 (2003)
9. Brameier, M., Banzhaf, W.: A Comparison of Linear Genetic Programming and Neural Networks in Medical Data Mining. IEEE Transactions on Evolutionary Computation 5, 17–26 (2001)
10. Brameier, M., Banzhaf, W.: Explicit Control of Diversity and Effective Variation Distance in Linear Genetic Programming. In: Proceedings of the Fourth European Conference on Genetic Programming, pp. 37–49 (2002)
11. Oltean, M., Grosan, C.: Evolving Evolutionary Algorithms by using Multi Expression Programming. In: Proceedings of the Seventh European Conference on Artificial Life, pp. 651–658 (2003)
12. Ryan, C., Collins, J.J., O'Neill, M.: Grammatical Evolution: Evolving Programs for an Arbitrary Language. In: Proceedings of the First European Workshop on Genetic Programming, pp. 83–96 (1998)
13. Ferreira, C.: Gene Expression Programming: A New Adaptive Algorithm for Solving Problems. Complex Systems 13, 87–129 (2001)

Verification Questions

1. Explain the main process of ES.
2. What is the general form of an ES individual?
3. Define and explain the four types of ES.

4. Define the main components of an EP chromosome.
5. Enumerate EP variants.
6. Explain the differences between GP and other evolutionary algorithms.
7. Enumerate and explain some of GP variants.
8. Enumerate the advantages and disadvantages of each of the four evolutionary computation techniques.
9. Present some of the applications domains of the evolutionary computation.

Chapter 16
Swarm Intelligence

16.1 Introduction

Swarm behavior can be seen in bird flocks, fish schools, as well as in insects like mosquitoes and midges. Many animal groups such as fish schools and bird flocks clearly display structural order, with the behavior of the organisms so integrated that even though they may change shape and direction, they appear to move as a single coherent entity [6]. The main principles of the collective behavior are:

- *homogeneity*: every bird in flock has the same behavior model. The flock moves without a leader, even though temporary leaders seem to appear.
- *locality*: the motion of each bird is only influenced by its nearest flock mates. Vision is considered to be the most important senses for flock organization.
- *collision avoidance*: avoid collision with nearby flock mates.
- *velocity matching*: attempt to match velocity with nearby flock mates.
- *flock centering*: attempt to stay close to nearby flock mates.

Individuals attempt to maintain a minimum distance between themselves and others at all times. This rule has the highest priority and corresponds to a frequently observed behavior of animals in nature [12]. If individuals are not performing an avoidance manoeuvre, they tend to be attracted towards other individuals (to avoid being isolated) and to align themselves with neighbors [9], [10].

Couzin et al. [6] identified four collective dynamical behaviors as illustrated in Figure 16.1:

- *torus*: individuals perpetually rotate around an empty core (milling). The direction of rotation is random.
- *dynamic parallel group*: the individuals are polarized and move as a coherent group, but individuals can move throughout the group and density and group form can fluctuate [11], [9].
- *swarm* : an aggregate with cohesion, but a low level of polarization (parallel alignment) among members

C. Grosan and A. Abraham: Intelligent Systems, ISRL 17, pp. 409–422.
springerlink.com

- *highly parallel group*: much more static in terms of exchange of spatial positions within the group than the dynamic parallel group and the variation in density and form is minimal.

As mentioned in [7], at a high-level, a swarm can be viewed as a group of agents cooperating to achieve some purposeful behavior and achieve some goal. This collective intelligence seems to emerge from what are often large groups of relatively simple agents. The agents use simple local rules to govern their actions and via the interactions of the entire group, the swarm achieves its objectives. A type of self-organization emerges from the collection of actions of the group.

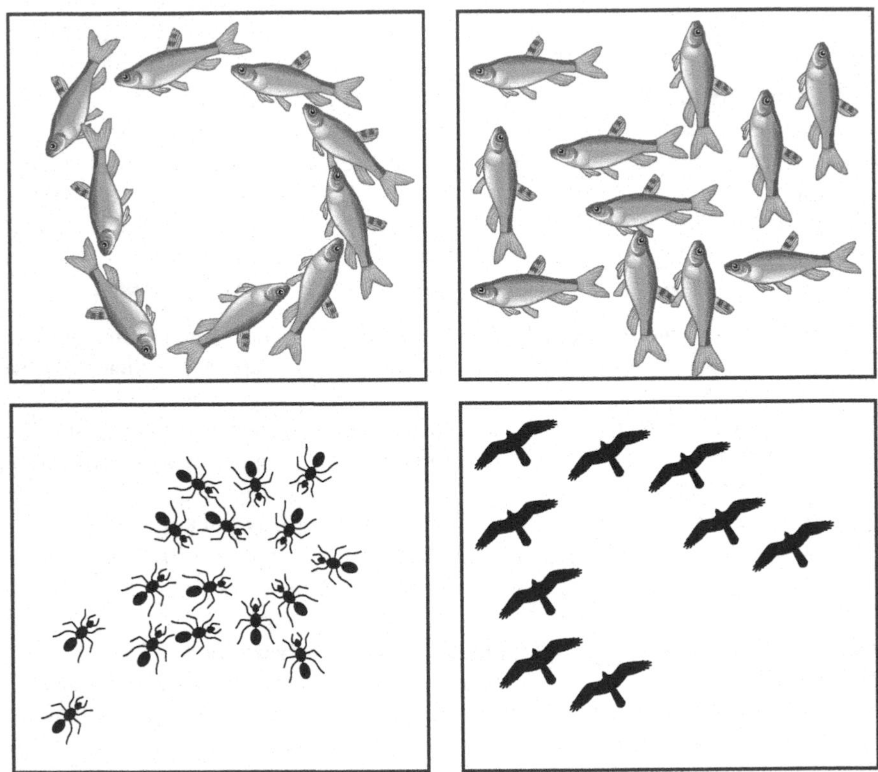

Fig. 16.1 Several models of collective behavior: torus (top-left) dynamic parallel group (top-right), swarm (bottom-left), and highly parallel group (bottom-right).

An autonomous agent is a subsystem that interacts with its environment, which probably consists of other agents, but acts relatively independently from all other agents [7]. The autonomous agent does not follow commands from a leader, or some global plan [8]. For example, for a bird to participate in a flock, it only adjusts its movements to coordinate with the movements of its flock mates, typically its neighbors that are close to it in the flock. A bird in a flock simply tries to stay close to its neighbors, but avoid collisions with them. Each bird does not take commands from any leader bird since there is no lead bird. Any bird can be in the front, center and back of the swarm. Swarm behavior helps birds take advantage of several things including protection from predators (especially for birds in the middle of the flock), and searching for food (essentially each bird is exploiting the eyes of every other bird) [7].

Since 1990, several collective behavior (like social insects, bird flocking) inspired algorithms have been proposed. The application areas of these algorithms refer to well studied optimization problems like NP-hard problems (Traveling Salesman Problem, Quadratic Assignment Problem, Graph problems), network routing, clustering, data mining, job scheduling, bioinformatics, etc. [1][2][3][4].

Particle Swarm Optimization (PSO) and Ant Colonies Optimization (ACO) are currently the most popular algorithms in the swarm intelligence domain.

16.2 Particle Swarm Optimization

The Particle Swarm Optimization (PSO) model [13][14] consists of a swarm of particles, which are initialized with a population of random candidate solutions. They move iteratively through the *d*-dimension problem space to search the new solutions, where the fitness, *f*, can be calculated as the certain qualities measure. Each particle has a position represented by a position-vector x_i (i is the index of the particle), and a velocity represented by a velocity-vector v_j. . Each particle remembers its own best position so far in a vector $x_i^{\#}$, and its j-th dimensional value is $x_{ij}^{\#}$. The best position-vector among the swarm so far is then stored in a vector x*, and its j-th dimensional value is x_j^*.

During the iteration time t, the update of the velocity from the previous velocity to the new velocity is determined by equation (16.1). The new position is then determined by the sum of the previous position and the new velocity by equation (16.2).

$$v_{ij}(t+1) = wv_{ij}(t) + c_1 r_1 \left(x_{ij}^{\#}(t) - x_{ij}(t)\right) + c_2 r_2 \left(x_j^{*}(t) - x_{ij}(t)\right) \tag{16.1}$$

$$x_{ij}(t+1) = x_{ij}(t) + v_{ij}(t+1) \tag{16.2}$$

where w is called as the inertia factor, r_1 and r_2 are the random numbers which are used to maintain the diversity of the population, and are uniformly distributed in the interval [0,1] for the j-th dimension of the i-th particle. c_1 is a positive constant, called as coefficient of the self-recognition component, c_2 is a positive constant, called as coefficient of the social component. A large inertia weight (w)

facilitates a global search while a small inertia weight facilitates a local search. By linearly decreasing the inertia weight from a relatively large value to a small value through the course of the PSO run gives the best PSO performance compared with fixed inertia weight settings.

From equation (16.1), a particle decides where to move next, considering its own experience, which is the memory of its best past position, and the experience of its most successful particle in the swarm. In the particle swarm model, the particle searches the solutions in the problem space with a range [-*s*; *s*] (if the range is not symmetrical, it can be translated to the corresponding symmetrical range.) In order to guide the particles effectively in the search space, the maximum moving distance during one iteration must be clamped in between the maximum velocity [- v_{max}, v_{max}] given in equation (16.3):

$$v_{ij} = \text{sign}\,(v_{ij})\,\min(|v_{ij}|,\, v_{max}) \tag{16.3}$$

The value of v_{max} is $p \times s$, with $0.1 \leq p \leq 1.0$ and is usually chosen to be *s*, i.e. $p = 1$. The pseudo-code for particle swarm optimization algorithm is illustrated below:

```
Step 1.   Initialize the size of the particle swarm n, and other
          parameters.
Step 2.   Initialize the positions and the velocities for all the
          particles randomly.
Step 3.   While (the end criterion is not met) do
       Step 3.1    t = t + 1;
       Step 3.2    Calculate the fitness value of each particle;
```

$$x^* = \arg\min_{i=1}^{n}\left(f\left(x^*(t-1)\right), f\left(x_1(t)\right), f\left(x_2(t)\right), \ldots, f\left(x_i(t)\right), \ldots, f\left(x_n(t)\right)\right)$$

```
       Step 3.3    For i= 1 to n
```

$$x_i^{\#}(t) = \arg\min_{i=1}^{n}\left(f\left(x_i^{\#}(t-1)\right), f\left(x_i(t)\right)\right)$$

```
            Step 3.3.1   For j = 1 to Dimension
                           Update the j-th dimension value of xi and vi
                          according to equations (16.1), (16.2), (16.3);
            Step 3.3.2   Next j
        Step 3.4   Next i
 End While.
```

The end criteria are usually one of the following:

- maximum number of iterations: the optimization process is terminated after a fixed number of iterations, for example, 1000 iterations;
- number of iterations without improvement: the optimization process is terminated after some fixed number of iterations without any improvement;
- minimum objective function error: the error between the obtained objective function value and the best fitness value is less than a prefixed anticipated threshold.

There are 3 terms that contribute to creating the new velocity:

1) *inertia term:*
   - this term forces the particle to move in the same direction;
   - audacious tendency, following own way using old velocity.
2) *cognitive term*
   - this term forces the particle to go back to the previous best position;
   - conservative tendency.
3) *social learning term*
   - this term forces the particle to move to the best previous position of its neighbors;
   - sheep like tendency, be a follower.

16.2.1 Parameters of PSO

The role of inertia weight w in equation (16.1) is considered critical for the convergence behavior of PSO. The inertia weight is employed to control the impact of the previous history of velocities on the current one. Accordingly, the parameter w regulates the trade-off between the global (wide-ranging) and local (nearby) exploration abilities of the swarm. A large inertia weight facilitates global exploration (searching new areas), while a small one tends to facilitate local exploration, i.e. fine-tuning the current search area. A suitable value for the inertia weight w usually provides balance between global and local exploration abilities and consequently results in a reduction of the number of iterations required to locate the optimum solution. Initially, the inertia weight is set as a constant. However, some experiment results indicates that it is better to initially set the inertia to a large value, in order to promote global exploration of the search space, and gradually decrease it to get more refined solutions [11]. Thus, an initial value around 1.2 and gradually reducing towards 0 can be considered as a good choice for w. A better method is to use some adaptive approaches (example: fuzzy controller), in which the parameters can be adaptively fine tuned according to the problems under consideration [5].

The parameters c_1 and c_2 in equation (16.1) are not critical for the convergence of PSO. However, proper fine-tuning may result in faster convergence and alleviation of local minima. As default values, usually, $c_1 = c_2 = 2$ are used, but some experiment results indicate that $c_1 = c_2 = 1.49$ might provide even better results. Recent work reports that it might be even better to choose a larger cognitive parameter, c_1, than a social parameter, c_2, but with $c_1+c_2 \leq 4$ [16].

The particle swarm algorithm can be described generally as a population of vectors whose trajectories oscillate around a region which is defined by each individual's previous best success and the success of some other particle. Various methods have been used to identify some other particle to influence the individual. Eberhart and Kennedy called the two basic methods as "*gbest model*" and "*lbest model*" [13]. In the lbest model, particles have information only of their own and their nearest array neighbors' best (lbest), rather than that of the entire group.

Namely, in Eq.(16.4), gbest is replaced by lbest in the model. So a new neighborhood relation is defined for the swarm:

$$v_{id}(t+1) = w \cdot v_{id}(t) + c_1 r_1 \left(p_{id}(t) - x_{id}(t)\right) + c_2 r_2 \left(p_{ld}(t) - x_{id}(t)\right) \quad (16.4)$$

$$x_{id}(t+1) = x_{id}(t) + v_{id}(t+1) \quad (16.5)$$

In the gbest model, the trajectory for each particle's search is influenced by the best point found by any member of the entire population. The best particle acts as an attractor, pulling all the particles towards it. Eventually all particles will converge to this position. The lbest model allows each individual to be influenced by some smaller number of adjacent members of the population array. The particles selected to be in one subset of the swarm have no direct relationship to the other particles in the other neighborhood. Typically lbest neighborhoods comprise exactly two neighbors. When the number of neighbors increases to all but itself in the lbest model, the case is equivalent to the gbest model. Some experiment results testified that gbest model converges quickly on problem solutions but has a weakness for becoming trapped in local optima, while lbest model converges slowly on problem solutions but is able to "flow around" local optima, as the individuals explore different regions. The gbest model is recommended strongly for unimodal objective functions, while a variable neighborhood model is recommended for multimodal objective functions.

Kennedy and Mendes [15] studied the various population topologies on the PSO performance. Different concepts for neighborhoods could be envisaged. It can be observed as a spatial neighborhood when it is determined by the Euclidean distance between the positions of two particles, or as a sociometric neighborhood (e.g. the index position in the storing array). The different concepts for neighborhood leads to different neighborhood topologies. Different topologies primarily affect the communication abilities and thus the group's performance. Different topologies are illustrated in Fig. 16.2. In the case of a global neighborhood, the structure is a fully connected network where every particle has access to the others' best position (Figure 16.2 (a)). But in local neighborhoods there are more possible variants. In the von Neumann topology (Figure 16.2 (b)), neighbors above, below, and each side on a two dimensional lattice are connected. Figure 16.2 (e) illustrates the von Neumann topology with one section flattened out. In a pyramid topology, three dimensional wire frame triangles are formulated as illustrated in Figure 16.2 (c). As shown in Figure 16.2 (d), one common structure for a local neighborhood is the circle topology where individuals are far away from others (in terms of graph structure, not necessarily distance) and are independent of each other but neighbors are closely connected. Another structure is called wheel (star) topology and has a more hierarchical structure, because all members of the neighborhood are connected to a 'leader' individual as shown in Fig. 16.2 (f). Thus, all information has to be communicated though this 'leader', which then compares the performances of all others.

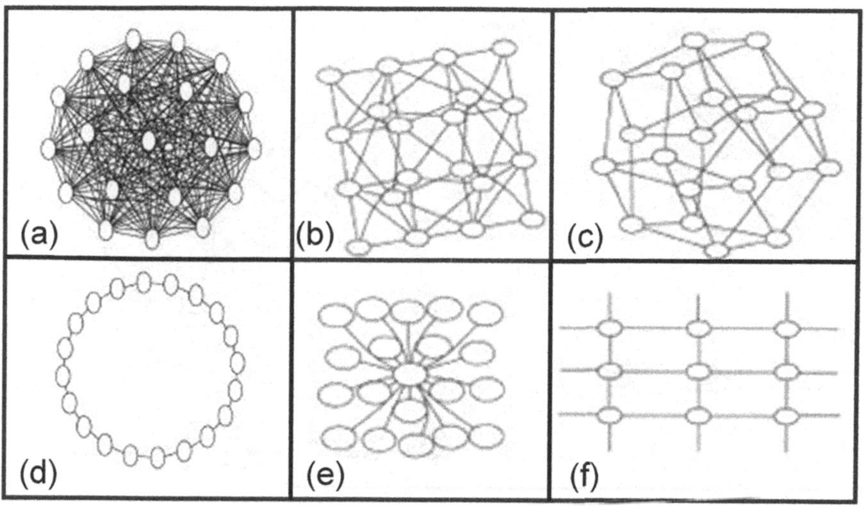

Fig. 16.2 Some neighborhood topologies (adapted from [15]).

16.3 Ant Colonies Optimization

When searching for food, ants initially explore the area surrounding their nest in a random manner. While moving, ants leave a chemical pheromone trail on the ground. Ants are guided by pheromone smell. Ants tend to choose the paths marked by the strongest pheromone concentration. When an ant finds a food source, it evaluates the quantity and the quality of the food and carries some of it back to the nest. During the return trip, the quantity of pheromone that an ant leaves on the ground may depend on the quantity and quality of the food. The pheromone trails will guide other ants to the food source.

The indirect communication between the ants via pheromone trails enables them to find shortest paths between their nest and food sources as illustrated in Figure 16.3.

Ants' ability to collectively find the shortest path to the best food source was studied by Jean-Louis Deneubourg [18][19][20]. He demonstrated how the Argentine ant was able to successfully choose the shortest between the two paths to a food source. Deneubourg was initially interested in self organization, a concept which until then had been the fare of chemists and physicists seeking to explain the natural order occurring in physical structures such as sand dunes and animal patterns. Deneubourg saw the potential for this concept, which by 1989 had turned into a sizeable research project amongst Physicists, to be applied to Biology. In his experiments, a group of ants are offered two branches leading to the same food source, one longer than the other. Initially, there is a 50% chance of an ant choosing either branch, but gradually more and more journeys are completed on the shorter branch than the longer one, causing a denser pheromone trail to be laid. This consequently tips the balance and the ants begin to concentrate on the shorter

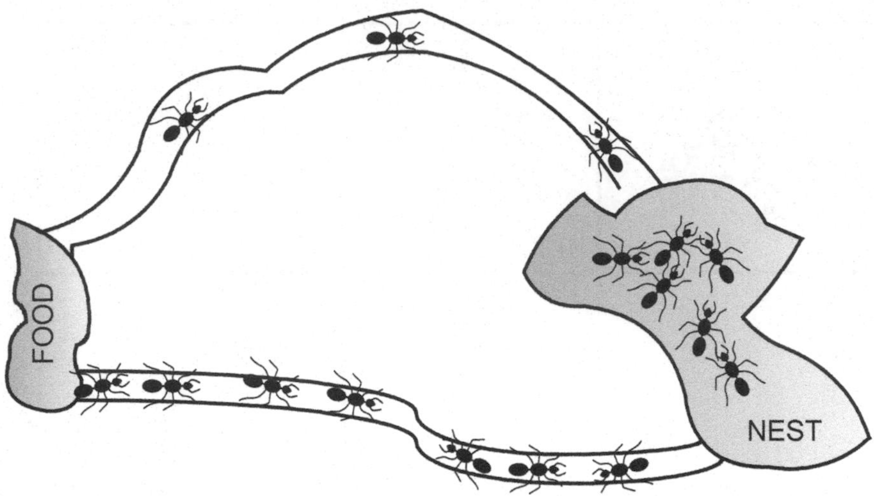

Fig. 16.3 The ants taking the shortest path can perform a greater number of trips between nest and food; implicitly the pheromone trail will be more than the one released by the ants following the longest path.

route, discarding the longer one. This is precisely the mechanism underpinning an ant colony's ability to efficiently exploit food sources in sequential order: strong trails will be established to the nearest source first, and then when it is depleted and the ants lose interest, the trails leading to the next nearest source will build up [4].

16.3.1 Ant System

Although an individual ant is quite small (measuring only 2.2 to 2.6 mm in length) and wanders quite aimlessly in isolation, a group of many ants exhibits extraordinarily intelligent behavior, recognizable to humans as meaningful pathways to food sources. This emergent intelligence can be summarized as:

1) At the outset of the foraging process, the ants move more or less randomly – this "random" movement is actually executed such that a considerable amount of surface area is covered, emanating outward from the nest.
2) If it is not carrying food, the ant "deposits" a nest pheromone and will prefer to walk in the direction of sensed food pheromone.
3) If it is carrying food, the ant deposits a food pheromone and will prefer to walk in the direction of sensed nest pheromone.
4) The ant will transport food from the source to the nest.

As a pheromone "trail" becomes stronger, the more ants follow it, leaving more pheromone along the way, which makes more ants follow it, and so on. ACO is implemented as a team of intelligent agents which simulate the ants' behavior, walking around the graph representing the problem to solve using mechanisms of cooperation and adaptation. ACO algorithm requires defining the following [23]:

- The problem needs to be represented appropriately, which would allow the ants to incrementally update the solutions through the use of a probabilistic transition rules, based on the amount of pheromone in the trail and other problem specific knowledge. It is also important to enforce a strategy to construct only valid solutions corresponding to the problem definition.
- A problem-dependent heuristic function h that measures the quality of components that can be added to the current partial solution.
- A rule set for pheromone updating, which specifies how to modify the pheromone value t.
- A probabilistic transition rule based on the value of the heuristic function h and the pheromone value t that is used to iteratively construct a solution.

According to [22], the main steps of the ACO algorithm are given below:

1) *pheromone trail initialization*;
2) *solution construction using pheromone trail*;
   Each ant constructs a complete solution to the problem according to a probabilistic rule;
3) *state transition rule;*
   The state transition rule depends mainly on the state of the pheromone;
4) *pheromone trail update;*
   A global pheromone updating rule is applied in two phases. First, an evaporation phase where a fraction of the pheromone evaporates, and then a reinforcement phase where each ant deposits an amount of pheromone which is proportional to the fitness of its solution [21]. This process is iterated until a termination condition is reached.

ACO was first introduced using the Traveling Salesman Problem. Starting from its start node, an ant iteratively moves from one node to another. When being at a node, an ant chooses to go to an unvisited node at time t with a probability given by:

$$p_{i,j}^{k}(t) = \begin{cases} \dfrac{\left[\tau_{i,j}(t)\right]^{\alpha}\left[\eta_{i,j}(t)\right]^{\beta}}{\sum_{k \in N_i^k}\left[\tau_{i,k}(t)\right]^{\alpha}\left[\eta_{i,k}(t)\right]^{\beta}}, & \text{if } j \in N_i^k \\ 0 & \text{otherwise} \end{cases} \tag{16.6}$$

where N_i^k is the feasible neighborhood of the ant_k, that is, the set of cities which ant_k has not yet visited; $\tau_{i,j}(t)$ is the pheromone value on the edge (i, j) at the time t,

α is the weight of pheromone; $\eta_{i,j}(t)$ is a priori available heuristic information on the edge (i, j) at the time t, β is the weight of heuristic information. Two parameters α and β determine the relative influence of pheromone trail and heuristic information. $\tau_{i,j}(t)$ is determined by:

$$\tau_{i,j}(t) = \rho\tau_{i,j}(t-1) + \sum_{k=1}^{n} \Delta\tau_{i,j}^{k}(t), \ \forall(i, j) \tag{16.7}$$

$$\Delta\tau_{i,j}^{k} = \begin{cases} \dfrac{Q}{L_k}, & if\ (i, j) \in tour\ described\ by\ tabu_k \\ 0 & \text{otherwise} \end{cases} \tag{16.8}$$

where ρ is the pheromone trail evaporation rate ($0 < \rho < 1$), n is the number of ants, Q is a constant for pheromone updating, L_k is the length of that path. A generalized version of the pseudocode for the ACO algorithm with reference to the TSP is illustrated below:

```
Step 1. Initialize the number of ants n, and other parameters.
Step 2.While (the end criterion is not met)
      Step 2.1 t = t + 1;
      Step 2.2 For k= 1 to n
            Step 2.2.1 antk is positioned on a starting node;
            Step 2.2.2 For m= 2 to problem size
                        1. Choose the state to move into according to
                        the probabilistic transition rules;
                        2. Append the chosen move into tabuk(t) for
                        antk;
      Step 2.3 Compute the length Lk(t) of the tour Tk(t) chosen by the
               antk;
      Step 2.4 Compute Δτi,j(t) (t) for every edge (i, j) in Tk(t)
             according to equations (16.8);
Step 4. Update the trail pheromone intensity for every edge (i, j)
        according to equation (16.7);
Step 5. Compare and update the best solution;
End While.
```

Other applications of the ACO algorithm include: sequential ordering problem, quadratic assignment problem, vehicle routing problem, scheduling problems, graph coloring, partitioning problems, timetabling, shortest subsequence problem, constraint satisfaction problems, maximum clique problem, edge-disjoint paths problem.

Summaries

This chapter introduced some of the theoretical foundations of swarm intelligence. We focus on the design and implementation of the Particle Swarm Optimization and Ant Colony Optimization algorithms. PSO is a population-based search algorithm and is initialized with a population of random solutions, called particles. Unlike in the other evolutionary computation techniques, each particle in PSO is also associated with a velocity. Particles fly through the search space with velocities, which are dynamically adjusted according to their historical behaviors. Therefore, the particles have the tendency to fly towards the better and better search area over the course of search process. The PSO was first designed to simulate birds seeking food, which is defined as a 'cornfield vector'.

In PSO, each single solution is like a 'bird' in the search space, which is called 'particle'. All particles have fitness values, which are evaluated by the fitness function to be optimized, and have velocities, which direct the flying of the particles. (The particles fly through the problem space by following the particles with the best solutions so far). PSO is initialized with a group of random particles (solutions) and then searches for optima by updating each generation.

There are some drawbacks, which the standard PSO algorithm encounters:

- particles tend to cluster, i.e., converge too fast and get stuck at local optimum;
- movement of particle carried it into infeasible region;
- inappropriate mapping of particle space into solution space.

Some of them are overcome by newer versions of PSO:

- PSO with multiple social learning terms;
- Measurement Indices for PSO:
  - Two measurement indices are defined for observing the dynamic behavior of the swarm:
    - *dispersion index*: It measures how particles are spreading around the best particle in the swarm, and is defined as the average absolute distance of each dimension from the best particle.

      It explains the coverage searching area of the swarm. A swarm with higher dispersion index has relatively wider coverage of searching area than the one with lower dispersion index.
    - *velocity index*: It measures how fast the swarm moves in certain iteration, and is defined as the average of absolute velocity.

      It shows the moving behavior of the swarm: higher index means the swarm move more aggressively in moving through the problem space than the swarm with lower index.
- Heterogeneous Particles;
- Hierarchical PSO.

ACO algorithms were inspired by the behavior of ant colonies. Ants are social insects, being interested mainly in the colony survival rather than individual survival. Of interests is ants' ability to find the shortest path from their nest to food. This idea was the source of the algorithms inspired from ants' behavior.

When searching for food, ants initially explore the area surrounding their nest in a random manner. While moving, ants leave a chemical pheromone trail on the ground. Ants are guided by pheromone smell and tend to choose the paths marked by the strongest pheromone concentration. When an ant finds a food source, it evaluates the quantity and the quality of the food and carries some of it back to the nest. During the return trip, the quantity of pheromone that an ant leaves on the ground may depend on the quantity and quality of the food. The pheromone trails will guide other ants to the food source. The indirect communication between the ants via pheromone trails enables them to find shortest paths between their nest and food sources.

Some advantages and disadvantages of the ACO system are:

- *Advantages*:
  - positive feedback accounts for rapid discovery of good solutions;
  - distributed computation avoids premature convergence;
  - the greedy heuristic helps find acceptable solution in the early solution in the early stages of the search process;
  - the collective interaction of a population of agents.
- *Disadvantages:*
  - it has a slower convergence than other heuristics;
  - it performed poorly for TSP problems larger than 75 cities;
  - no centralized processor to guide the AS towards good solutions.

The subject of copying, imitating, and learning from biology was coined *Biomimetics* by Otto H. Schmitt in 1969 [17]. This field is increasingly involved with emerging subjects of science and engineering and it represents the studies and imitation of nature's methods, designs and processes. Nature has produced effective solutions to innumerable complex real-world problems. Even though there are several computational nature inspired models, there is still a lot of room for more research, at least in the form of finding some collaborations and interactions between the existing systems as well as developing new systems by borrowing ideas from nature. Butler [16] suggests some potential research areas:

1) Spiders spin silk that is stronger than synthetic substances developed by man but require only insects as inputs.
2) Diatoms, microscopic phytoplankton responsible for a quarter of all the photosynthesis on Earth, make glass using silicon dissolved in seawater.
3) Abalone, a type of shellfish, produces a crack-resistant shell twice as tough as ceramic from calcium found in seawater using a process known as biomineralization.
4) Trees "turn sunlight, water, and air into cellulose, a sugar stiffer and stronger than nylon, and bind it into wood, a natural compo-

site with a higher bending strength and stiffness than concrete or steel," as noted by Paul Hawken, Amory and L. Hunter Lovins in *Natural Capitalism*.

5) Countless plants generate compounds that fight off infection from fungi, insects, and other pests [4].

References

1. Grosan, C., Abraham, A., Chis, M.: Swarm intelligence in data mining, Swarm Intelligence and Data Mining. In: Abraham, A., Grosan, C., Ramos, V. (eds.) Studies in Computational Intelligence, vol. 34, pp. 1–20. Springer, Germany (2006)
2. Das, S., Abraham, A., Konar, A.: Swarm Intelligence Algorithms in Bioinformatics. In: Kelemen, A., et al. (eds.) Computational Intelligence in Bioinformatics, pp. 113–147. Springer, Germany (2008)
3. Abraham, A., Das, S., Roy, S.: Swarm Intelligence Algorithms for Data Clustering. In: Maimon, O., Rokach, L. (eds.) Soft Computing for Knowledge Discovery and Data Mining, pp. 279–313. Springer, Heidelberg (2007)
4. Grosan, C., Abraham, A.: Stigmergic Optimization: Inspiration, Technologies and Perspectives. In: Abraham, A., Grosan, C., Ramos, V. (eds.) Studies in Computational Intelligence, pp. 1–24. Springer, Germany (2006)
5. Abraham, A., Guo, H.: H. Liu, Swarm Intelligence: Foundations, Perspectives and Applications. In: Nedjah, N., Mourelle, L. (eds.) Swarm Intelligent Systems, pp. 3–25. Springer, Germany (2006)
6. Couzin, I.D., Krause, J., James, R., Ruxton, G.D., Franks, N.R.: Collective Memory and Spatial Sorting in Animal Groups. Journal of Theoretical Biology 218, 1–11 (2002)
7. Fayyad, U., Piatestku-Shapio, G., Smyth, P., Uthurusamy, R.: Advances in knowledge discovery and data mining. AAAI/MIT Press (1996)
8. Flake, G.: The computational beauty of nature. MIT Press, Cambridge (1999)
9. Partridge, B.L., Pitcher, T.J.: The sensory basis of fish schools: relative role of lateral line and vision. Journal of Comparative Physiology 135, 315–325 (1980)
10. Partridge, B.L.: The structure and function of fish schools. Science American 245, 90–99 (1982)
11. Major, P.F., Dill, L.M.: The three-dimensional structure of airborne bird flocks. Behavioral Ecology and Sociobiology 4, 111–122 (1978)
12. Krause, J., Ruxton, G.D.: Living in groups. Oxford University Press, Oxford (2002)
13. Kennedy, J., Eberhart, R.: Swarm intelligence. Morgan Kaufmann, Academic Press (2001)
14. Clerc, M., Kennedy, J.: The particle swarm-explosion, stability, and convergence in a multidimensional complex space. IEEE Transactions on Evolutionary Computation 6(1), 58–73 (2002)
15. Kennedy, J., Mendes, R.: Population structure and particle swarm performance. In: Proceeding of IEEE Conference on Evolutionary Computation, pp. 1671–1676 (2002)
16. Butler, R.: Biomimetics, technology that mimics nature, available online at `http://mongabay.com`
17. Cohen, Y.B.: Biomimetics: Biologically Inspired Technologies. CRC Press, Boca Raton (2005)

18. Deneubourg, J.L., Aron, S., Goss, S., Pasteels, J.M., Duerinck, G.: Random behaviour, amplification processes and number of participants: how they contribute to the foraging properties of ants. Physica D 22, 176–186 (1986)
19. Deneubourg, J.L., Aron, S., Goss, S., Pasteels, J.M.: Error, communication and learning in ant societies. European Journal of Operational Research 30, 168–172 (1987)
20. Deneubourg, L.J., Goss, S., Pasteels, J.M., Fresneau, D., Lachaud, J.P.: Self-organization mechanisms in ant societies (II): learning in foraging and division of labor. In: From Individual to Collective Behavior in Social Insects. Experientia Supplementum, vol. 54, pp. 177–196 (1984)
21. Toksari, M.D.: Ant colony optimization for finding the global minimum. Applied Mathematics and Computation 176(1), 308–316 (2006)
22. Dorigo, M., Gambardella, L.M.: Ant Colonies for the Traveling Salesman Problem. BioSystems 43, 73–81 (1997)
23. Dorigo, M., Bonaneau, E., Theraulaz, G.: Ant algorithms and stigmergy. Future Generation Computer Systems 16, 851–871 (2000)

Verification Questions

1. Present some models of collective behavior.
2. What are the main principles of collective behavior?
3. Explain in detail the mechanism behind PSO.
4. How to set inertia weight in PSO?
5. How is the new velocity created?
6. Explain the main PSO models and the differences between them.
7. Define and explain neighborhood topologies in PSO.
8. What is the nature association with the artificial Ant Colonies Systems?
9. What are the main steps of ACO Algorithm?
10. Which problems are suitable to be approached with ACO?
11. Enumerate known ones or find by yourself some possible nature inspired research ideas.

Exercises

Use PSO for all the optimization problems given in Chapter 13. Some of them might not have a direct real representation; thus, try to adapt PSO (whenever it is possible) for these problems.

Chapter 17
Hybrid Intelligent Systems

17.1 Introduction

Computational intelligence is an innovative framework for constructing intelligent hybrid architectures involving Neural Networks (NN), Fuzzy Inference Systems (FIS), Probabilistic Reasoning (PR), Evolutionary Computation (EC) and Swarm Intelligence (SI). Most of these hybridization approaches, however, follow an ad hoc design methodology, justified by success in certain application domains. Due to the lack of a common framework it often remains difficult to compare the various hybrid systems conceptually and to evaluate their performance comparatively.

Several adaptive hybrid computational intelligence frameworks have been developed. Many of these approaches use a combination of different knowledge representation schemes, decision making models and learning strategies to solve a computational task. This integration aims at overcoming the limitations of individual techniques through hybridization or the fusion of various techniques.

To achieve a highly intelligent system, a synthesis of various techniques is required. Figure 17.1 shows the synthesis of NN, FIS and EC and their mutual interactions leading to different architectures. Each technique plays a very important role in the development of different hybrid soft computing architectures. Experience has shown that it is crucial, in the design of hybrid systems, to focus primarily on the integration and interaction of different techniques rather than to merge different methods to create ever-new techniques. Techniques already well understood should be applied to solve specific domain problems within the system. Their weaknesses must be addressed by combining them with complementary methods [1].

Neural networks offer a highly structured architecture with learning and generalization capabilities, which attempts to mimic the neurological mechanisms of the brain. NN stores knowledge in a distributive manner within its weights, which have been determined by learning from known samples. The generalization ability of new inputs is then based on the inherent algebraic structure of the NN. However it is very hard to incorporate human *a priori* knowledge into a NN mainly because the connectionist paradigm gains most of its strength from a distributed knowledge representation.

C. Grosan and A. Abraham: Intelligent Systems, ISRL 17, pp. 423–450.
springerlink.com

By contrast, fuzzy inference systems exhibit complementary characteristics, offering a very powerful framework for approximate reasoning, which attempts to model the human reasoning process at a cognitive level [8]. FIS acquires knowledge from domain experts which is encoded within the algorithm in terms of the set of *if-then* rules. FIS employ this rule-based approach and interpolative reasoning to respond to new inputs [5]. The incorporation and interpretation of knowledge is straightforward, whereas learning and adaptation constitute major problems.

Probabilistic reasoning such as Bayesian belief networks gives us a mechanism for evaluating the outcome of systems affected by randomness or other types of probabilistic uncertainty. An important advantage of probabilistic reasoning is its ability to update previous outcome estimates by conditioning them with newly available evidence [7].

Global optimization involves finding the absolutely best set of parameters to optimize an objective function. In general, it may be possible to have solutions that are locally but not globally optimal. Consequently, global optimization problems are typically quite difficult to solve exactly: in the context of combinatorial problems, they are often NP-hard. Evolutionary Computation works by simulating evolution on a computer by iterative generations and alteration processes operating on a set of candidate solutions that form a population. The entire population evolves towards better candidate solutions via the selection operation and genetic operators such as crossover and mutation. The selection operator decides which candidate solutions move on into the next generation and thus limits the search space [1][6].

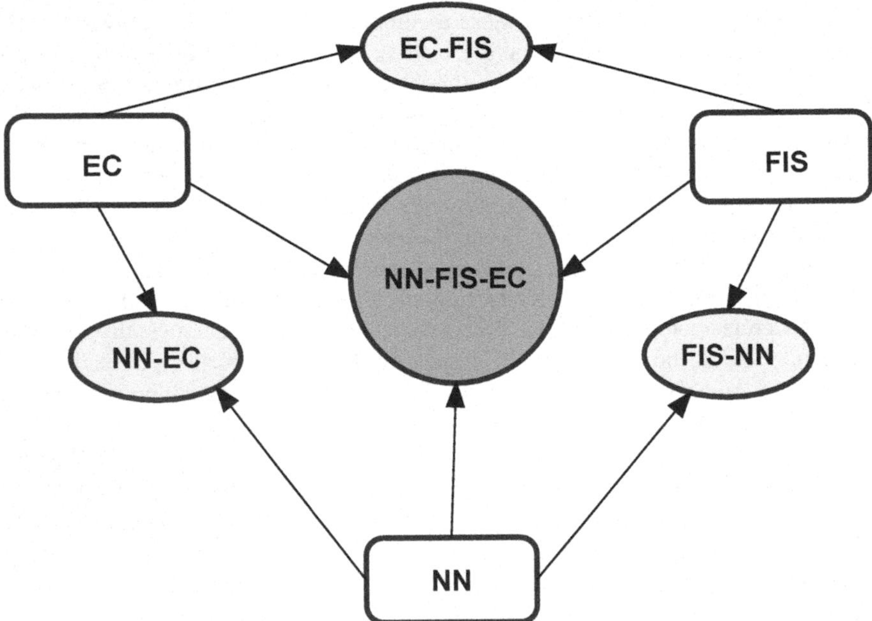

Fig. 17.1 General framework for hybrid architectures

17.2 Models of Hybrid Computational Intelligence Architectures

The hybrid intelligent architectures can be classified into 4 different categories based on the system's overall architecture [1]:

1) Stand-alone
2) Transformational
3) Hierarchical hybrid
4) Integrated hybrid.

The following sections discuss each of these strategies.

17.2.1 Stand-Alone Systems

Stand-alone models consist of independent software components, which do not interact in any way. Developing stand-alone systems can have several purposes: first, they provide a direct means of comparing the problem solving capabilities of different techniques with reference to a certain application [1]. Running different techniques in a parallel environment permits a loose approximation of integration. Stand-alone models are often used to develop a quick initial prototype, while a more time-consuming application is developed. Figure 17.2 displays a stand-alone system where a neural network and a fuzzy system are used separately.

Some of the benefits are simplicity and ease of development by using commercially available software packages. On the other hand, stand-alone techniques are not transferable: neither can support the weakness of the other technique.

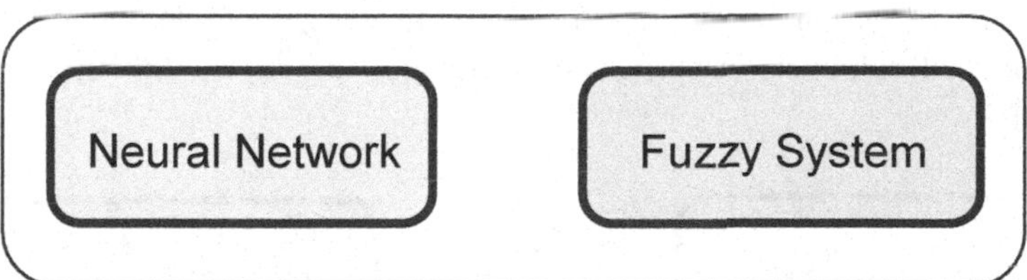

Fig. 17.2 Stand-alone system.

17.2.2 Transformational Hybrid Intelligent System

In a transformational hybrid model, the system begins as one type and ends up as the other. Determining, which technique is used for development and which is used for delivery is based on the desirable features that the technique offers. Figure 17.3 shows the interaction between a neural network and an expert system in a transformational hybrid model [9]. Obviously, either the expert system is incapable of adequately solving the problem, or the speed, adaptability, and robustness of neural network is required. Knowledge from the expert system is used to determine the initial conditions and the training set for the artificial neural network.

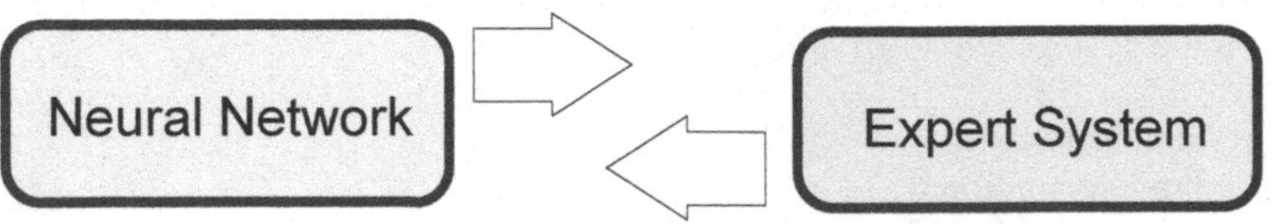

Fig. 17.3 Transformational hybrid architecture.

Transformational hybrid models are often quick to develop and ultimately require maintenance on only one system. They can be developed to suit the environment and offer many operational benefits. Unfortunately, transformational models are significantly limited: most are just application-oriented. For a different application, a totally new development effort might be required such as a fully automated means of transforming an expert system to a neural network and vice versa.

17.2.3 Hierarchical Hybrid Intelligent System

This architecture is built in a hierarchical fashion, associating a different functionality with each layer. The overall functioning of the model depends on the correct functioning of all the layers. Figure 17.4 demonstrates a hierarchical hybrid architecture involving a neural network, an evolutionary algorithm and a fuzzy system. The neural network uses an evolutionary algorithm to optimize its performance and the network output acts as a pre-processor to a fuzzy system, which then produces the final output. Poor performance in one of the layers directly affects the final output.

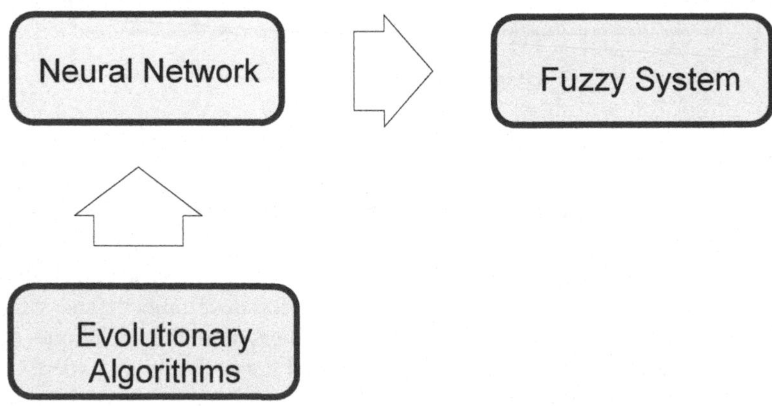

Fig. 17.4 Hierarchical hybrid architectures.

17.2.4 Integrated Intelligent System

Fused architectures are the first true form of integrated intelligent systems. They include systems, which combine different techniques into one single computational model. They share data structures and knowledge representations. Another approach is to put the various techniques side-by-side and focus on their interaction in a problem-solving task. This method can allow for integrating alternative techniques and exploiting their mutuality. Furthermore, the conceptual view of the agent allows one to abstract from the individual techniques and focus on the global system behavior, as well as to study the individual contribution of each component [10].

The benefits of integrated models include robustness, improved performance and increased problem-solving capabilities. Finally, fully integrated models can provide a full range of capabilities such as adaptation, generalization, noise tolerance and justification. Fused systems have limitations caused by the increased complexity of the inter-module interactions and specifying, designing, and building fully integrated models is complex. Some examples are neuro-fuzzy systems, evolutionary neural networks, evolutionary fuzzy systems etc.

17.3 Neuro-fuzzy Systems

A feedforward neural network could approximate any fuzzy-rule-based system and any feedforward neural network may be approximated by a rule-based fuzzy inference system [12]. A fusion of artificial neural networks and fuzzy inference systems has attracted growing interest among researchers in various scientific and engineering areas due to the growing need for adaptive intelligent systems to solve real world problems. The advantages of a combination of neural networks and fuzzy inference systems are obvious. An analysis reveals that the drawbacks pertaining to these approaches seem complementary and therefore, it is natural to consider building an integrated system combining the concepts. While the learning capability is an advantage from the viewpoint of a fuzzy inference system, the automatic formation of a linguistic rule base is an advantage from the viewpoint of neural networks. Neural network learning techniques could be used to learn the fuzzy inference system in a cooperative and an integrated environment [1].

17.3.1 Cooperative and Concurrent Neuro-fuzzy Systems

In the simplest way, a cooperative model can be considered as a preprocessor wherein ANN learning mechanism determines the FIS membership functions or fuzzy rules from the training data. Once the FIS parameters are determined, ANN goes to the background. The rule based is usually determined by a clustering approach (self organizing maps) or fuzzy clustering algorithms. Membership functions are usually approximated by neural network from the training data.

Kosko's fuzzy associative memories [11], Pedryz's (*et al)* fuzzy rule extraction using self organizing maps [14] and Nomura's. (*et al)* systems capable of learning of fuzzy set parameters [13] are some good examples of cooperative neuro-fuzzy systems.

In a concurrent model, ANN assists the FIS continuously to determine the required parameters especially if the input variables of the controller cannot be measured directly. In some cases the FIS outputs might not be directly applicable to the process. In that case ANN can act as a postprocessor of FIS outputs. Figures 17.5 -6 depict the cooperative and concurrent NF models.

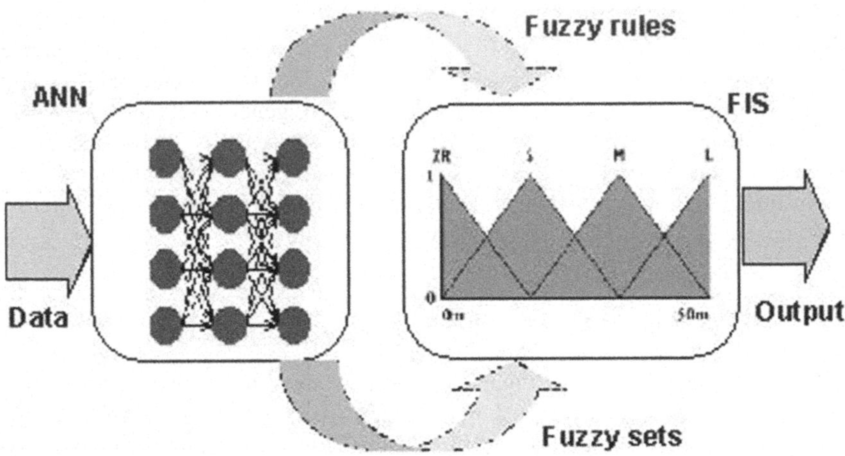

Fig. 17.5 Cooperative NF model

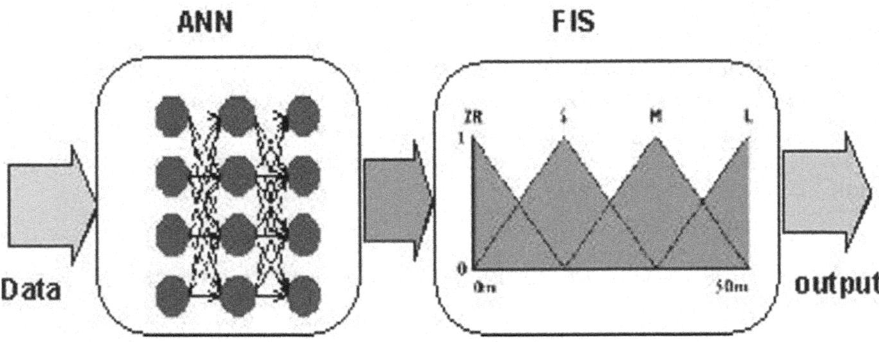

Fig. 17.6 Concurrent NF model

17.3.2 Fused Neuro Fuzzy Systems

In a fused NF architecture, ANN learning algorithms are used to determine the parameters of FIS. Fused NF systems share data structures and knowledge representations. A common way to apply a learning algorithm to a fuzzy system is to represent it in a special ANN like architecture. However the conventional ANN learning algorithms (gradient descent) cannot be applied directly to such a system as the functions used in the inference process are usually non differentiable. This problem can be tackled by using differentiable functions in the inference system or by not using the standard neural learning algorithm. Some of the major woks in this area are GARIC [27], FALCON [26], ANFIS [19], NEFCON [25], FUN [21], SONFIN [20], FINEST [22], EFuNN [23], dmEFuNN [23] and many others[24].

- **Fuzzy Adaptive learning Control Network (FALCON)**

FALCON [26] has a five-layered architecture as shown in Figure 17.7. There are two linguistic nodes for each output variable. One is for training data (desired output) and the other is for the actual output of FALCON. The first hidden layer is responsible for the fuzzification of each input variable. Each node can be a single node representing a simple membership function (MF) or composed of multilayer nodes that compute a complex MF. The Second hidden layer defines the preconditions of the rule followed by rule consequents in the third hidden layer. FALCON uses a hybrid-learning algorithm comprising of unsupervised learning to locate initial membership functions/ rule base and a gradient descent learning to optimally adjust the parameters of the MF to produce the desired outputs.

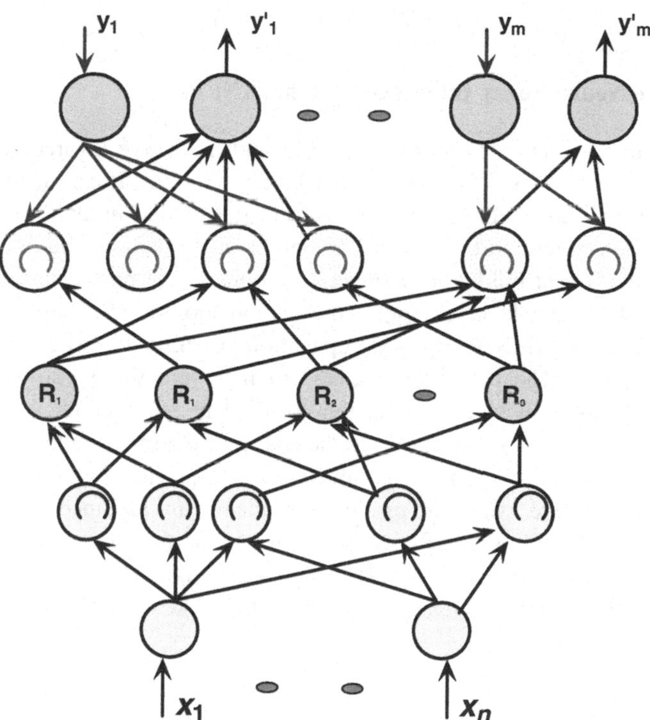

Fig. 17.7 Architecture of FALCON

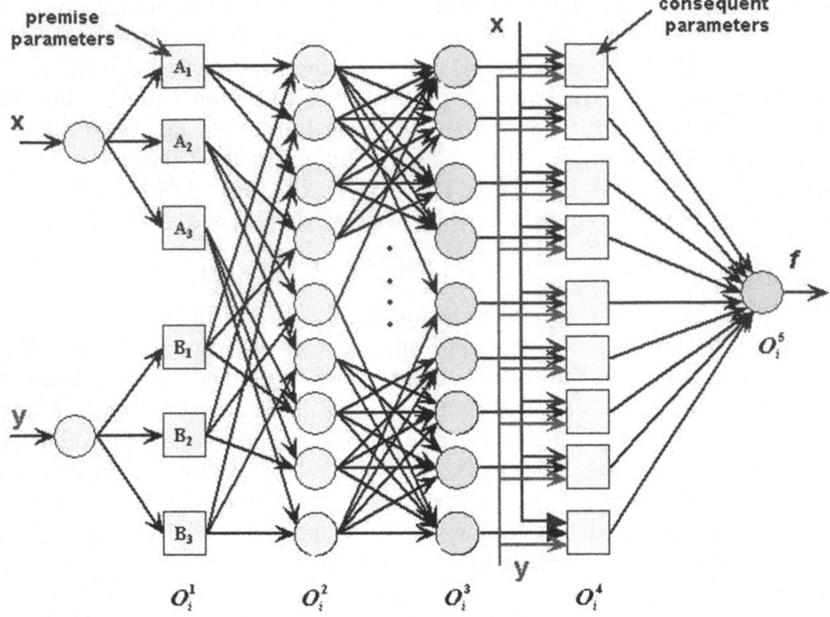

Fig. 17.8 Structure of ANFIS

- **Adaptive Neuro Fuzzy Inference System (ANFIS)**

ANFIS [19] implements a Takagi Sugeno FIS and has a five layered architecture as shown in Figure 17.8. The first hidden layer is for fuzzification of the input variables and T-norm operators are deployed in the second hidden layer to compute the rule antecedent part. The third hidden layer normalizes the rule strengths followed by the fourth hidden layer where the consequent parameters of the rule are determined. Output layer computes the overall input as the summation of all incoming signals. ANFIS uses backpropagation learning to determine premise parameters (to learn the parameters related to membership functions) and least mean square estimation to determine the consequent parameters. A step in the learning procedure has got two parts: In the first part the input patterns are propagated, and the optimal consequent parameters are estimated by an iterative least mean square procedure, while the premise parameters are assumed to be fixed for the current cycle through the training set. In the second part the patterns are propagated again, and in this epoch, backpropagation is used to modify the premise parameters, while the consequent parameters remain fixed. This procedure is then iterated.

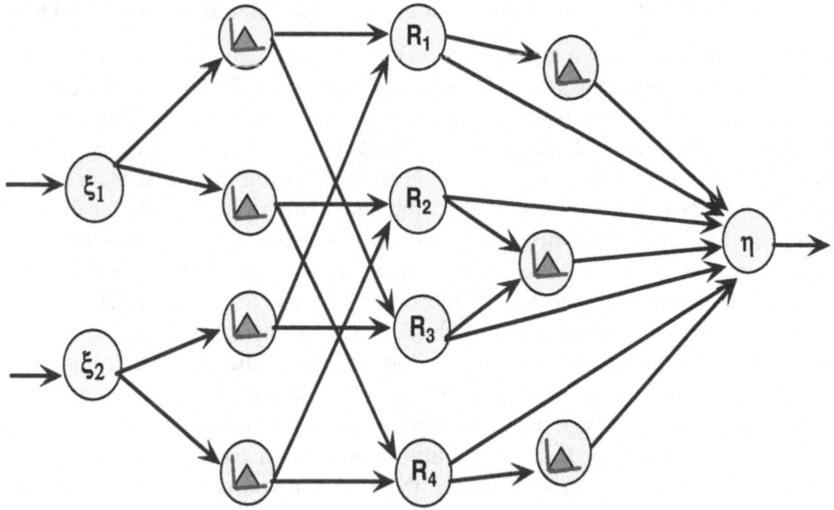

Fig. 17.9 ASN of GARIC

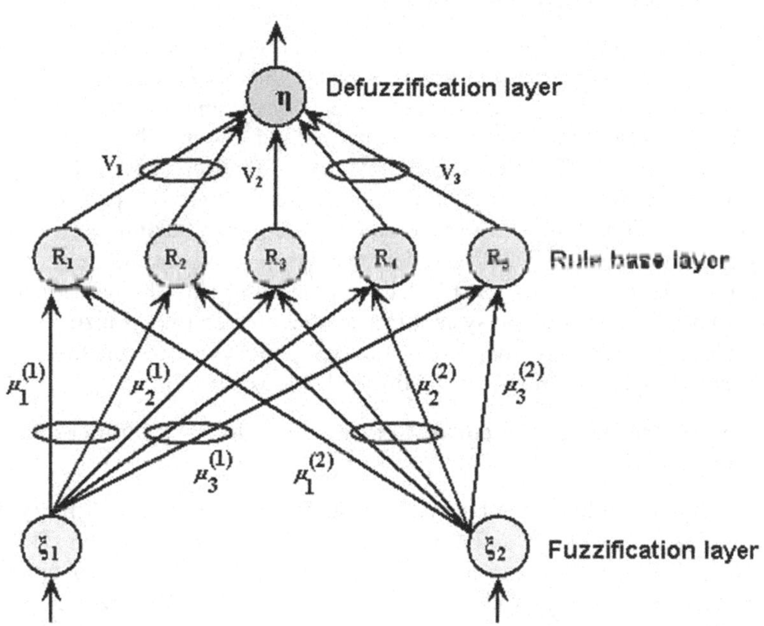

Fig. 17.10 Architecture of NEFCON

- **Generalized Approximate Reasoning based Intelligent Control (GARIC)**

GARIC [27] implements a neuro-fuzzy controller by using two neural network modules, the ASN (Action Selection Network) and the AEN (Action State Evaluation Network). The AEN is an adaptive critic that evaluates the actions of the ASN. ASN of GARIC is feedforward network with five layers. Figure 17.9 illustrates the structure of GARIC – ASN. The connections between layers are not weighted. The first hidden layer stores the linguistic values of all the input variables. Each input unit is only connected to those units of the first hidden layer, which represent its associated linguistic values. The second hidden layer represents the fuzzy rules nodes, which determine the degree of fulfillment of a rule using a *softmin* operation. The third hidden layer represents the linguistic values of the control output variable η. Conclusions of the rule are computed depending on the strength of the rule antecedents computed by the rule node layer. GARIC makes use of local mean-of-maximum method for computing the rule outputs. This method needs a crisp output value from each rule. Therefore the conclusions must be defuzzified before they are accumulated to the final output value of the controller. GARIC uses a mixture of gradient descent and reinforcement learning to fine-tune the node parameters.

- **Neuro-Fuzzy Control (NEFCON)**

NEFCON [25] is designed to implement Mamdani type FIS and is illustrated in Figure 6. Connections in NEFCON are weighted with fuzzy sets and rules (μ_r, v_r are the fuzzy sets describing the antecedents and consequents) with the same antecedent use so-called shared weights, which are represented by ellipses drawn around the connections. They ensure the integrity of the rule base. The input units assume the task of fuzzification interface, the inference logic is represented by the propagation functions, and the output unit is the defuzzification interface. The learning process of the NEFCON model is based on a mixture of reinforcement and backpropagation learning. NEFCON can be used to learn an initial rule base, if no prior knowledge about the system is available or even to optimize a manually defined rule base. NEFCON has two variants: NEFPROX (for function approximation) and NEFCLASS (for classification tasks) [25].

- **Fuzzy Inference and Neural Network in Fuzzy Inference Software (FINEST)**

FINEST [22] is capable of two kinds of tuning process, the tuning of fuzzy predicates, combination functions and the tuning of an implication function. The generalized modus ponens is improved in the following four ways (1) Aggregation operators that have synergy and cancellation nature (2) A parameterized implication function (3) A combination function that can reduce fuzziness (4) Backward chaining based on generalized modus ponens. FINEST make use of a backpropagation algorithm for the fine-tuning of the parameters. Figure 17.11 shows the layered architecture of FINEST and the calculation process of the fuzzy inference. FINEST provides a framework to tune any parameter, which appears in the nodes of the network representing the calculation process of the fuzzy data if the derivative function with respect to the parameters is given.

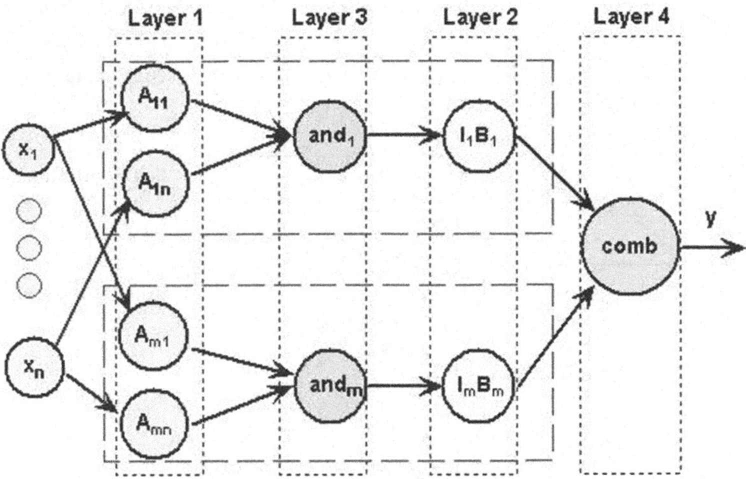

Fig. 17.11 Architecture of FINEST

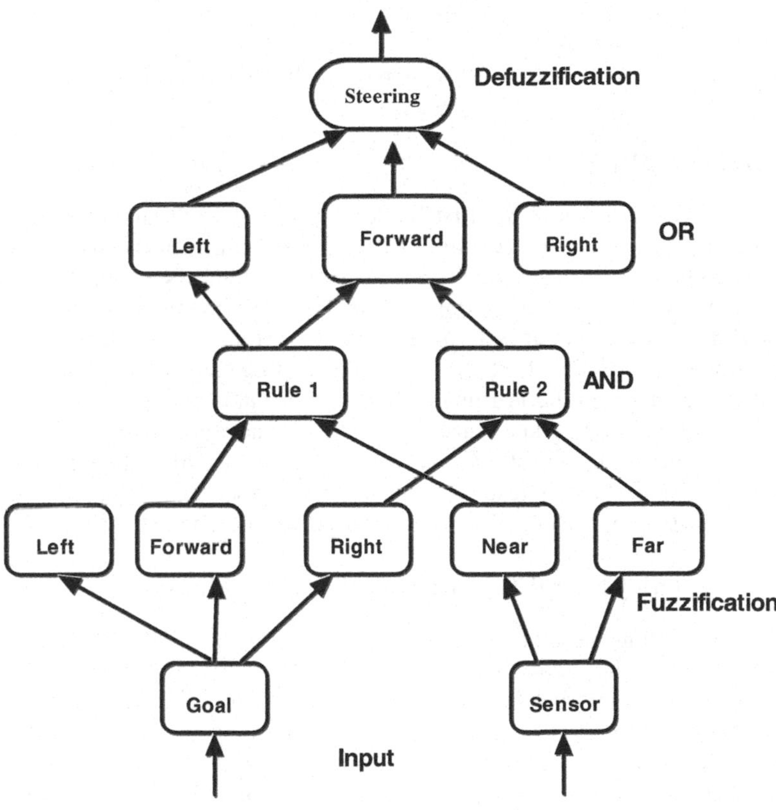

Fig. 17.12. Architecture of FUN

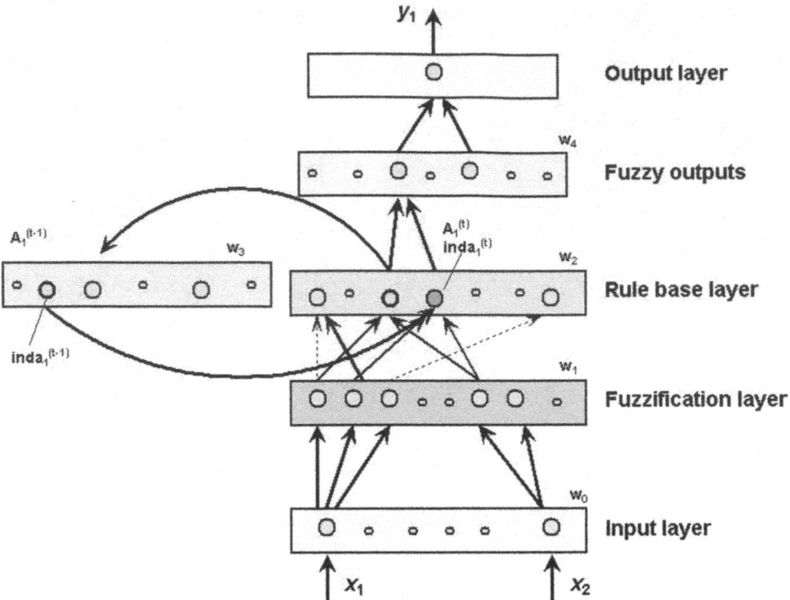

Fig. 17.13 Architecture of EFuNN

- **FUzzy Net (FUN)**

In FUN [21] the neurons in the first hidden layer contain the membership functions and this performs a fuzzification of the input values. In the second hidden layer, the conjunctions (fuzzy-*AND*) are calculated. Membership functions of the output variables are stored in the third hidden layer. Their activation function is a fuzzy-*OR*. Finally the output neuron performs the defuzzification. The network is initialized with a fuzzy rule base and the corresponding membership functions and there after uses a stochastic learning technique that randomly changes parameters of membership functions and connections within the network structure. The learning process is driven by a cost function, which is evaluated after the random modification. If the modification resulted in an improved performance the modification is kept, otherwise it is undone. The architecture is illustrated in Figure 17.12.

- **Evolving Fuzzy Neural Network (EFuNN)**

In EFuNN [23] all nodes are created during learning (Figure 17.13). The input layer passes the data to the second layer, which calculates the fuzzy membership degrees to which the input values belong to predefined fuzzy membership functions. The third layer contains fuzzy rule nodes representing prototypes of input-output data as an association of hyper-spheres from the fuzzy input and fuzzy output spaces. Each rule node is defined by 2 vectors of connection weights, which are adjusted through the hybrid learning technique. The fourth layer calculates the

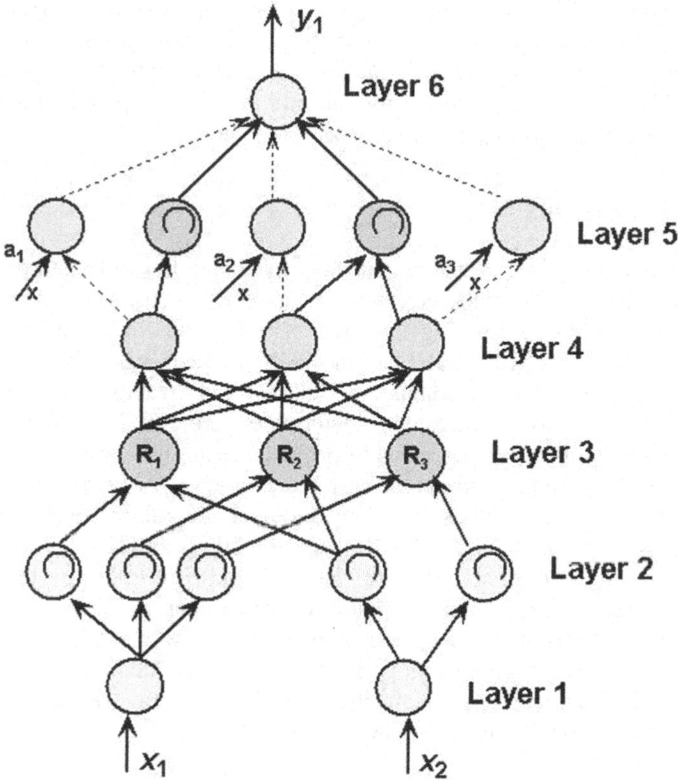

Fig. 17.14 Architecture of SONFIN

degrees to which output membership functions are matched by the input data, and the fifth layer does defuzzification and calculates exact values for the output variables. Dynamic Evolving Fuzzy Neural Network (dmEFuNN) [23] is a modified version of EFuNN with the idea that not just the winning rule node's activation is propagated but a group of rule nodes is dynamically selected for every new input vector and their activation values are used to calculate the dynamical parameters of the output function. While EFuNN implements fuzzy rules of Mamdani type, dmEFuNN estimates the Takagi-Sugeno fuzzy rules based on a least squares algorithm.

- **Self Constructing Neural Fuzzy Inference Network (SONFIN)**

SONFIN [20] implements a modified Takagi-Sugeno FIS and is illustrated in Figure 17.14. In the structure identification of the precondition part, the input space is partitioned in a flexible way according to an aligned clustering based algorithm. As to the structure identification of the consequent part, only a singleton value selected by a clustering method is assigned to each rule initially. Afterwards, some additional significant terms (input variables) selected via a projection-based correlation measure for each rule are added to the consequent part (forming a linear equation of input variables) incrementally as learning proceeds. For parameter identification, the consequent parameters are tuned optimally by either least mean squares or recursive least squares algorithms and the precondition parameters are tuned by backpropagation algorithm.

17.3.3 Discussions

As evident, both cooperative and concurrent models are not fully interpretable due to the presence of ANN (black box concept), whereas a fused NF model is interpretable and capable of learning in a supervised mode. In FALCON, GARIC, ANFIS, NEFCON, SONFIN, FINEST and FUN the learning process is only concerned with parameter level adaptation within fixed structures. For large-scale problems, it will be too complicated to determine the optimal premise-consequent structures, rule numbers etc. User has to provide the architecture details (type and quantity of MF's for input and output variables), type of fuzzy operators etc. FINEST provides a mechanism based on the improved generalized modus ponens for fine tuning of fuzzy predicates & combination functions and tuning of an implication function. An important feature of EFuNN and dmEFuNN is the one pass (epoch) training, which is highly capable for online learning. Since FUN system uses a stochastic learning procedure, it is questionable to call FUN a NF system. Table 17.1 provides a comparative performance of some neuro fuzzy systems for predicting the Mackey-Glass chaotic time series. Training was done using 500 data sets and NF models were tested with another 500 data sets.

Table 17.1 Performance of NF systems and ANN

System	Epochs	RMSE
ANFIS	75	0.0017
NEFPROX	216	0.0332
EFuNN	1	0.0140
dmEFuNN	1	0.0042
SONFIN	-	0.0180

17.4 Evolutionary Fuzzy Systems

Fuzzy logic has been successfully used to capture heuristic control laws obtained from human experience or engineering practice in automated algorithm. These control laws are defined by means of linguistic rules. As man-machine interaction increases, the need to find a common framework to represent key elements in these two worlds becomes essential. Adaptive fuzzy systems provide such a framework. For the fuzzy controller to be fully adaptive, shape of membership functions, number of rules, reasoning method used to aggregate multiple actions, output actions associated with each partition etc. are to be decided automatically.

It is known that the performance of a fuzzy control system may be significantly improved if the fuzzy reasoning model is supplemented by a evolutionary learning mechanism.

Several researchers are busy exploring the integration of evolutionary algorithms with fuzzy logic. Majority of the works are concerned with the automatic design or optimization of fuzzy logic controllers either by adapting the fuzzy membership functions or by learning the fuzzy *if-then* rules. The first method results in a self-tuning controller in which an evolutionary algorithm adapts the fuzzy membership functions. The genome encodes parameters of trapezoidal, triangle, logistic, Laplace, hyperbolic-tangent or Gaussian membership functions etc. This approach requires a previous defined rule base and is primarily useful in order to optimize the performance of an already existing controller.

Evolutionary search of fuzzy rules can be implemented using two approaches. In the first approach the fuzzy knowledge base is adapted as a result of antagonistic roles of competition and cooperation of fuzzy rules. Each genotype represents a single fuzzy rule and the entire population represents a solution. A classifier rule triggers whenever its condition part matches the current input, in which case the proposed action is sent to the process to be controlled. Referring to Figure 17.15, a Mamdani or TSK rule may be formed as:

If **input-1** is *medium* and **input-2** is *large* then rule $\mathbf{R_8}$ is fired.

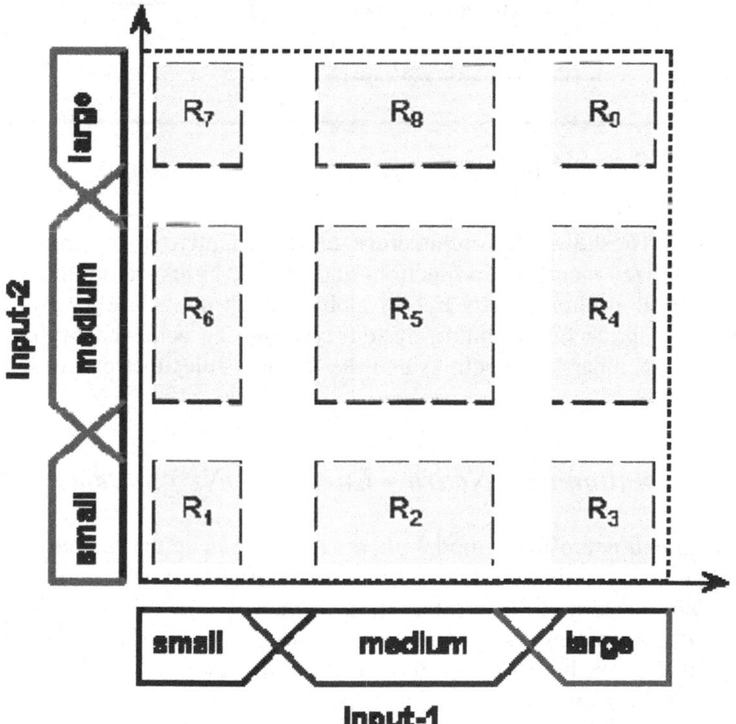

Fig. 17.15 Two-dimensional space partitioned using 3 trapezoidal membership functions.

The evolutionary algorithm generates new classifier rules based on the rule strengths acquired during the entire process. The fuzzy behavior is created by an activation sequence of mutually collaborating fuzzy rules. The entire knowledge base is build up by a cooperation of competing multiple fuzzy rules. The second approach evolves a population of knowledge bases rather than individual fuzzy rules. The disadvantage is the increased complexity of search space and additional computational burden especially for online learning. The size of the genotype depends on the number of input/output variables and fuzzy sets.

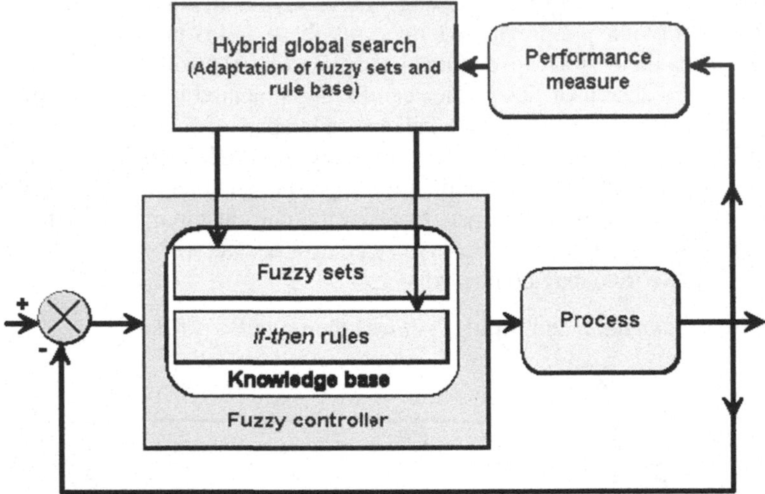

Fig. 17.16 Adaptive fuzzy control system architecture.

Figure 17.16 shows the architecture of the adaptive fuzzy control system wherein the fuzzy membership functions and the rule bases are optimized using an evolutionary algorithm (or any hybrid global search procedure). An optimal design of an adaptive fuzzy control system can only be achieved by the adaptive evolution of membership functions and the learning rules that progress on different time scales.

17.4.1 Evolutionary – Neuro – Fuzzy (EvoNF) Systems

In an integrated neuro-fuzzy model, there is no guarantee that the neural network learning algorithm will converge and the tuning of fuzzy inference system is successful. Optimization of fuzzy inference systems could be further improved using neural network learning algorithm and evolutionary algorithms. This could be considered as a methodology to integrate neural networks, fuzzy inference systems and evolutionary search procedures [2] [3] [4].

The EvoNF framework could adapt to Mamdani, Takagi-Sugeno or other fuzzy inference systems. The architecture and the evolving mechanism could be considered as a general framework for adaptive fuzzy systems that is a fuzzy model that can change membership functions (quantity and shape), rule base (architecture), fuzzy operators and learning parameters according to different environments without human intervention. Solving multi-objective scientific and engineering problems is, generally, a very difficult goal. In these particular optimization problems, the objectives often conflict across a high-dimension problem space and may also require extensive computational resources.

Figure 17.17 illustrates the interaction of various evolutionary search procedures and shows that for every fuzzy inference system, there exists a global search of learning algorithm parameters, an inference mechanism, a rule base and membership functions in an environment decided by the problem. Thus, the evolution of the fuzzy inference system evolves at the slowest time scale while the evolution of the quantity and type of membership functions evolves at the fastest rate. The function of the other layers could be derived similarly [1].

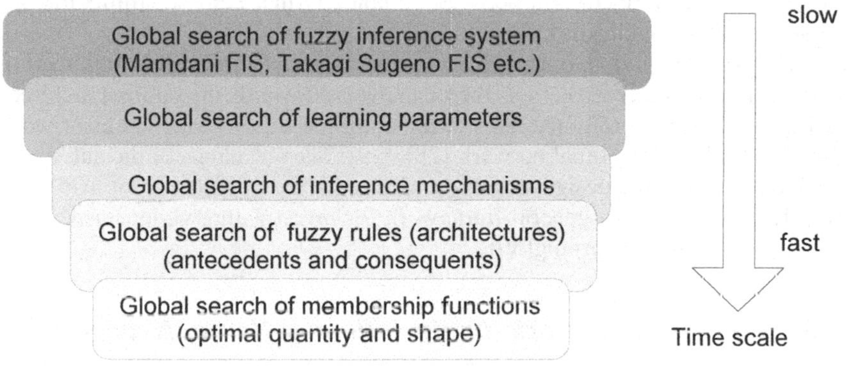

Fig. 17.17 General computational framework for EvoNF

17.5 Evolutionary Neural Networks (EANN)

Even though artificial neural networks are capable of performing a wide variety of tasks, yet in practice sometimes they deliver only marginal performance. Inappropriate topology selection and learning algorithm are frequently blamed. There is little reason to expect that one can find a uniformly best algorithm for selecting the weights in a feed-forward artificial neural network [31][32]. This is in accordance with the no free lunch theorem, which explains that for any algorithm, any elevated performance over one class of problems is exactly paid for in performance over another class. In sum, one should be skeptical of claims in the literature on training algorithms that one being proposed is substantially better than most others. Such claims are often defended through some simulations based on applications in which the proposed algorithm performed better than some familiar alternative.

At present, neural network design relies heavily on human experts who have sufficient knowledge about the different aspects of the network and the problem domain. As the complexity of the problem domain increases, manual design becomes more difficult and unmanageable. Evolutionary design of artificial neural networks has been widely explored. Evolutionary algorithms are used to adapt the connection weights, network architecture and learning rules according to the problem environment. A distinct feature of evolutionary neural networks is their adaptability to a dynamic environment. In other words, such neural networks can adapt to an environment as well as changes in the environment. The two forms of adaptation: evolution and learning in evolutionary artificial neural networks make their adaptation to a dynamic environment much more effective and efficient than the conventional learning approach.

Many of the conventional ANNs now being designed are statistically quite accurate but they still leave a bad taste with users who expect computers to solve their problems accurately. The important drawback is that the designer has to specify the number of neurons, their distribution over several layers and interconnection between them. Several methods have been proposed to automatically construct ANNs for reduction in network complexity that is to determine the appropriate number of hidden units, layers, etc.

The interest in evolutionary search procedures for designing ANN architecture has been growing in recent years as they can evolve towards the optimal architecture without outside interference, thus eliminating the tedious trial and error work of manually finding an optimal network [28], [29]. The advantage of the automatic design over the manual design becomes clearer as the complexity of ANN increases. EANNs provide a general framework for investigating various aspects of simulated evolution and learning[30].

17.5.1 General Framework for Evolutionary Neural Networks

In an Evolutionary Artificial Neural Network (EANN), evolution can be introduced at various levels. At the lowest level, evolution can be introduced into weight training, where ANN weights are evolved. At the next higher level, evolution can be introduced into neural network architecture adaptation, where the architecture (number of hidden layers, no of hidden neurons and node transfer functions) is evolved. At the highest level, evolution can be introduced into the learning mechanism. A general framework of EANNs which includes the above three levels of evolution is given in Figure 17.18.

From the point of view of engineering, the decision on the level of evolution depends on what kind of prior knowledge is available. If there is more prior knowledge about EANN's architectures than that about their learning rules or a particular class of architectures is pursued, it is better to implement the evolution of architectures at the highest level because such knowledge can be used to reduce the search space and the lower level evolution of learning rules can be more biased towards this kind of architectures. On the other hand, the evolution of learning rules should be at the highest level if there is more prior knowledge about them available or there is a special interest in certain type of learning rules.

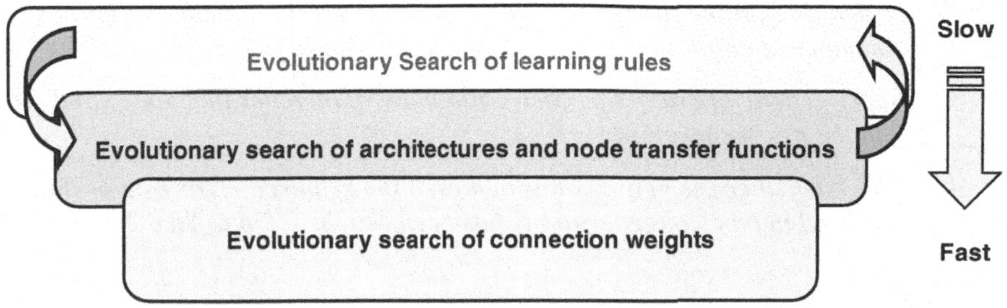

Fig. 17.18 A General Framework for EANNs

17.5.2 Evolutionary Search of Connection Weights

The shortcomings of the backpropagation algorithm could be overcome if the training process is formulated as a global search of connection weights towards an optimal set defined by the evolutionary algorithm.. Optimal connection weights can be formulated as a global search problem wherein the architecture of the neural network is pre-defined and fixed during the evolution. Connection weights may be represented as binary strings represented by a certain length. The whole network is encoded by concatenation of all the connection weights of the network in the chromosome. A heuristic concerning the order of the concatenation is to put connection weights to the same node together. Figure 17.19 illustrates the binary representation of connection weights wherein each weight is represented by 4 bits.

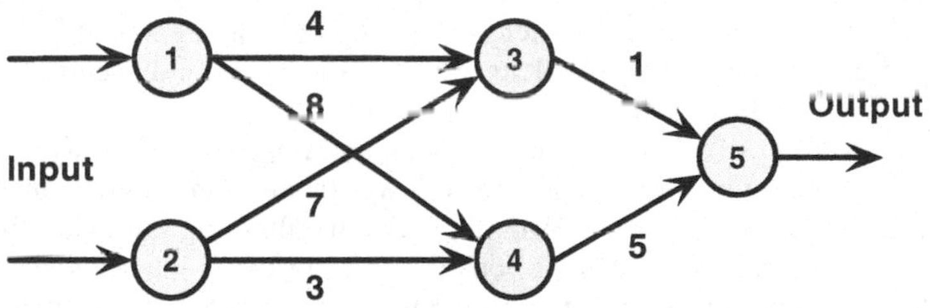

Fig. 17.19 Connection weight chromosome encoding using binary representation

Real numbers have been proposed to represent connection weights directly. A representation of the ANN could be (2.0, 6.0, 5.0, 1.0, 4.0, 10.0). However proper genetic operators are to be chosen depending upon the representation used.

Evolutionary search of connection weights can be formulated as follows:

1) *Generate an initial population of N weight chromosomes. Evaluate the fitness of each EANN depending on the problem.*

2) *Depending on the fitness and using suitable selection methods reproduce a number of children for each individual in the current generation.*

3) *Apply genetic operators to each child individual generated above and obtain the next generation.*

4) *Check whether the network has achieved the required error rate or the specified number of generations has been reached. Go to Step 2.*

5) *End*

While gradient based techniques are very much dependant on the initial setting of weights, this algorithm can be considered generally much less sensitive to initial conditions. When compared to any gradient descent or second order optimization technique that can only find local optimum in a neighborhood of the initial solution, evolutionary algorithms always try to search for a global optimal solution. Performance by using the above approach will directly depend on the problem.

17.5.3 Evolutionary Search of Architectures

Evolutionary architecture adaptation can be achieved by constructive and destructive algorithms. Constructive algorithms, add complexity to the network starting from a very simple architecture until the entire network is able to learn the task. Destructive algorithms start with large architectures and remove nodes and interconnections until the ANN is no longer able to perform its task. Then the last removal is undone. Figure 17.20 demonstrates how typical neural network architecture could be directly encoded and how the genotype is represented. For an optimal network, the required node transfer function (Gaussian, sigmoidal, etc.) can be formulated as a global search problem, which is evolved simultaneously with the search for architectures.

To minimize the size of the genotype string and improve scalability, when priori knowledge of the architecture is known it will be efficient to use some indirect coding (high level) schemes. For example, if two neighboring layers are fully connected then the architecture can be coded by simply using the number of layers and nodes. The blueprint representation is a popular indirect coding scheme where it assumes architecture consists of various segments or areas. Each segment or area will define a set of neurons, their spatial arrangement and their efferent connectivity.

Global search of transfer function and the connectivity of the ANN using evolutionary algorithms can be formulated as follows:

1) *The evolution of architectures has to be implemented such that the evolution of weight chromosomes are evolved at a faster rate i.e. for every architecture chromosome, there will be several weight chromosomes evolving at a faster time scale.*

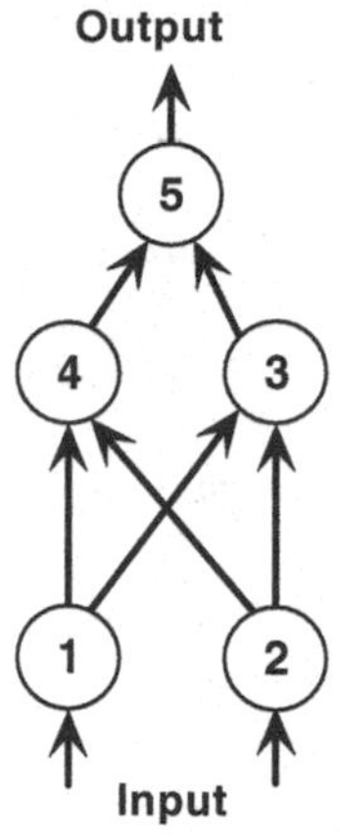

From / To	1	2	3	4	5	Bias	Gene
1	0	0	0	0	0	0	000000
2	0	0	0	0	0	0	000000
3	1	1	0	0	0	1	110001
4	1	1	0	0	0	1	110001
5	0	0	1	1	0	1	001101

Genotype: 000000 **000000** 110001 110001 **001101**

Fig. 17.20 Architecture chromosome using binary coding

2) *Generate an initial population of N architecture chromosomes. Evaluate the fitness of each EANN depending on the problem.*
3) *Depending on the fitness and using suitable selection methods reproduce a number of children for each individual in the current generation.*
4) *Apply genetic operators to each child individual generated above and obtain the next generation.*
5) *Check whether the network has achieved the required error rate or the specified number of generations has been reached. Go to Step 3.*
6) *End*

17.5.4 Evolutionary Search of Learning Rules

For the neural network to be fully optimal the learning rules are to be adapted dynamically according to its architecture and the given problem. Deciding the learning rate and momentum can be considered as the first attempt of learning rules. The basic learning rule can be generalized by the function

$$\Delta w(t) = \sum_{k=1}^{n} \sum_{i_1, i_2, \ldots, i_k = 1}^{n} \left(\theta_{i_1, i_2, \ldots, i_k} \prod_{j=1}^{k} x_{ij}(t-1) \right)$$

where t is the time, Δw is the weight change, x_1, x_2,..... x_n are local variables and the θ*s* are the real values coefficients which will be determined by the global search algorithm. In the above equation different values of θ*s* determine different learning rules. The above equation is arrived based on the assumption that the same rule is applicable at every node of the network and the weight updating is only dependent on the input/output activations and the connection weights on a particular node. Genotypes (θ*s*) can be encoded as real-valued coefficients and the global search for learning rules using the hybrid algorithm can be formulated as follows:

1. *The evolution of learning rules has to be implemented such that the evolution of architecture chromosomes are evolved at a faster rate i.e. for every learning rule chromosome, there will be several architecture chromosomes evolving at a faster time scale*
2. *Generate an initial population of N learning rules. Evaluate the fitness of each EANN depending on the problem.*
3. *Depending on the fitness and using suitable selection methods reproduce a number of children for each individual in the current generation.*
4. *Apply genetic operators to each child individual generated above and obtain the next generation.*
5. *Check whether the network has achieved the required error rate or the specified number of generations has been reached. Go to Step 3.*
6. *End*

17.5.5 Meta Learning Evolutionary Artificial Neural Networks

Experimental evidence had indicated cases where evolutionary algorithms are inefficient at fine tuning solutions, but better at finding global basins of attraction [28]. The efficiency of evolutionary training can be improved significantly by incorporating a local search procedure into the evolution. Evolutionary algorithms are used to first locate a good region in the space and then a local search procedure is used to find a near optimal solution in this region. It is interesting to consider finding good initial weights as locating a good region in the space. Defining that the basin of attraction of a local minimum is composed of all the points, sets of weights in this case, which can converge to the local minimum through a local search algorithm, then a global minimum can easily be found by the local search algorithm if the evolutionary algorithm can locate any point, i.e, a set of initial weights, in the basin of attraction of the global minimum. Referring to Figure 17.21, G_1 and G_2 could be considered as the initial weights as located by the evolutionary search and W_A and W_B the corresponding final weights fine-tuned by the meta-learning technique.

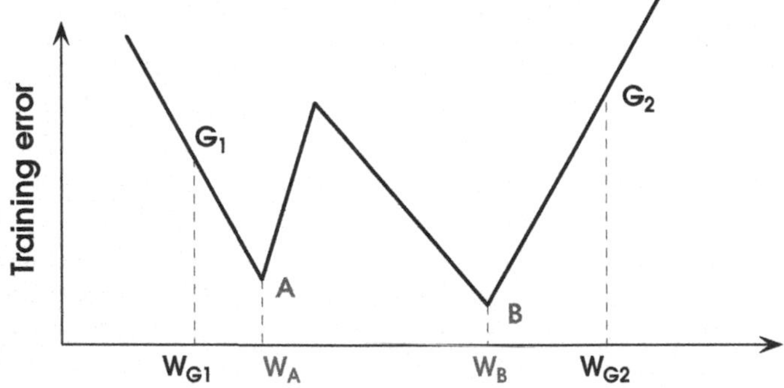

Fig. 17. 21 Fine tuning of weights using meta-learning

Figure 17.22 illustrates the general interaction mechanism with the learning mechanism of the EANN evolving at the highest level on the slowest time scale. All the randomly generated architecture of the initial population are trained by four different learning algorithms (backpropagation-BP, scaled conjugate gradient-SCG, quasi-Newton algorithm-QNA and Levenberg-Marquardt-LM) and evolved in a parallel environment. Parameters controlling the performance of the learning algorithm are adapted (example, learning rate and momentum for BP) according to the problem [31]. Figure 17.23 depicts the basic algorithm of proposed meta-learning EANN. Architecture of the chromosome is depicted in Figure 17.24.

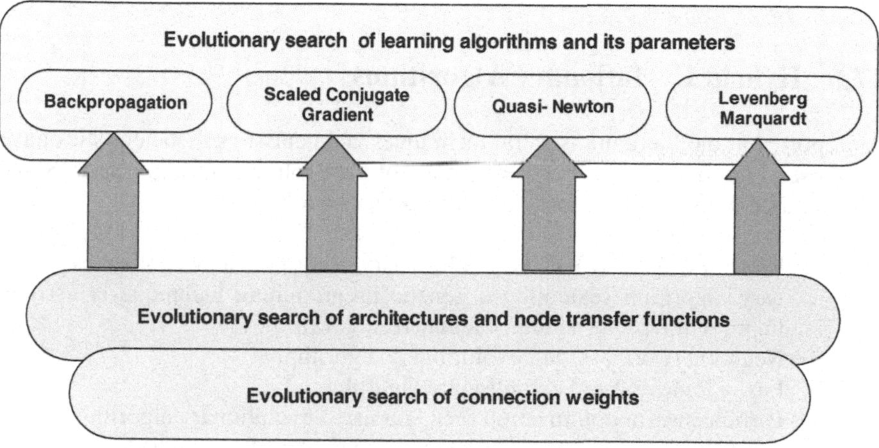

Fig. 17. 22 Interaction of various evolutionary search mechanisms

1. *Set t=0 and randomly generate an initial population of neural networks with architectures, node transfer functions and connection weights assigned at random.*
2. *In a parallel mode, evaluate fitness of each ANN using BP/SCG/QNA and LM*
3. *Based on fitness value, select parents for reproduction*
4. *Apply mutation to the parents and produce offspring (s) for next generation. Refill the population back to the defined size.*
5. *Repeat step 2*
6. *STOP when the required solution is found or number of iterations has reached the required limit.*

Fig. 17.23 Meta-learning algorithm for EANNs

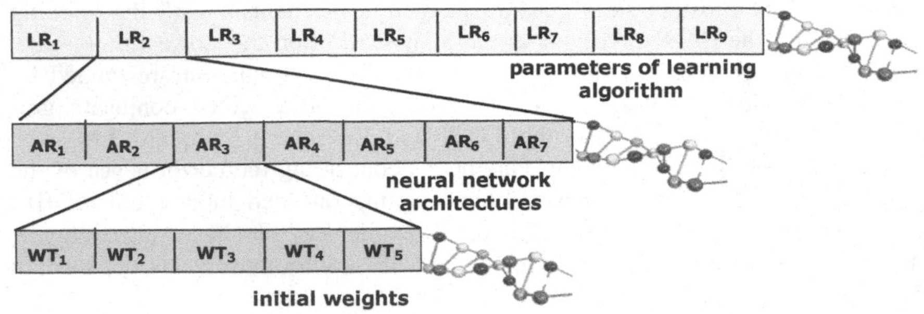

Fig. 17.24 Chromosome representation of the proposed MLEANN framework

17.6 Hybrid Evolutionary Algorithms

As reported in the literature, several techniques and heuristics/metaheuristics have been used to improve the general efficiency of the evolutionary algorithms. Some of most used hybrid architectures are summarized as follows:

1) hybridization between an evolutionary algorithm and another evolutionary algorithm (example: a genetic programming technique is used to improve the performance of a genetic algorithm);
2) Neural network assisted evolutionary algorithm;
3) Fuzzy logic assisted evolutionary algorithm;
4) Particle swarm optimization (PSO) assisted evolutionary algorithm;
5) Ant colony optimization (ACO) assisted evolutionary algorithm;
6) Bacterial foraging optimization assisted evolutionary algorithm;
7) hybridization between evolutionary algorithm and other heuristics (such as local search, tabu search, simulated annealing, hill climbing, dynamic programming, greedy random adaptive search procedure, etc).

Figure 17.25 represents a concurrent architecture where all the components are required for the proper functioning of the model. As depicted in Figure17.25 (a), evolutionary algorithm acts as a preprocessor and the intelligent paradigm is used to fine tune the solutions formulated by the evolutionary algorithm.

In Figure 17.25 (b), intelligent paradigm acts as a preprocessor and the evolutionary algorithm is used to fine tune the solutions formulated by the intelligent paradigm.

Figure 17.25 (c), represents a transformational hybrid system in which the evolutionary algorithm is used to fine tune the performance of the intelligent paradigm and at the same time, the intelligent paradigm is used to optimize the performance of the evolutionary algorithm. Required information is exchanged between the two techniques during the search (problem solving) process. In a cooperative model the intelligent paradigm is used only for initialization or for determining some parameters of the evolutionary algorithm. As depicted in Figure17.25 (d), thereafter, the intelligent paradigm is not required for the proper functioning of the system. Also, there are several ways to hybridize two or more techniques [15].

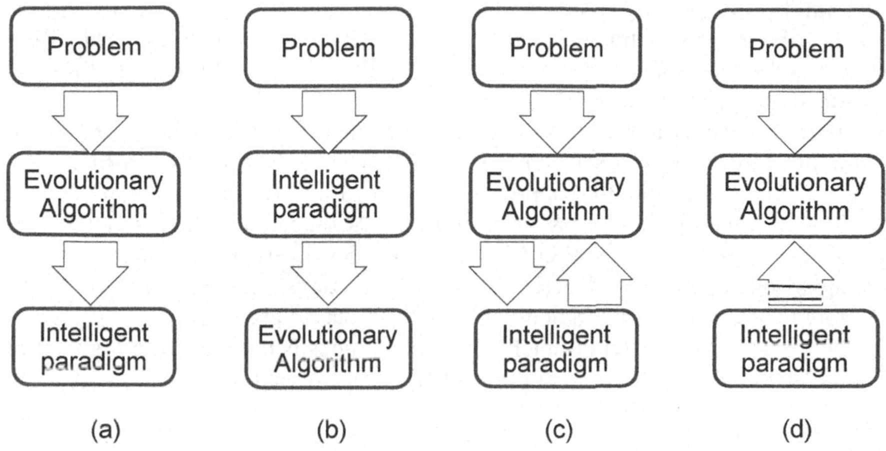

Fig. 17.25 Hybrid evolutionary algorithms: generic architecture.

The integration of different learning and adaptation techniques, to overcome individual limitations and achieve synergetic effects through hybridization or fusion of these techniques, has in recent years contributed to a large number of new hybrid evolutionary systems. Most of these approaches, however, follow an ad hoc design methodology, further justified by success in certain application domains. Due to the lack of a common framework it remains often difficult to compare the various hybrid systems conceptually and evaluate their performance comparatively. There are several ways to hybridize a conventional evolutionary algorithm for solving optimization problems. Some of them are summarized below [15][18]:

- The solutions of the initial population of EA may be created by problem-specific heuristics;
- Some or all the solutions obtained by the EA may be improved by local search. These kinds of algorithms are known as memetic algorithms [16][17].
- Solutions may be represented in an indirect way and a decoding algorithm maps any genotype to a corresponding phenotypic solution. In this mapping, the decoder can exploit problem-specific characteristics and apply heuristics etc.
- Variation operators may exploit problem knowledge. For example, in recombination more promising properties of one parent solution may be inherited with higher probabilities than the corresponding properties of the other parent(s). Also mutation may be biased to include in solutions promising properties with higher probabilities than others.

Summaries

The integration of different learning and adaptation techniques to overcome individual limitations and to achieve synergetic effects through the hybridization or fusion of these techniques has, in recent years, contributed to a large number of new intelligent system designs. Computational intelligence is an innovative framework for constructing intelligent hybrid architectures involving Neural Networks (NN), Fuzzy Inference Systems (FIS), Probabilistic Reasoning (PR) and Evolutionary Computation (EC)

Different generic architectures for integrating intelligent systems can be found in the literature such as: NN-FIS, EC-FIS, EC-NN, FIS-PR and NN-FIS-EC.

The hybrid soft computing approach has many important practical applications in science, technology, business and commercial. Compared to the individual constituents (NN, FIS, EC and PR) hybrid approaches are relatively young. As the strengths and weakness of different hybrid architectures are understood, it will be possible to use them more efficiently to solve real world problems.

The integration of different intelligent technologies is the most exciting fruit of modern artificial intelligence and is an active area of research.

References

1. Abraham, A.: Intelligent Systems: Architectures and Perspectives, Recent Advances in Intelligent Paradigms and Applications. In: Abraham, A., Jain, L., Kacprzyk, J. (eds.) Studies in Fuzziness and Soft Computing, pp. 1–35. Springer, Germany (2002)
2. Abraham, A.: EvoNF: A Framework for Optimization of Fuzzy Inference Systems Using Neural Network Learning and Evolutionary Computation. In: 2002 IEEE International Symposium on Intelligent Control (ISIC 2002). IEEE Press, Canada (2002)
3. Abraham, A.: How Important is Meta-Learning in Evolutionary Fuzzy Systems? In: Proceedings of Sixth International Conference on Cognitive and Neural Systems, ICCNS 2002. Boston University Press, USA (2002)
4. Abraham, A., Nath, B.: Evolutionary Design of Neuro-Fuzzy Systems - A Generic Framework. In: Namatame, A., et al. (eds.) Proceedings of The 4-th Japan-Australia Joint Workshop on Intelligent and Evolutionary Systems, Japan, pp. 106–113 (2000)
5. Cherkassky, V.: Fuzzy Inference Systems: A Critical Review. In: Kayak, O., et al. (eds.) Computational Intelligence: Soft Computing and Fuzzy-Neuro Integration with Applications, pp. 177–197. Springer, Heidelberg (1998)
6. Fogel, D.B.: Evolutionary Computation: Towards a New Philosophy of Machine Intelligence, 2nd edn. IEEE Press, Los Alamitos (2000)
7. Judea, P.: Probabilistic Reasoning in Intelligent Systems: Networks of Plausible Inference. Morgan Kaufmann Publishers, USA (1997)
8. Kosko, B.: Fuzzy Engineering. Prentice Hall, Upper Saddle River (1997)
9. Medsker, L.R.: Hybrid Intelligent Systems. Kluwer Academic Publishers, Dordrecht (1995)
10. Jacobsen, H.A.: A Generic Architecture for Hybrid Intelligent Systems. In: Proceedings of The IEEE World Congress on Computational Intelligence (FUZZ IEEE), USA, vol. 1, pp. 709–714 (1998)

11. Kosko, B.: Neural Networks and Fuzzy Systems: A Dynamical Systems Approach to Machine Intelligence. Prentice Hall, Englewood Cliffs (1992)
12. Li, X.H., Chen, C.L.P.: The Equivalence Between Fuzzy Logic Systems and Feedforward Neural Networks. IEEE Transactions on Neural Networks 11(2), 356–365 (2000)
13. Nomura, H., Hayashi, I., Wakami, N.: A Learning Method of Fuzzy Inference Systems by Descent Method. In: Proceedings of the First IEEE International conference on Fuzzy Systems, San Diego, USA, pp. 203–210 (1992)
14. Pedrycz, W., Card, H.C.: Linguistic Interpretation of Self Organizing Maps. In: Proceedings of the IEEE International Conference on Fuzzy Systems, San Diego, pp. 371–378 (1992)
15. Grosan, C., Abraham, A.: Hybrid Evolutionary Algorithms: Methodologies, Architectures and Reviews, Hybrid Evolutionary Algorithms. In: Grosan, C., et al. (eds.) Studies in Computational Intelligence, vol. 75, pp. 1–17. Springer, Germany (2007)
16. Hart, W.E., Krasnogor, N., Smith, J.E. (eds.): Recent Advances in Memetic Algorithms. Studies in Fuzziness and Soft Computing, vol. 166 (2005)
17. Moscato, P.: Memetic algorithms: A short introduction. In: Corne, D., et al. (eds.) New Ideas in Optimisation, pp. 219–234 (1999)
18. Swain, A.K., Morris, A.S.: A novel hybrid evolutionary programming method for function optimization. In: Proceedings of the Congress on Evolutionary Computation (CEC2000), pp. 1369–1376 (2000)
19. Jang, R.: Neuro-Fuzzy Modeling: Architectures, Analyses and Applications, PhD Thesis, University of California, Berkeley (July 1992)
20. Feng, J.C., Teng, L.C.: An Online Self Constructing Neural Fuzzy Inference Network and its Applications. IEEE Transactions on Fuzzy Systems 6(1), 12–32 (1998)
21. Sulzberger, S.M., Tschicholg-Gurman, N.N., Vestli, S.J.: FUN: Optimization of Fuzzy Rule Based Systems Using Neural Networks. In: Proceedings of IEEE Conference on Neural Networks, San Francisco, pp. 312–316 (March 1993)
22. Tano, S., Oyama, T., Arnould, T.: Deep combination of Fuzzy Inference and Neural Network in Fuzzy Inference. Fuzzy Sets and Systems 82(2), 151–160 (1996)
23. Kasabov, N., Song, Q.: Dynamic Evolving Fuzzy Neural Networks with 'm-out-of-n' Activation Nodes for On-line Adaptive Systems, Technical Report TR99/04, Department of information science, University of Otago (1999)
24. Mackey, M.C., Glass, L.: Oscillation and Chaos in Physiological Control Systems. Science 197, 287–289 (1977)
25. Nauck, D., Kruse, R.: Neuro-Fuzzy Systems for Function Approximation. In: 4th International Workshop Fuzzy-Neuro Systems (1997)
26. Lin, C.T., Lee, C.S.G.: Neural Network based Fuzzy Logic Control and Decision System. IEEE Transactions on Comput. 40(12), 1320–1336 (1991)
27. Bherenji, H.R., Khedkar, P.: Learning and Tuning Fuzzy Logic Controllers through Reinforcements. IEEE Transactions on Neural Networks (3), 724–740 (1992)
28. Abraham, A.: Neuro-Fuzzy Systems: State-of-the-Art Modeling Techniques. In: Mira, J., Prieto, A. (eds.) Connectionist Models of Neurons, Learning Processes, and Artificial Intelligence, Granada, Spain, pp. 269–276. Springer, Germany (2001)
29. Abraham, A.: Optimization of Evolutionary Neural Networks Using Hybrid Learning Algorithms. In: IEEE 2002 Joint International Conference on Neural Networks, World Congress on Computational Intelligence (WCCI 2002), Hawaii, May 12-17 (2002)
30. Abraham, A.: ALEC -An Adaptive Learning Framework for Optimizing Artificial Neural Networks. In: Alexandrov, V.N., et al. (eds.) Computational Science, San Francisco, USA, pp. 171–180. Springer, Germany (2001)

31. Abraham, A.: Meta-Learning Evolutionary Artificial Neural Networks. Neurocomputing Journal 56c, 1–38 (2004)
32. Yao, X.: Evolving Artificial Neural Networks. Proceedings of the IEEE 87(9), 1, 423–1447 (1999)

Verification Questions

1. Why do we need to hybridize intelligent techniques?
2. What are the main hybrid architectures? Describe each of them with an example.
3. What is a neuro-fuzzy system? What are the different types of neuro-fuzzy systems?
4. Describe with illustrative diagrams how to design an evolutionary neural network involving architecture and weight adaptation?
5. Describe with illustrative diagrams how to design an evolutionary fuzzy system involving tuning the membership functions and learning the rules?
6. What are hybrid evolutionary algorithms? How they are useful in practice?

Exercises

Find a problem (or more) for which a couple of techniques do not work very well, Try to find a hybridization between two or more of them in such a way that the hybrid works better than each of the techniques independently.